AI for Legal Practice

A Practical Guide to Tools, Risk, and Professional Responsibility

Carlos A. Baradat

Copyright

AI for Legal Practice

A Practical Guide to Tools, Risk, and Professional Responsibility

Disclaimer. This book is intended for educational and informational purposes only. It does not constitute legal advice and should not be relied upon as a substitute for consultation with a qualified attorney regarding the reader's specific circumstances. Nothing in this book creates an attorney-client relationship between the author and the reader. Readers should independently verify all information, rules, and authorities cited in this book against the law and professional responsibility requirements of their own jurisdiction before relying on them in any professional matter.

The law, court orders, and AI tools referenced in this book reflect information available through April 2026. This field evolves rapidly; readers should verify that cited authorities, regulations, and tools remain current before relying on them in practice.

Accuracy. The author has made reasonable efforts to ensure the accuracy of the information presented in this book. However, the field of artificial intelligence and its application to legal practice is evolving rapidly. The author makes no warranty, express or implied, regarding the completeness or currency of the information contained herein and assumes no liability for any errors or omissions or for any outcomes resulting from the use of this information.

Trademarks. Product names, service names, and company names mentioned in this book may be trademarks or registered trademarks of their respective owners. Their use in this book is for identification and reference purposes only and does not imply endorsement.

First Edition, 2026

ISBN 979-8-9958263-0-9 (paperback)

ISBN 979-8-9958263-1-6 (ebook)

Published by The Virtual Jurist

An imprint of Legal Tech Dynamics, LLC
Boca Raton, Florida
www.ltdynamics.com

Dedication

To Liz, my vision and my guiding light.

This book bears my name, but it simply would not exist without you. Thank you for your endless patience during the long nights, the early mornings, and the endless monologues about technology and law. You are the operational heart of everything I build, the anchor that keeps me grounded, and the love that makes the work worth doing. Every page of this achievement is as much yours as it is mine.

To my mother, Marisol.

You are quite literally my world. Everything I am, everything I have achieved, and every boundary I have ever crossed is because of your unconditional love, your sacrifice, and your unshakeable faith in me. You believed in my future long before I could see it myself. This milestone, and every clean victory in my life, belongs entirely to you. I love you.

To Karen, my sister and my rock.

Thank you for your lifelong support, your unshakeable strength, and for always being the steady ground beneath my feet through every single chapter of life.

To my little sister, Beby (Marisol).

Watching you grow, pour yourself into forming a family, and reach your goals has been one of the greatest joys of my life. I have always looked at you with a protective heart, and I could not be prouder of the incredible, dedicated woman you have become. You will always be my Beby.

For Chuck

You gave me my start in this profession, but what came afterward mattered far more. What I have learned from you over the years is beyond measure. You became a mentor when I needed direction, and a true friend in the most unlikable places. The foundation you helped me build holds up everything that followed.

For My Friends, Colleagues, and Mentors

For the questions that made me think harder, the long conversations that refined emerging ideas, and the insights that appear throughout these pages even when you do not realize it. No book is written alone. This one certainly was not. Your fingerprints are everywhere, and I am better for every exchange we have shared.

For My Students

You inspire me every single day. Your curiosity, your persistence, and your willingness to grow remind me why I teach. I am proud of every one of you. Keep showing up. The profession is better because you are entering it.

Table of Contents

Preface

Artificial intelligence is already changing legal practice. It is changing how lawyers research, draft, communicate with clients, manage information, and think about professional judgment. It is also changing how mistakes happen. Some of those mistakes are obvious. Others are subtle, persuasive, and dangerous precisely because they look competent on first read. That combination of power and risk is what motivated this book.

I did not write this book to celebrate AI, and I did not write it to warn lawyers away from it. I wrote it because too much of the current conversation falls into one of two camps. One camp treats AI as a miracle tool that will transform practice if lawyers simply get out of its way. The other treats AI as a reckless shortcut that serious professionals should avoid. Neither view is good enough for the real world lawyers inhabit. Most lawyers do not have the luxury of ignoring technology that clients, courts, counterparties, vendors, and colleagues are already using. At the same time, no responsible lawyer should outsource judgment, accuracy, confidentiality, or professional responsibility to a machine that predicts language but does not understand the law the way a lawyer must.

This book takes the middle path, which I believe is the only sustainable one. AI can make lawyers faster, more organized, and in many contexts more effective. It can also make them sloppier, overconfident, and more vulnerable to mistakes if used without structure. The answer is not blind adoption or blanket resistance. The answer is disciplined integration.

My goal has been to create a practical guide for lawyers, law students, legal educators, and legal professionals who want to use AI competently, ethically, and strategically. I wanted a book that would be useful on a desk, not just interesting on a shelf. That meant moving beyond abstract commentary and building something operational: a book that explains what these tools are, where they help, where they fail, how to supervise their output, how to protect client information, and how to fit them into real legal workflows without losing the human judgment that defines excellent practice.

You will see that approach throughout the chapters. The early chapters establish the foundation: what large language models do, why they produce the outputs they produce, and what the core risks are. From there, the book moves into prompt engineering, professional responsibility, confidentiality, tool selection, and workflow design. The later chapters focus on actual legal work: research, drafting, litigation, client communication, practice management, governance, and the rapidly developing regulatory environment. The structure is intentional. I wanted the reader first to understand the technology, then to understand the constraints, and only then to apply it in practice.

I have also tried to write this book in a way that respects the reality of legal work. Lawyers are busy. They do not need hype. They do not need vague encouragement to innovate. They need frameworks they can trust, examples that feel real, and warnings that are specific enough to change behavior. They also need permission to be skeptical. One of the recurring themes in this book is that verification is not paranoia. It is professionalism. The fact that AI can be useful does not reduce the duty to check what it produces. In many contexts, it increases that duty.

Another recurring theme is that the most valuable parts of lawyering are not disappearing. If anything, they are becoming more important. Judgment, empathy, credibility, strategic thinking, ethical reasoning, and the ability to explain difficult realities to clients in a steady and humane way remain central to the profession. AI can accelerate many tasks. It cannot replace the lawyer's responsibility to choose, to prioritize, to advise, and to stand behind the final work product.

This book is also shaped by the belief that legal education and legal practice should not be separated when it comes to AI. Law students are entering a profession that is already being reshaped by these tools. Practicing lawyers are being asked questions by clients and courts right now. Educators are trying to prepare students for a professional world that is moving faster than most curricula. I hope this book helps bridge those groups by providing a shared language and a shared set of practical standards.

If you are new to AI, this book will help you build a solid and responsible foundation. If you have already experimented with AI tools, it will help you tighten your process, spot risks earlier, and use the technology more deliberately. If you lead a practice, a law firm, a legal department, or a classroom, it will help you think beyond individual prompts and individual tasks toward systems, standards, and governance.

No book on AI can freeze the technology in place. Tools will change. Regulations will develop. New capabilities will emerge. Specific product names, interfaces, and feature sets will evolve. That is unavoidable. What I have tried to provide instead are durable principles and workflows that will remain useful even as the surrounding landscape changes. A lawyer who understands verification, supervision, confidentiality, structured prompting, workflow design, and governance will be better prepared for whatever the next generation of tools looks like than a lawyer who only learns how to use one product well.

My hope is that this book helps you approach AI neither with fear nor with overconfidence, but with seriousness, curiosity, and professional discipline. Used well, these tools can help lawyers serve clients better. Used badly, they can magnify errors and compromise professional judgment. The difference lies not in the tool alone, but in the lawyer using it.

That is the premise of this book, and it is the standard I hope it helps you build into your own practice.

Acknowledgments

No serious book gets written in isolation, even when much of the actual drafting happens alone.

This project was shaped by years of legal practice, teaching, experimentation, frustration, revision, and the steady accumulation of questions about how lawyers should respond to a changing professional landscape. I am grateful for the many clients, students, colleagues, and professional contacts whose questions, challenges, and conversations sharpened the thinking that ultimately became this book. Even where no specific example appears on these pages, the real-world problems people brought to me over time helped define what this book needed to be: practical, candid, and grounded in the realities of legal work rather than hype.

I am also grateful to the broader legal and legal-technology communities. Lawyers, judges, educators, technologists, and bar leaders are all trying, in different ways, to understand how artificial intelligence should fit into a profession built on judgment, trust, and responsibility. Their writing, presentations, debates, and examples have helped move the conversation beyond novelty and toward standards that are more serious, more useful, and more worthy of the profession.

My thanks also go to the students I have taught over the years. Teaching has a way of forcing clarity. Questions from students often expose gaps in explanation faster than anything else, and that pressure to explain difficult ideas plainly has helped shape both the structure and the tone of this book. If parts of this manuscript feel practical and teachable rather than abstract and distant, that is in no small part because so much of the underlying material was tested in classrooms before it was tested on the page.

Finally, I am grateful for the opportunity to do work at the intersection of law, technology, and education. It is demanding work, and at times chaotic work, but it is also deeply meaningful. This book is my effort to contribute something useful to lawyers trying to navigate that intersection with competence, integrity, and confidence.

Any errors, of course, are mine alone.

A Note on Jurisdiction and Professional Rules

This book is written primarily for U.S. lawyers, law students, and legal professionals working in or studying American legal practice. Throughout the book, references to professional responsibility generally use the ABA Model Rules of Professional Conduct and related federal rules as the baseline point of reference because they provide the closest thing to a common vocabulary across jurisdictions. They are a useful starting point, but they are not the final word in any given matter.

Every U.S. jurisdiction adopts its own professional rules, comments, court rules, ethics opinions, and local practices. In some jurisdictions, the language tracks the ABA Model Rules closely. In others, the variations are meaningful. The same is true of court standing orders, local rules, and judge-specific requirements relating to artificial intelligence, disclosure, verification, confidentiality, and filing practices. Those rules are developing quickly, and in some places they are changing faster than any printed book can track.

For that reason, you should treat the rules, standards, and workflows discussed in this book as a professional framework rather than as a substitute for checking the rules that actually govern your work. Before applying any rule-based guidance in a live matter, verify the current version of the governing rule in your jurisdiction, review any applicable ethics opinions or local orders, and confirm that no more recent authority changes the analysis.

That point is especially important in areas where AI use intersects with duties of competence, confidentiality, supervision, candor to the tribunal, billing, and disclosure. The principles in this book are meant to help you think and work more responsibly. The final responsibility for determining what applies in your jurisdiction, in your court, and in your client's matter remains yours.

Readers outside the United States may still find the frameworks in this book useful, especially the discussions of workflow design, verification, confidentiality, supervision, governance, and client communication. But the rule citations and procedural examples are U.S.-focused and should be translated carefully into the professional and regulatory framework that governs your own practice.

How to Use This Book

This book is designed to work in more than one way.

You can read it straight through from beginning to end, which is the best approach if you want a full framework for understanding how artificial intelligence fits into legal practice. The early chapters explain how these tools work, where they are useful, where they fail, and what professional obligations govern their use. The later chapters apply those principles to actual workflows such as research, drafting, litigation, client communication, practice management, and governance.

You can also use the book selectively, depending on what you need most right now.

If you are new to AI in legal practice, start with Chapters 1 through 3. Those chapters explain the core concepts, the major risks, and the prompt-engineering principles that make the rest of the book easier to use well.

If your main concern is ethics and professional responsibility, go directly to Chapters 4 and 5, then Chapter 12. Those chapters address competence, supervision, confidentiality, governance, audit trails, and incident response.

If your main interest is applying AI in real legal work, focus on Chapters 7 through 11. Those chapters cover research, drafting, litigation and discovery, client communications, and practice management workflows.

If you lead a firm, legal department, or practice group, Chapters 6, 11, 12, and 13 will likely be especially important. Those chapters address tool selection, organizational systems, governance, training, and future-proofing.

Each chapter is built to be practical. The callout boxes highlight the points most worth remembering in active practice. The chapter-end Key Terms sections give you quick reference points without interrupting the body text. The Mini-Labs are there to help you translate concepts into habits. You do not need to complete every exercise to benefit from the book, but the more of them you actually use, the more quickly the material becomes operational.

This book is not organized around the idea that AI is either magic or poison. It is organized around a more useful question: how do you use these tools in ways that are efficient, defensible, and professionally sound? That is why so much of the book emphasizes verification, judgment, confidentiality, and workflow design rather than novelty for its own sake.

A final note: do not treat any single tool, prompt, or workflow in this book as permanent. The tools will change. Interfaces will change. Capabilities will change. The underlying habits matter more than the products. If you take away a strong process for supervision, verification, and responsible use, the book will remain useful even as the surrounding technology evolves.

Use the book in the way that serves your practice best. Read it as a guide, return to it as a desk reference, and adapt its frameworks to the specific realities of your work.

Companion Resources

This book has a companion resource hub at **thevirtualjurist.com/books/aiforlegalpractice**, where readers can access downloadable tools and ongoing updates built to extend the material in this book into daily practice.

Available to readers at no cost:

- **The full prompt library from Appendix B**, as a searchable web page and downloadable file, so readers can copy prompts directly into their AI tools rather than retyping them from the page.
- **Fillable versions of the Appendix C verification checklist and workflow log**, formatted as Word and PDF templates ready to use in court-facing work.
- **Editable tool cards** based on the Appendix F templates, so firms can build and maintain their own living tool card library without starting from scratch.
- **Sample AI disclosure language** for client engagement letters, practice-area variations, and matter-specific disclosures.
- **Errata and clarifications** as the book is read and refined by working practitioners.

The field of AI in legal practice evolves faster than any printed book can track. The companion site is where the living edition of this material is maintained between print editions — new prompts, new tool cards, new governance frameworks, and new commentary as the professional responsibility landscape develops.

The companion site also hosts practitioner resource bundles for firms and practice groups that want to go deeper, including policy templates, training materials, and implementation frameworks. Details and current offerings are available at the site.

Readers can subscribe to practitioner updates at the companion site. Corrections, questions, and practitioner conversations are welcome at **editorial@thevirtualjurist.com**. Reader feedback directly shapes future editions and the ongoing resources published online.

List of Figures

Part I: Foundations and the Prompt System

Chapter 1: The New Baseline: What AI Changes in Legal Work

Before generative AI became widely available, drafting a motion to compel could take an attorney hours of manual writing and record matching. Today, a competent first draft can be generated in minutes. That shift is not a distant prediction. It is an active reality, and it is rapidly separating practitioners who use these tools from those who ignore them.

This chapter establishes the foundation for everything that follows. Before we get into prompts, workflows, and risk management, we need a shared understanding of what AI actually changes in legal work and, just as importantly, what it does not change. The tools are new. The professional obligations are not.

What This Chapter Covers

- How AI shifts the lawyer's role from primary drafter to editor and supervisor
- Where AI tools fit across the lifecycle of a legal matter
- The five task types AI handles, and how their reliability differs
- The three core risks that create professional responsibility exposure
- High-risk categories where AI should stay out of your workflow
- How to establish your own rules of use before you need them

1.1 From "Research Tool" to "Drafting Engine"

For decades, legal technology has helped lawyers work faster without changing what lawyers actually did. Westlaw and Lexis changed how you found authorities. Document management changed how you stored files. E-discovery changed how you processed volumes. But the work product still came from you. You still read the cases yourself. You still wrote the memo yourself. The technology touched specific phases. Everything that mattered — the analysis, the drafting, the judgment — remained yours.

Generative AI is different in kind, not just degree. Previous tools helped you find authorities, organize files, or process volumes of documents. Generative AI can produce drafts. Give it a set of facts and a legal question, and it returns a draft analysis. Give it a contract and revision instructions, and it returns a redlined version. Give it a deposition transcript and a request for impeachment opportunities, and it returns a draft impeachment outline. The drafts vary in quality. Some are usable starting points. Others are confident-sounding mistakes. But the output is no longer a list of sources for you to read — it is something that resembles work product, which is why supervising it is a different skill than producing it.

This distinction matters because it changes your role. When technology helps you research, you are still the drafter. You are just a better-informed drafter. When technology helps you draft, you become something else: an editor, a supervisor, a quality-control function. The first draft exists before you have written a word. Your job is to evaluate whether that draft is any good.

That is a genuinely different skill. Drafting from scratch requires you to organize your thoughts, structure an argument, and translate analysis into prose. Supervising a draft requires you to spot what is missing, catch what is wrong, and recognize when confident-sounding prose is covering for thin reasoning. Many lawyers find the second task harder than the first, not because it is more intellectually demanding, but because it is less familiar. We trained for years to produce work product. We did not train to audit it.

Consider a routine task: drafting a motion for summary judgment on a straightforward contract dispute. In the traditional workflow, you would review the record, identify the key facts, research the governing standard, outline your argument, write a first draft, revise it, and polish the final version. That process might take eight to twelve hours of attorney time, depending on complexity.

In an AI-assisted workflow, the middle steps compress dramatically. You feed the AI the relevant facts, the contract, and the key authorities. You instruct it to draft a motion arguing that no genuine dispute of material fact exists. Ten minutes later, you have a fifteen-page draft with a statement of facts, a legal standard section, an argument section, and a conclusion. The draft might even include citations to the cases you provided.

Here is the problem: that draft is not reliable. It might be excellent. It might be competent. It might contain a fabricated case citation that will earn you a sanctions motion. It might state the legal standard correctly but apply it to facts you did not provide. It might omit the best argument entirely because you did not think to mention it in your prompt. You will not know which of these is true until you review the draft with the same rigor you would apply to work from a first-year associate who has never handled this type of motion before.

> **Key Point** AI is changing how lawyers spend their time. Less time on raw drafting and research, more time on supervision, judgment, and strategic choices. The shift is real, but it doesn't replace the lawyer; it reshapes what the lawyer's day looks like. Most lawyers have not been trained for that reshaped role.

The lawyers who struggle with AI-assisted drafting tend to fall into two camps. The first camp does not review carefully enough. They see a polished-looking draft, spot-check a few citations, and file it. When something goes wrong, a hallucinated case, a misstated holding, a fact pulled from nowhere, they are caught off guard. They trusted the output because it looked trustworthy.

The second camp reviews so heavily that they lose the efficiency gains entirely. They rewrite every sentence, re-research every point, and effectively draft the document from scratch using the AI output as a vaguely useful outline. They are not wrong to be cautious, but they have turned a twelve-minute draft into a ten-hour revision. The tool has not saved them time. It has added a step.

The goal is neither blind trust nor obsessive rework. The goal is calibrated supervision, a review process scaled to the actual risk of the task and the known reliability of the tool for that task type. A low-stakes internal memo needs less scrutiny than a brief going to a federal appellate court. A document summarization task, which AI does well, needs less scrutiny than a novel legal analysis task, which AI does inconsistently. Calibration is the skill. We will spend much of this book developing it.

The shift from research tool to drafting engine also changes how you think about time. Traditional legal work involved long stretches of production punctuated by moments of review. AI-assisted work inverts that ratio: short bursts of prompting followed by extended periods of evaluation. A partner who expects to see six hours of drafting time on a motion may not understand why the bill shows one hour of prompting and four hours of verification. But if the verification was rigorous and the output is sound, the client got a better result at lower cost. The economics have changed, and billing practices will eventually follow. In the meantime, you will need to explain what you are actually doing, to partners, to clients, and sometimes to courts.

None of this means AI makes legal work easy. It means AI makes legal work different. The production burden drops. The supervision burden rises. The time per task drops. The judgment per task rises. You can process more matters, but each matter still requires a lawyer's eye. The technology is not practicing law. You are, just with a very fast, very confident, and occasionally very wrong assistant.

1.2 Where AI Fits in the Matter Lifecycle

A legal matter moves through phases: intake, investigation, strategy, production, negotiation or litigation, and resolution. AI does not help equally at every stage. Understanding where it adds value, and where it introduces risk, lets you deploy it strategically rather than reflexively.

To see how these phases work in practice, consider a straightforward breach-of-contract matter. A small business client calls to report that a supplier failed to deliver goods specified in their agreement (intake). You review the contract, communications between the parties, and the client's records of the failed delivery to understand what happened (investigation). You assess the strength of the claim, the available remedies, and whether the matter is better resolved through negotiation, mediation, or litigation (strategy). You draft a demand letter outlining the breach and the remedy sought (production). When the supplier's counsel responds, you negotiate terms — or, if talks fail, prepare and file suit (negotiation or litigation). Eventually, you finalize a settlement agreement or obtain a judgment, close out the file, and deliver a final summary to the client (resolution). Every phase of this matter involves tasks that AI can potentially support, but the supervision required, and the risks involved, differ sharply from phase to phase. Understanding that difference is what this section is about.

Start with intake. A new client calls with a problem. They talk for twenty minutes, mixing relevant facts with irrelevant background, legal conclusions with emotional reactions, and chronological narrative with stream-of-consciousness tangents. Traditionally, you would take notes, ask clarifying questions, and then spend another twenty minutes writing up a summary of the matter for your file. AI handles this well. Feed it a transcript or your rough notes, instruct it to extract the key facts, identify the potential legal issues, and flag gaps in the information, and you will get a structured intake memo in two minutes. The task is low risk because you are not relying on the AI for legal judgment. You are using it to organize information you already have. You will still read the output and correct any errors, but the errors are easy to catch because you just heard the client tell you the same story.

Investigation works similarly. When you are reviewing a stack of documents to understand what happened, AI can accelerate the process dramatically. It can summarize contracts, extract key dates into a timeline, identify the parties involved and their relationships, and flag documents that mention specific topics. These are extraction and classification tasks, areas where AI performs reliably. You are not asking it to evaluate whether the facts support a claim. You are asking it to surface the facts so you can evaluate them yourself.

> **Risk Note** Never let AI draft a communication that goes directly to opposing counsel or a court without full human review. "I'll just send what it wrote" is how lawyers end up citing cases that do not exist, misstating their own client's position, or making concessions they did not intend.

Strategy is where the calculus shifts. Developing a case theory, choosing which claims to pursue, deciding whether to file in state or federal court, assessing settlement value, these tasks require judgment that integrates legal knowledge with client goals, risk tolerance, factual nuance, and practical constraints. AI can support strategic thinking, but it cannot replace it. You might ask an AI to outline the elements of each potential claim, identify the strongest and weakest facts for each, or list the typical defenses a defendant might raise. That output can sharpen your thinking. But the decision itself, which theory to pursue, how aggressively to litigate, when to push and when to settle, remains yours. Lawyers who outsource strategic judgment to AI are making a category error, asking the tool to do something it is not designed for and should not be trusted with.

Production is AI's natural habitat. Once you have decided what to create, a demand letter, a motion, a contract, a discovery request, AI can generate a first draft quickly. This is the "drafting engine" function discussed in the previous section. The draft needs supervision, but the production phase is where AI delivers its most obvious efficiency gains. Routine documents benefit most: discovery responses, standard contract clauses, correspondence, memoranda on well-settled legal questions. Complex or novel work benefits less, because the supervision burden rises faster than the production time drops.

Negotiation and litigation present a mixed picture. AI can help you prepare, drafting talking points, outlining negotiation positions, identifying concessions you might offer or extract, generating deposition questions, summarizing testimony, preparing cross-examination outlines. But the live interaction remains human. You are the one reading the room, adjusting on the fly, deciding when to press and when to let something go. AI can hand you a script. It cannot perform the play. Lawyers who over-rely on AI-generated outlines sometimes find themselves flat-footed when the conversation departs from the predicted path. The preparation was thorough, but it was also brittle.

Resolution and close-out tasks, finalizing settlement agreements, preparing closing documents, generating matter summaries for the file, are generally well suited to AI. These tasks tend to be structured, to involve synthesis of existing information rather than novel analysis, and to carry lower risk than mid-matter strategic decisions. A settlement agreement drafted by AI still needs careful review, particularly for terms that interact in non-obvious ways, but the review is checking for errors in execution rather than errors in judgment.

1.3 Task Types: Generate, Transform, Extract, Classify, Critique

Not every task you hand to an AI is the same kind of task. A request to "draft a motion" is fundamentally different from a request to "summarize this contract," even though both produce text output. Understanding the category of work you are asking for helps you predict how reliable the output will be and how much supervision it needs.

Five task types cover most of what lawyers ask AI to do: generate, transform, extract, classify, and critique. Each has a different reliability profile.

Generate means creating new content from instructions. You provide a prompt describing what you want, and the AI produces text that did not exist before. Drafting a demand letter, writing contract clauses, composing a research memo, creating deposition questions. Generation is the most powerful application and the most dangerous. The AI is not retrieving information or reorganizing existing material. It is constructing something new. That construction draws on patterns learned during training, which means the output reflects statistical regularities in the training data rather than verified legal accuracy. Generated content can be excellent. It can also be fabricated, outdated, or subtly wrong in ways that look right. Generation tasks require the heaviest supervision.

Transform means converting existing content from one form to another. You provide source material and instructions for changing its format, length, tone, or structure. Turning a formal brief into a plain-language client summary. Converting bullet points into flowing paragraphs. Rewriting a clause to favor your client's position. Translating a technical document for a lay audience. Transformation is more constrained than generation because the source material anchors the output. The AI is not inventing. It is reshaping. Errors still occur, particularly when the transformation requires judgment about what to emphasize or omit, but they are easier to catch because you can compare the output against the input. Transformation tasks need moderate supervision.

Extract means pulling specific information out of a larger body of material. You provide documents and instructions about what to find. Identifying all payment terms in a set of contracts. Pulling every date mentioned in a deposition transcript. Listing the parties referenced in a collection of emails. Extraction is highly reliable because the task is concrete and verifiable. Either the extracted information appears in the source or it does not. The main failure modes are omission and hallucination. Both are catchable with reasonable review. Extraction tasks need lighter supervision, focused on completeness and accuracy rather than judgment.

Classify means sorting materials into categories. You provide items and a classification scheme, and the AI assigns each item to the appropriate bucket. Sorting documents by relevance to specific issues. Categorizing contract clauses by type. Identifying which emails discuss a particular topic. Flagging privileged communications in a document set. Classification reliability depends heavily on how clear your categories are. Well-defined categories with objective criteria produce consistent results. Fuzzy categories that require judgment produce inconsistent results. A request to classify documents as "relevant" or "not relevant" will struggle because relevance depends on context the AI may not fully grasp. A request to classify documents as "mentioning the June 15 meeting" or "not mentioning the June 15 meeting" will succeed because the criterion is concrete. Classification tasks need supervision calibrated to category clarity.

Critique means evaluating existing work and identifying problems. You provide a document and ask the AI to find weaknesses, gaps, errors, or areas for improvement. Reviewing a contract for missing provisions. Checking a brief for logical inconsistencies. Identifying ambiguities in a settlement agreement. Spotting factual assertions that lack record support. Critique is underused by most lawyers but remarkably valuable. The AI approaches your work without the cognitive biases you developed while creating it. It does not know which arguments you are proud of or which sections you rushed through. It just reads what is there and flags what is missing or broken. Critique tasks are moderately reliable. The AI will catch many genuine problems, but it will also raise false alarms and miss issues that require deeper contextual understanding. Treat critique output as a checklist to consider, not a definitive assessment.

The categories are not always pure. A single prompt might combine types. "Review this contract and draft a summary of the key risks" involves extraction, classification, and generation. When tasks combine, reliability tends to follow the least reliable component. A prompt that is mostly extraction but includes some generation will need supervision appropriate for generation.

The practical application is supervision budgeting. You have limited time for review. Allocating that time based on task type makes your review more efficient and more effective. Spend less time verifying a document extraction. Spend more time verifying a generated brief. Spend less time checking a concrete classification. Spend more time checking a judgment-dependent one. The goal is not to review everything equally but to review everything appropriately.

1.4 The Three Risks: Accuracy, Confidentiality, Accountability

Every AI-assisted legal task carries three risks. They travel together, and managing one without attending to the others leaves you exposed. Understanding how they interact is essential before you build AI into your workflows.

Accuracy risk is the most discussed and the easiest to grasp. The AI might produce output that is wrong. The wrongness can take many forms: a case citation that does not exist, a holding that is misstated, a fact that was never in the record, a legal standard from the wrong jurisdiction, an analysis that sounds persuasive but does not hold up under scrutiny. Accuracy failures happen because generative AI does not retrieve verified information. It predicts plausible text based on patterns. Plausible and correct are not the same thing.

Figure 1.1 — The Three Risks of AI in Legal Practice

The severity of accuracy risk varies by task type. Extraction tasks carry lower accuracy risk than generation tasks. But no task type is immune. Even a straightforward extraction can miss items or report items that were not there. Even a simple transformation can introduce errors when the AI misunderstands what to preserve and what to change. Accuracy risk is always present. Only its magnitude varies.

What makes accuracy risk particularly dangerous in legal work is that AI errors often do not look like errors. A hallucinated case citation comes complete with a party name, a volume and reporter, a page number, and a parenthetical summary of the holding. It looks exactly like a real citation. A misstated legal standard is written in confident, professional prose indistinguishable from an accurate statement. The errors do not announce themselves. They hide in plain sight until someone checks.

> **Risk Note** Confident tone is not a signal of accuracy. AI produces the same polished prose whether the underlying content is correct, partially correct, or completely fabricated. Never use fluency as a proxy for reliability.

Confidentiality risk runs in the opposite direction. Accuracy risk concerns what comes out of the AI. Confidentiality risk concerns what goes in. When you enter information into an AI system, that information leaves your control. Depending on the system's architecture, data handling policies, and terms of service, your input might be stored, logged, used for model training, or accessible to the system's operators. For lawyers, this creates a direct collision with the duty to protect client confidences.

The confidentiality calculus depends heavily on which tool you are using and how it is deployed. Consumer-facing AI products typically offer weaker confidentiality protections than enterprise deployments. Some systems explicitly state that user inputs may be used to improve the model, which means your client's sensitive information could influence outputs that other users see. Other systems offer data isolation, no-training guarantees, or on-premises deployment that keeps information within your firm's infrastructure. Later chapters address these distinctions in detail. For now, the key point is that confidentiality risk exists every time you enter client information into an AI system, and managing that risk requires understanding what happens to your data after you hit enter.

The interaction between accuracy and confidentiality creates a practical tension. To get accurate, useful output, you often need to provide detailed input: the specific facts of your case, the exact contract language at issue, the particular client circumstances that shape the analysis. But detailed input means greater confidentiality exposure. Vague prompts protect confidentiality but produce generic, less useful outputs. Specific prompts produce better outputs but put more client information into the system. Navigating this tension is a core skill in responsible AI use.

> **Practice Point** Before entering any information into an AI tool, ask: If this input became public, would I have a problem? If the answer is yes, either use a more secure tool, anonymize the information, or handle the task without AI assistance.

Accountability risk concerns the human element: whether a lawyer has maintained sufficient oversight to take professional responsibility for the work product. The rules of professional conduct do not care whether a document was drafted by an associate, a contract attorney, a paralegal, or an AI system. They care whether the supervising lawyer ensured that the work meets professional standards. When you sign a brief, you are certifying that the factual contentions have evidentiary support and the legal contentions are warranted by existing law or nonfrivolous arguments for changing it. That certification does not come with an asterisk for AI-assisted work.

Accountability risk materializes when the supervision is inadequate to the task. If you generate a motion using AI, skim it quickly, and file it without checking the citations, you have not supervised the work. You have laundered it. The AI produced something. You added your signature. No lawyer actually reviewed it with the care the rules require. When the fabricated citation surfaces, "the AI wrote it" is not a defense. You filed it. You are responsible.

The courts have started to address this directly. Federal courts have begun imposing sanctions on lawyers who filed briefs containing AI-generated citations to nonexistent cases. The opinions are pointed: the lawyers failed in their duty of candor to the tribunal and their basic obligation to verify the accuracy of their submissions. Some courts now require disclosure of AI assistance in filings precisely because they want lawyers to think twice before submitting unverified AI output as their own work. Judge Brantley Starr of the Northern District of Texas issued one of the earliest and most widely cited of these standing orders in May 2023, requiring attorneys to certify either that no portion of a filing was drafted by generative AI or that any AI-generated language was verified for accuracy by a human being. Other courts have since issued similar orders.

> **Authority Note** In *Mata v. Avianca, Inc.*, 678 F. Supp. 3d 443 (S.D.N.Y. 2023), a federal court sanctioned lawyers who submitted a brief containing fabricated case citations generated by ChatGPT. The court found that the lawyers "abandoned their responsibilities" when they submitted nonexistent authorities created by the tool. The case has become the cautionary standard for AI use in litigation.

Accountability risk also extends beyond court filings. Every piece of legal work carries an implicit representation that a competent lawyer stands behind it. A contract you send to opposing counsel. A memo you deliver to a client. Advice you give in a meeting. If that work was produced by AI and you did not review it adequately, you have made a representation you cannot support. The client trusted your judgment. You gave them an algorithm's output dressed up as lawyering.

The three risks compound each other in practice. Low-accuracy output that reaches a client or court is an accountability failure and may also reflect confidentiality failures in what was entered. A confidentiality breach that exposes client data creates malpractice exposure regardless of whether the underlying output was accurate. A lawyer who does not understand accuracy limitations is more likely to commit accountability failures, which makes any confidentiality exposure more damaging. The risks do not exist in isolation. They form a system, and responsible AI use requires managing the system as a whole.

The framework for managing all three risks comes down to two commitments. First, use tools whose data handling practices you understand and that match the sensitivity of the information you are entering. Second, review AI output with rigor proportionate to the stakes and the task type before that output reaches anyone outside your control. The first commitment addresses confidentiality. The second addresses accuracy. Together, they satisfy accountability. Every workflow in this book builds on these two commitments.

1.5 When Not to Use AI (High-Risk Categories)

The previous sections might suggest that AI belongs everywhere in legal practice, provided you manage the risks appropriately. That is not quite right. Some categories of work are poor candidates for AI assistance no matter how rigorous your supervision process. Recognizing these categories upfront saves you from learning the hard way.

The common thread in high-risk categories is that the consequences of error are severe, the errors are difficult to detect, or both. When those conditions combine, the supervision burden overwhelms any efficiency gain. You spend so much time verifying that you might as well have done the work yourself. Worse, you might miss something despite your best efforts, because some errors only become visible in hindsight.

Novel legal questions sit at the top of the list. When you are arguing for an extension of existing doctrine, interpreting a statute with no case law on point, or navigating an area where courts have reached conflicting conclusions, AI provides little value and considerable danger. Generative AI draws on patterns in its training data. Novel questions, by definition, lack established patterns. The AI will produce something, because it always produces something. But that something will be extrapolation dressed up as analysis, and you will not be able to verify it against authorities that do not exist. For cutting-edge legal work, you need a lawyer's judgment, not a statistical prediction of what plausible legal text looks like.

Matters involving imminent deadlines with high stakes present a different problem. The time pressure that makes AI attractive is precisely what makes AI dangerous. Adequate supervision takes time. When you are filing an emergency motion at 11 p.m. or closing a deal at midnight, you do not have the hours required to verify AI output thoroughly. The temptation is to trust the draft because you need to trust the draft. That is when accountability failures happen. If the stakes are high and the clock is short, do the work yourself or get human help. Speed without verification is not efficiency. It is recklessness.

> **Risk Note** Time pressure is a risk multiplier, not a reason to rely on AI more heavily. The situations where you most want to skip verification are exactly the situations where skipping verification is most likely to cause harm.

Confidential or privileged information that cannot be adequately protected creates an absolute bar. If your available AI tools do not offer data protections sufficient for the sensitivity of the information, the information does not go into those tools. This is not a risk to be managed. It is a line not to be crossed. Some matters involve trade secrets, grand jury material, sealed court records, national security information, or communications so sensitive that any external processing is unacceptable. For those matters, AI assistance using cloud-based consumer tools is simply off the table. You either use an air-gapped system with appropriate security clearances, or you proceed without AI.

Work requiring empathy, judgment about human dynamics, or sensitive client communication resists AI involvement for reasons that have nothing to do with accuracy. Telling a client that their case is weaker than they hoped. Counseling a business owner through a difficult negotiation. Navigating a family law matter with emotional complexity. These situations require you to read the room, adjust your tone in real time, and bring genuine human presence to the interaction. AI can help you prepare talking points, but the conversation itself needs to be yours. Clients who sense they are receiving algorithmic output in moments that call for human connection will lose trust in you, and they will be right to.

Criminal defense work warrants particular caution. The stakes are liberty and sometimes life. The right to effective assistance of counsel is constitutional. An error in a criminal case can result in wrongful conviction or, in capital cases, irreversible harm. Some criminal defense lawyers use AI for discrete tasks like summarizing discovery or organizing case files, but the core work of defending a client, including developing theory, preparing cross-examination, and making strategic judgments, should remain human. The margin for error is too thin, and the consequences of AI-induced mistakes are too grave.

Tasks requiring authentication or personal knowledge cannot involve AI at all for the portions that require attestation. When you sign a declaration stating that facts are true to your personal knowledge, those facts must actually be within your personal knowledge. When you authenticate a document, you must have a basis for that authentication. AI can help you organize information, but it cannot provide you with knowledge you do not have. Lawyers who use AI to draft declarations and then sign them without independently verifying every factual assertion are flirting with perjury.

Matters where the client has prohibited AI use create a contractual or fiduciary bar. Some clients, particularly in regulated industries or government contracting, have policies restricting the use of AI tools on their matters. Some engagement letters now include provisions addressing AI. If your client has told you not to use AI, that instruction controls regardless of your own assessment of the risks. Violating a client instruction to serve your own convenience is a breach of the relationship, even if you believe your AI use would have been safe.

AI should stay out of certain categories of work altogether. Those categories include:

- novel legal questions
- high-stakes work under severe time pressure
- matters involving information too sensitive for your available tools
- work requiring human empathy and presence
- criminal defense at its core
- tasks requiring personal attestation
- any matter where the client has prohibited AI use

Lawyers sometimes push back on these limitations by pointing to the potential for AI to improve. As the tools get better, will these categories shrink? Perhaps. But the standard for adopting AI in your practice today is not what the technology might do in three years. It is what the technology can do reliably now, in your hands, with the supervision resources you actually have. Future capabilities do not reduce present risks. When the tools improve, reassess. Until then, respect the boundaries.

1.6 Your Personal AI Rules of Use

Policies imposed from above rarely stick. Lawyers find workarounds, make exceptions, or simply ignore guidelines that do not match how they actually work. The rules that govern your AI use most effectively will be the ones you develop yourself, based on your own practice, your own risk tolerance, and your own honest assessment of where AI helps and where it does not.

This section is not about firm-wide governance. Later chapters address that. This section is about the framework you carry in your head every time you sit down to work on a matter and consider whether AI belongs in the process.

Start with a threshold question: Does this task even need AI? The answer is not always yes. AI has become so accessible that lawyers sometimes reach for it reflexively, prompting a chatbot to do work they could have done faster themselves. If the task takes five minutes without AI and requires ten minutes of verification with AI, you have not gained efficiency. You have added a step. The goal is not to maximize AI use. The goal is to maximize effective work. Sometimes that means closing the AI tab and just writing the email yourself.

When AI does make sense, your personal rules should address four questions: What goes in? What comes out? How do I verify? How do I document?

What goes in is a confidentiality decision. Before you enter anything into an AI tool, you need to know what kind of tool you are using and what protections it offers. Consumer tools with broad data-use policies get different treatment than enterprise tools with contractual confidentiality guarantees. Your rule might be: "Consumer AI tools receive only public information, anonymized hypotheticals, or information the client has authorized me to share. Client-specific details go only into firm-approved tools with appropriate data protections." The exact line depends on your firm's infrastructure and your client's expectations, but you need to draw a line somewhere and hold it consistently.

What comes out is a scope decision. AI will produce whatever you ask for, but not everything it produces belongs in your workflow. Your rule might be: "AI can produce first drafts, but not final versions. It can suggest arguments, but I choose which arguments to make. It can generate research starting points, but I verify every authority myself." These limits keep you in control of the work product. You are using AI as a tool, not delegating your judgment to it.

Some lawyers find it helpful to define their scope rules by output type. Court filings receive the heaviest scrutiny and the most limited AI involvement. Client communications receive careful review with attention to tone and accuracy. Internal work product receives lighter review because the audience is only yourself or close colleagues. Informal research receives the lightest touch because it is a starting point, not an endpoint. Your categories might differ, but the principle holds: different outputs warrant different levels of AI involvement and different intensities of verification.

How do I verify is a supervision decision. Knowing that you should verify AI output is not the same as having a process for doing so. Your rules should specify what verification actually looks like for different task types. For generated legal analysis, verification means checking every citation, confirming every stated holding, and tracing the reasoning to ensure it actually supports the conclusion. For extracted information, verification means spot-checking a sample of extractions against the source material and scanning for obvious gaps. For transformed content, verification means comparing the output to the input to ensure nothing was lost or distorted in translation.

The depth of verification scales with stakes. A summary of a contract for your own internal reference needs a quick read-through. A summary of a contract going into a due diligence report needs line-by-line comparison against the source. A motion going to federal court needs verification of every factual assertion and legal citation, full stop. Your rules should acknowledge these gradations rather than applying the same process to everything.

> **Risk Note** "I'll review it carefully" is not a verification process. It is a vague intention. Specify what careful review means for each type of output: which elements do you check, against what sources, and how do you know the check occurred?

How do I document is an accountability decision. When something goes wrong, you want a record showing that you used AI responsibly. That record protects you professionally and helps you learn from mistakes. Your documentation might include what tool you used, what prompt you entered, with confidential information redacted if necessary, what output you received, what verification steps you performed, and what changes you made before using the work product. You do not need a bureaucratic log for every interaction, but you should be able to reconstruct your process for any work product that matters.

Some lawyers keep a simple note in the matter file: "Initial draft of motion generated using [Tool] on [Date]. All citations verified against Westlaw. Argument structure revised to add Section III. Final version reflects attorney work product." That is enough to demonstrate a responsible process without creating undue administrative burden.

Beyond these four questions, your personal rules should include a commitment to learning. AI tools change rapidly. Your understanding of their capabilities and limitations should change with them. Build in periodic reassessment. Every few months, ask yourself: What have I learned about where AI helps and where it does not? Have I encountered failure modes I did not anticipate? Are there tasks I have been avoiding with AI that I should try, or tasks I have been using AI for that I should handle differently? The lawyers who use AI well are not the ones who figured it out once and stopped thinking. They are the ones who keep paying attention.

Finally, write your rules down. Not for anyone else to see, but for yourself. The act of articulating your framework forces clarity. Vague intentions like "I'll be careful" collapse under time pressure. Specific commitments like "I verify every citation before filing" hold firm because you know exactly what you promised yourself. Your rules will evolve as you gain experience, but starting with a written framework gives you something concrete to test against reality.

The Mini-Lab at the end of this chapter walks you through developing your initial rules. Before you do that exercise, take stock of where you are. Have you used AI for legal work before? If so, what went well and what did not? If not, what concerns are holding you back? Your answers will shape the rules you create.

1.7 Mini-Lab: Build Your Personal AI Rules of Use

Section 1.6 argued that the rules governing your AI use most effectively will be the ones you develop yourself. This Mini-Lab walks you through doing exactly that. By the end, you will have a one-page Personal AI Rules of Use drawn from the frameworks in Chapter 1, tested against a real matter on your desk.

This is not a compliance exercise. It is a working tool. Write it in plain language. Keep it short enough that you will actually read it. If it takes more than one page, you are overthinking it.

Objective: Produce a one-page Personal AI Rules of Use using the frameworks from Chapter 1, then pressure-test it against a real matter on your desk to find the questions you still need to answer.

Time Required: 30 to 40 minutes

What You Will Need:

- The task type framework from Section 1.3 (generate, transform, extract, classify, critique)
- The three-risk framework from Section 1.4 (accuracy, confidentiality, accountability)
- The high-risk categories from Section 1.5
- One matter currently on your desk that involves at least some written work
- A single sheet of paper, a document, or a note in whatever tool you use for practice notes

Part 1: Draft Your Rules of Use (20 minutes)

Your Rules of Use answer four questions, drawn directly from Section 1.6. For each question, write two to four sentences. Do not write a treatise. Write the kind of statement you would actually hold yourself to under time pressure.

Question 1: What tasks am I willing to use AI for?

Using the five task types from Section 1.3, identify which types you are comfortable using AI for at this stage. A reasonable answer might read: "I will use AI for extraction, classification, and transformation tasks, but not for generating court filings or novel legal analysis until I have a reliable verification process."

Your answer should name specific task types. Vague commitments like "I will use AI carefully" are not commitments. They are intentions, and Section 1.6 already explained why intentions collapse under pressure.

Question 2: What are my hard red lines?

Using the high-risk categories from Section 1.5, write down the categories of work where you will not use AI under any circumstances. These are not risks you manage. They are lines you do not cross. Be specific. "I will not use AI for criminal defense work on matters involving liberty" is better than "I will be careful with criminal cases." Include at least three hard red lines.

Include any client-specific prohibitions you are aware of. If a client has told you not to use AI on their matters, that is a red line regardless of your own assessment.

Question 3: How will I protect client information?

Based on Section 1.4's discussion of confidentiality risk, make a starting commitment about what information you will and will not enter into AI tools. A reasonable answer might read: "I will not enter client-identifying information into consumer AI tools. For matters requiring AI assistance, I will use only firm-approved tools with confidentiality protections, or anonymize the material before entering it."

Write this as a rule you can follow today, with the tools you have access to right now.

Question 4: How will I maintain accountability?

This is where you commit to supervision and documentation. Your answer should address both. A reasonable answer might read: "I will verify every factual assertion and legal citation in AI-assisted work product against primary sources before it leaves my desk. I will note in the matter file what tool

I used, what task I performed, and what verification I completed. If I cannot verify, I will not rely on the output."

When you have answered all four questions, you have a working Rules of Use statement. Write the date at the top. Keep it somewhere you will actually find it when you need it.

Part 2: Pressure-Test Your Rules Against a Real Matter (10 to 20 minutes)

Now take one matter currently on your desk. Not a hypothetical. A real file with real facts and a real client. Pick one that involves at least some written work — a motion to be drafted, a contract to be reviewed, a discovery response to be prepared, a research memo to be written. Any matter will work as long as it is real.

Look at the work that matter requires over the next week. Identify two or three discrete tasks. For each task, walk through your Rules of Use and ask:

1. Do my rules tell me whether I can use AI for this task? Apply Question 1. If the task falls within the task types you committed to, move on. If it falls in the types you said you would not use AI for, you have your answer. If you are uncertain which task type applies, your rules have a gap.

2. Do my rules tell me how to protect the information involved? Apply Question 3. Think about what inputs you would need to give the AI to get useful output. Does your rule cover that information? If the matter involves sensitive client information and your rule only addresses "consumer AI tools," you may have a gap.

3. Do my rules tell me how to verify the output? Apply Question 4. Think about what the AI would produce for this task. Do your rules tell you how you would verify it? If the task is a research memo and your rule only says "verify citations," but you also need to verify that the analysis is sound, you have a gap.

4. Do my rules tell me whether this task is off-limits? Apply Question 2. Check whether the task touches any of your hard red lines. If it does, your answer is clear. If it does not touch a red line but still feels risky, ask why. Your instinct may be telling you that your red lines are incomplete.

Keep your notes on the gaps short. You are not solving the gaps today. You are identifying what you still need to learn.

Deliverable

At the end of this Mini-Lab, you should have:

1. **A one-page Personal AI Rules of Use** — a working document answering the four questions from Section 1.6, written in your own voice, ready to adopt as your standard operating procedure.
2. **A short gap list** — the specific questions your rules could not answer when tested against a real matter.

Keep both. The rules are your working framework. The gap list tells you exactly what you still need to learn.

Bonus Challenge

Share your Rules of Use with one colleague whose judgment you trust. Not to ask for edits. Just to see what questions they ask. If they push back on any of your commitments, that is useful information —

either your rules need revision, or you need to be able to explain why you drew the line where you did. Both outcomes strengthen the rules.

If you are working through this book as part of a firm or study group, compare rules with two or three colleagues. You will find that reasonable practitioners land in different places on the same questions. That variation is informative and worth discussing.

Key Takeaways

1. AI has shifted the lawyer's role from drafter to supervisor. The primary skill is no longer producing work product from scratch. It is evaluating work product that already exists and catching errors before they cause harm.
2. AI fits best at the bookends and production middle of the matter lifecycle. Intake, investigation, drafting, and close-out tasks offer strong returns with manageable risk. Strategy and live interaction remain fundamentally human.
3. Five task types define what you are asking AI to do. Generate, transform, extract, classify, and critique each carry different reliability profiles, with extraction and classification most reliable and generation requiring the heaviest supervision.
4. Three risks travel together on every AI-assisted task. Accuracy risk concerns whether the output is correct. Confidentiality risk concerns whether the input is protected. Accountability risk concerns whether a lawyer has actually supervised the work.
5. Some categories of work should remain off-limits to AI. Novel legal questions, high-stakes work under time pressure, overly sensitive information, tasks requiring human presence, and client-prohibited matters are boundaries to respect, not risks to manage.

Chapter Checkpoint

Before moving on, make sure you can:

- ☐ Explain why supervising AI output requires different skills than drafting from scratch, and identify which skills you need to develop
- ☐ Map a typical matter in your practice to the lifecycle framework and identify which phases are strong candidates for AI assistance
- ☐ Classify any legal task by type and predict its relative reliability
- ☐ Assess any proposed AI use against the three risks and articulate what safeguards would be required to proceed responsibly

What's Next

Understanding when and where to use AI is the first step. Understanding how AI actually works is the second. Chapter 2 explains what large language models do, why they produce the outputs they produce, and why those outputs sometimes go wrong in predictable ways. You do not need a computer science degree to use these tools well, but you do need a mental model of what is happening under the hood. That mental model will make everything else in this book more intuitive.

Key Terms

Transformation — Converting content from one form to another while preserving its substance. The source material constrains the output, making transformation more reliable than open-ended generation.

Generate — Creating new content from instructions rather than reshaping existing material. Generation draws on patterns from training data rather than verified sources, which makes it the most powerful and the least reliable of the task types. Generation requires the heaviest supervision.

Extract — Pulling specific information out of a larger body of source material. Extraction is concrete and verifiable because the information either appears in the source or it does not. The main failure modes are omission and hallucination, both catchable with reasonable review.

Classify — Sorting items into defined categories according to a classification scheme. Reliability depends on category clarity: well-defined categories with objective criteria produce consistent results, while fuzzy categories that require judgment produce inconsistent ones.

Critique — Evaluating existing work to identify weaknesses, gaps, errors, or areas for improvement. Critique is useful because AI approaches the work without the cognitive biases its author developed while creating it, though it will also raise false alarms and miss issues requiring deeper contextual understanding.

Chapter 2: How Generative AI Works for Lawyers (Without the Math)

You do not need to understand internal combustion to drive a car, but you do need to know that cars require fuel, that brakes work differently on wet roads, and that the engine will seize if you ignore the oil light. The same principle applies to generative AI. You do not need a background in machine learning to use these tools effectively. But you do need a working mental model of what is happening when you enter a prompt and receive a response. Without that model, you cannot predict when the tools will fail, and you cannot design workflows that account for their limitations.

What This Chapter Covers

- What large language models actually do, explained without technical jargon
- Why context windows matter and how they affect what the AI can "remember"
- The critical difference between training and retrieval, and why citations are unreliable
- The common failure modes you will encounter and how to recognize them
- How bias and uneven training data affect output quality across different topics
- A reliability framework for matching AI capabilities to task risk levels
- A practical triage process for reviewing AI output efficiently and thoroughly

Chapter Summary

Large language models generate text by predicting what words are likely to come next, based on patterns learned from massive training datasets. They do not retrieve information from a database, search the internet in real time, or reason the way lawyers reason. Understanding this distinction explains most of their failure modes. Hallucinations occur because the model predicts plausible text rather than verified facts. Omissions occur because the model has no checklist ensuring completeness. Anchoring occurs because early predictions constrain later ones. Bias reflects the training data, which overrepresents some topics and perspectives while underrepresenting others. This chapter builds a mental model that helps you anticipate these failures before they reach your work product. It introduces a reliability ladder that maps task types to risk levels and provides a triage framework for reviewing AI output in a way that catches errors without consuming all the time you saved by using AI in the first place. The goal is not to make you a technologist. The goal is to make you a competent supervisor of tools you are already using.

2.1 What an LLM Is (and Isn't)

A large language model is a prediction engine. You give it a sequence of text, and it predicts what text should come next. That is it. Everything else these tools appear to do, from writing briefs to answering legal questions to analyzing contracts, emerges from that single capability: predicting the next word, then the next, then the next, thousands of times in succession until the response is complete.

This sounds simple because it is simple, at least conceptually. The complexity lies in how the predictions are made. During training, the model processed enormous quantities of text, far more than any human could read in a lifetime. It learned statistical patterns: which words tend to follow other words, which phrases cluster together, which structures appear in legal documents versus marketing copy versus scientific papers. When you prompt the model, it uses those patterns to generate text that resembles the kind of text that would plausibly follow your input.

The key word is "resembles." The model is not retrieving a correct answer from storage. It is not consulting a database of verified legal authorities. It is not reasoning through your problem the way you would reason through it. It is generating text that looks like the text a knowledgeable person might produce in response to your prompt. Sometimes that generated text is accurate. Sometimes it is not. The model does not know the difference, because the model does not "know" anything in the sense that lawyers mean when they use that word.

Consider an analogy. Imagine a lawyer who has read every legal document ever written but has never practiced law, never spoken with a client, and never learned which statements in those documents are true versus false. This lawyer can produce text that sounds exactly like a brief, a memo, or a contract, because they have seen millions of examples. But they have no way to verify whether the cases they cite exist, whether the facts they state are accurate, or whether the analysis they provide reflects current law. They are pattern-matching, not practicing.

That analogy is imperfect, but it captures something essential. The model's outputs sound authoritative because the model learned from authoritative-sounding text. The confidence in its tone reflects the confidence in its training data, not the accuracy of its claims. A fabricated case citation is delivered with the same fluency as a real one, because the model has no mechanism for distinguishing between them. It simply produces text that fits the pattern of what a case citation looks like.

This explains why generative AI excels at certain tasks and struggles with others. Tasks that depend on pattern replication work well: drafting a standard contract clause, converting a document from one format to another, writing in a particular style or tone. The patterns are clear, the training data is abundant, and deviations from accuracy are either minor or easily caught. Tasks that depend on factual accuracy or logical reasoning work less well: answering a novel legal question, verifying whether a statement is true, ensuring that an argument actually follows from its premises. These tasks require something beyond pattern prediction, and the model does not have that something.

> **Key Point** An LLM predicts plausible text based on patterns. It does not retrieve facts, verify accuracy, or reason logically. Outputs that look authoritative may still be wrong, because the model optimizes for plausibility, not truth.

The "large" in large language model refers to scale, both in training data and in model architecture. Current models are trained on enormous corpora—by recent estimates, trillions of words drawn from books, websites, articles, legal documents, and countless other sources. The models themselves contain billions of parameters, which are numerical values adjusted during training to improve prediction accuracy. The training corpus is much larger than the model that results from it; the model is a compressed representation of patterns found across the corpus, not a copy of it. This scale is what enables the models to produce coherent, contextually appropriate text across an extraordinary range of topics. But scale does not solve the fundamental limitation. A model trained on more data is better at predicting plausible text. It is not necessarily better at predicting true text. Plausibility and truth are different properties, and the training process optimizes for one, not the other.

Lawyers sometimes ask whether these models "understand" language. The question is contested among researchers, and the answer depends on what you mean by understanding. The models can perform tasks that seem to require understanding: following complex instructions, maintaining coherence across long passages, adjusting tone based on context. But they lack what we might call grounded understanding. They do not know what a courtroom looks like, what it feels like to lose a case, or what a client is actually worried about when they ask whether they will be okay. They manipulate symbols according to patterns. Whether that constitutes understanding is a philosophical question. For practical purposes, what matters is that the models behave as if they understand while lacking the verification mechanisms that human understanding provides.

One more distinction matters for legal work. Some AI tools combine large language models with other systems, such as databases of legal authorities or real-time search capabilities. These hybrid tools can partially address the accuracy problem by grounding the model's outputs in verified sources. Later chapters discuss these tools in detail. For now, recognize that a pure language model, the kind you access through a general-purpose chatbot, has no external verification built in. What you see is what the model predicted, nothing more. Any verification must come from you.

Understanding what an LLM is helps you understand what it is not. It is not a legal research platform with a conversational interface. It is not a junior associate with a photographic memory. It is not a calculator that produces reliable outputs given valid inputs. It is a sophisticated text predictor that generates plausible content, some of which happens to be accurate and some of which does not. Your job, as a supervising lawyer, is to tell the difference.

2.2 Tokens, Context Limits, and Why Details Disappear

When you paste a twenty-page contract into an AI tool and ask for a summary, something has to happen to that text before the model can process it. The model does not read words the way you do. It converts your input into tokens, which are chunks of text that the model treats as single units. Understanding tokens explains why AI tools sometimes lose track of details you provided, why long documents cause problems, and why you occasionally get responses that seem to ignore half of what you wrote.

A token is roughly three-quarters of a word in English, though the exact ratio varies. Common words like "the" or "contract" are usually single tokens. Longer or less common words get split into multiple tokens — for example, "interpretation" might be processed as three tokens: "inter", "pret", and "ation". This matters because models have limits on how many tokens they can process at once, and those limits constrain what you can do.

The context window is the total number of tokens a model can hold in working memory during a single interaction. Think of it as the model's attention span. Everything in the context window, your prompt, any documents you have provided, the conversation history, and the model's own response as it generates, must fit within this limit. Current models have context windows ranging from tens of thousands to over a million tokens, depending on the specific tool. That sounds like a lot, and for many tasks it is. But legal work regularly involves documents that push or exceed these limits.

Here is where problems emerge. When you provide more text than the context window can hold, something has to give. Different tools handle this differently. Some truncate your input silently, processing only the first portion and ignoring the rest. Some summarize or compress the input before processing. Some refuse to proceed and ask you to shorten your submission. The worst outcome is silent truncation, because you will not know that the model never saw the second half of the contract you asked it to review. Its analysis will be incomplete, but the output will not announce that incompleteness. You will get a confident summary that covers only the material the model actually processed.

Even within the context window, not all tokens receive equal attention. Models use mechanisms that allocate processing resources across the input, and those mechanisms can cause the model to focus on some portions while effectively ignoring others. Long documents are particularly vulnerable. The model might attend carefully to the beginning and end while skimming the middle. Important details buried on page fifteen of a twenty-page document may not influence the output as much as details on page one or page twenty. This is not a bug. It is a consequence of how attention mechanisms work at scale. But the practical effect is that comprehensiveness is not guaranteed, even when your entire document fits within the context window.

> **Risk Note** When working with long documents, never assume the model processed everything with equal care. Key provisions in the middle of a lengthy contract are more likely to be overlooked than provisions at the beginning or end. Verify that critical details actually appear in the output.

Context windows also explain why AI tools seem to forget things during extended conversations. Each time you send a new message, the entire conversation history must fit within the window. As the conversation grows, older exchanges may be dropped or compressed to make room for newer ones. You might reference something you discussed ten messages ago, only to find that the model no longer has access to that context. It is not that the model forgot. It is that the earlier exchange no longer fits in the window. The model is working with an incomplete record while appearing to engage in a continuous conversation.

For legal work, these constraints have practical implications. If you are reviewing a contract, consider breaking it into sections and analyzing each section separately rather than submitting the entire document at once. If you are having a long conversation about a complex matter, periodically restate key facts and context rather than assuming the model remembers everything. If you are asking the model to cross-reference multiple documents, recognize that its ability to hold all of them in mind simultaneously depends on their combined length relative to the context window.

> **Practice Point** For documents longer than ten pages, work in chunks. Analyze sections individually, then ask the model to synthesize the section-level analyses. This approach ensures that every portion receives adequate attention and reduces the risk of silent truncation or attention drift.

The token limit also affects output length. The model's response consumes tokens from the same context window as your input. If you have used most of the window on a long document, the model has less room to generate a thorough response. You might ask for a comprehensive analysis and receive something abbreviated, not because the model is being lazy, but because it literally ran out of space. Some tools handle this more gracefully than others, but the underlying constraint is always present.

One more wrinkle: tokens do not map perfectly to meaning. The model processes text as a sequence of tokens, but legal meaning often depends on relationships that span many tokens. A limitation of liability clause might reference definitions from fifteen pages earlier. A representation might be qualified by an exception buried in a schedule. The model can sometimes track these relationships, especially when they are explicit and clearly signaled. But implicit relationships, the kind that experienced lawyers recognize instantly, may not register. The model is processing tokens in sequence. It is not building the mental map of the document that you build as you read.

Understanding tokens and context windows demystifies some otherwise puzzling AI behavior. When the model seems to ignore something you told it, the explanation is often mechanical: that something fell outside the window or received insufficient attention. When the model's response seems truncated or superficial, the explanation may be that your input left too little room for output. These are not failures of intelligence. They are failures of capacity. Recognizing the difference helps you design prompts and workflows that work within the tool's actual constraints rather than the constraints you wish it had.

2.3 Training vs. Retrieval: Why "Citations" Can Be Unreliable

When you ask Westlaw for cases involving breach of fiduciary duty in Delaware, the system searches a database of actual court opinions and returns results that exist. When you ask a general-purpose AI the same question, something fundamentally different happens. The model does not search anything. It generates text that resembles what a response to your question would look like, based on patterns learned during training. The cases it produces may or may not exist. The model has no way to tell, and neither do you until you check.

This distinction between retrieval and generation is the single most important concept for lawyers using AI tools. Retrieval systems look things up. Generative systems make things up. That is an overstatement, since generative systems often produce accurate information, but it captures the essential difference. A retrieval system can only return what is in its database. A generative system can produce anything that fits the pattern, regardless of whether it corresponds to reality.

Training is the process by which the model learned its patterns. Months before you ever typed a prompt, the model processed billions of documents, including legal opinions, law review articles, briefs, and countless other texts. It learned what legal citations look like: a case name, a volume number, a reporter abbreviation, a page number, a court and year in parentheses. It learned what judicial opinions contain: procedural history, statements of fact, legal analysis, holdings. It learned the rhythm and structure of legal writing so thoroughly that it can produce text indistinguishable in form from the real thing.

But training did not teach the model which citations are real. The model learned patterns, not facts. It knows that "Smith v. Jones, 542 F.3d 867 (9th Cir. 2008)" looks like a valid citation, because it has the right structure. It does not know whether that citation corresponds to an actual opinion in the Federal Reporter. When you ask for cases on a topic, the model generates citations that fit the pattern of citations it has seen. Some of those generated citations happen to match real cases. Others are fabrications that look identical to real citations but point to nothing.

> **Risk Note** Never cite a case in any document based solely on AI output. Every citation must be verified against an authoritative source such as Westlaw, Lexis, or the court's own records. This is not optional caution. It is the minimum standard of professional competence. *Mata v. Avianca*, discussed in Chapter 1, showed what happens when lawyers skip this step.

The problem is worse than random error. Hallucinated citations are not obviously wrong. They do not contain typos or formatting mistakes that would signal fabrication. They often include plausible party names, appropriate reporters for the jurisdiction you asked about, and realistic date ranges. Some hallucinated citations even come with parenthetical descriptions of holdings that sound substantively correct for the legal issue you raised. The model is very good at generating text that looks right. It has no mechanism for ensuring that the text is right.

Consider what happens when you ask the model to find cases supporting a particular proposition. The model generates text predicting what a helpful response would look like. A helpful response would include case citations with supporting holdings. So the model produces citations and holdings that fit the pattern of supporting authorities. If real cases exist in its training data that match your request, the model may reproduce them accurately. If no perfect match exists, the model does not return an empty result the way a database search would. It generates something anyway, because that is what a helpful response looks like. The fabrication is not intentional. It is an artifact of a system optimized to produce plausible responses rather than verified ones.

The training data itself introduces another layer of unreliability. Models are trained on snapshots of text from a particular period. The training data has a cutoff date, after which the model knows nothing. Cases decided after that cutoff do not exist in the model's world. Law that changed after the cutoff is invisible. If you ask about a legal question where the controlling authority was decided recently, the model may give you an answer based on outdated law without any indication that the law has changed. The model does not know what it does not know.

Even within the training period, coverage is uneven. The model learned more about topics that appeared frequently in its training data and less about topics that appeared rarely. Federal circuit court opinions are well represented. Unpublished state trial court decisions are not. Mainstream legal doctrines have abundant training examples. Niche regulatory matters have fewer. The model's confidence does not track these coverage gaps. It produces authoritative-sounding text regardless of whether the underlying topic was well represented in training.

Authority-connected legal research tools improve this picture because they retrieve from verified databases rather than generating from training patterns alone. But even then, you still must verify what the cases say and whether the tool's characterization is accurate. The presence of retrieval reduces the risk of invented citations. It does not remove the need for legal judgment.

2.4 Common Failure Modes: Hallucinations, Omissions, Anchoring

Hallucinated citations get the headlines, but they are only one way AI output goes wrong. Understanding the full catalog of failure modes helps you design verification processes that catch problems before they cause harm. Each failure mode has a pattern, and once you learn to recognize these patterns, you will spot them more reliably in your own review. The failure modes in this chapter map onto the task types from Chapter 1: generate tasks are most vulnerable to hallucinations and anchoring, while extract and classify tasks are more vulnerable to omissions and attention drift.

Hallucinations extend beyond citations. The model can fabricate any kind of factual content. It might invent a statutory provision, create a regulatory requirement that does not exist, attribute a statement to a party who never said it, or manufacture procedural history for a case. The mechanism is always the same: the model generates text that fits the pattern of what accurate information would look like, without verifying that the information is actually accurate.

Omissions are the inverse of hallucinations: important information that should appear in the output but does not. You ask for a summary of a contract's termination provisions, and the model covers termination for convenience but misses termination for cause. You request an analysis of claims your client might bring, and the model discusses three theories while overlooking a fourth that might be the strongest. Omissions are dangerous precisely because you do not see what is missing. The output looks complete. It reads as thorough. Only someone who already knows what should be there can recognize what is not.

> **Risk Note** Comprehensive-sounding output is not necessarily comprehensive. When completeness matters, verify against a checklist, a source document, or your own independent analysis. Never assume the model covered everything just because what it did cover seems thorough.

Anchoring occurs when the model locks onto an early interpretation or approach and fails to consider alternatives. You provide a set of facts and ask for legal analysis. The model identifies the first plausible theory, develops that theory thoroughly, and presents it as the answer. Alternative theories that might be stronger, or that might represent what opposing counsel will argue, do not appear.

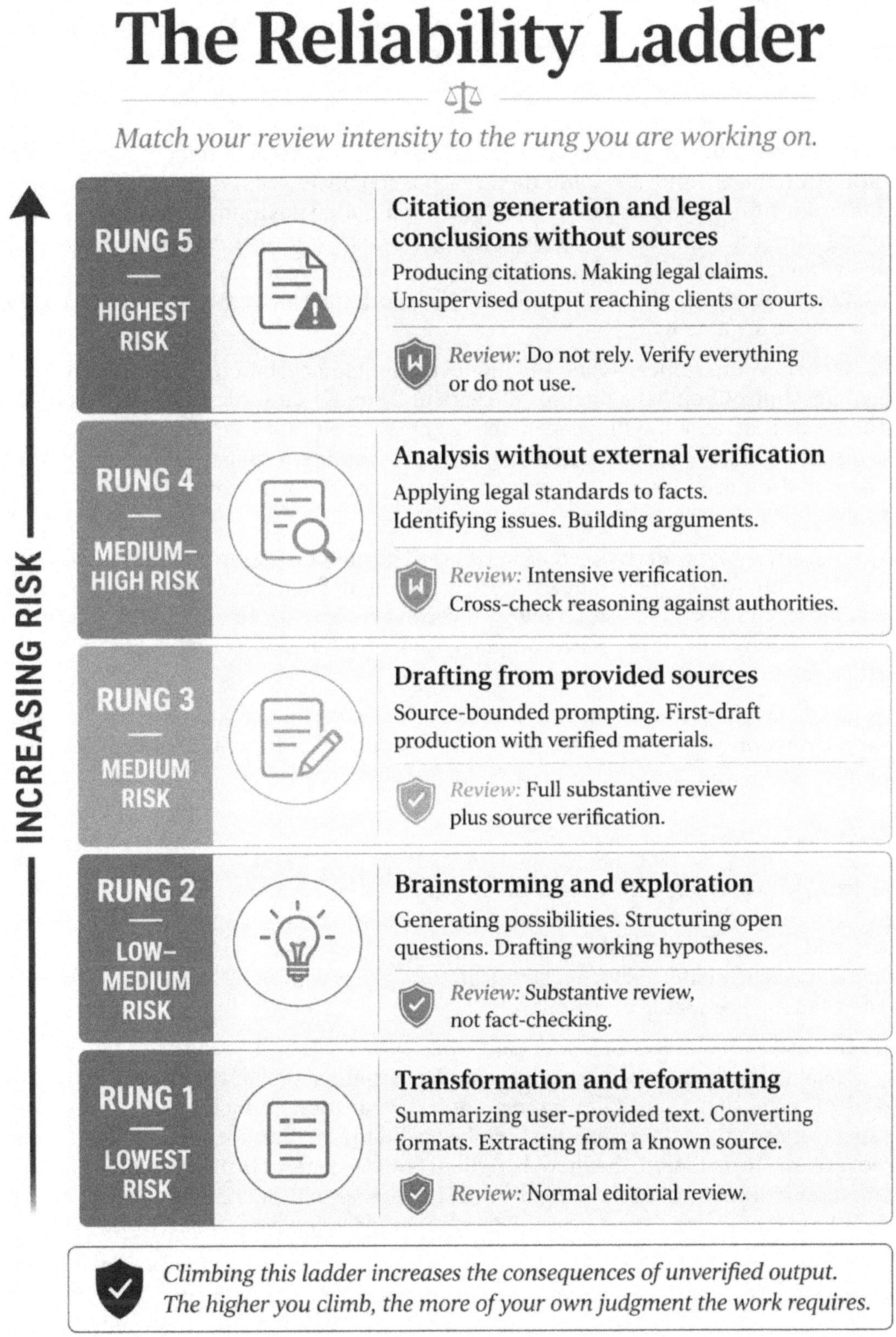

Figure 2.1 — The Reliability Ladder

The model is not suppressing alternatives deliberately. It is following a path of sequential prediction, where early outputs constrain later ones. Once the model commits to a direction, the patterns for continuing in that direction are stronger than the patterns for stopping, reconsidering, and going a different way.

Inconsistent reasoning appears when the model's conclusions do not follow from its analysis, or when different parts of the output contradict each other. The model might state a legal standard correctly, apply it to facts accurately, and then reach a conclusion that does not match the analysis. Or it might take one position in paragraph three and a conflicting position in paragraph seven without acknowledging the tension. These inconsistencies occur because the model generates text sequentially without maintaining a global check for logical coherence. Each sentence is optimized to follow plausibly from the preceding sentences, but nothing ensures that the full output hangs together as a unified argument.

Overconfident conclusions appear when the model presents uncertain or contested propositions as settled. Legal questions often have degrees of certainty. Some issues are black-letter law with clear answers. Others are unsettled, with reasonable arguments on multiple sides. The model does not track these distinctions well. It generates text in the same confident register regardless of underlying certainty. A well-established rule and a speculative prediction receive the same authoritative treatment.

Temporal confusion is another risk in legal work. The model might conflate current law with superseded law, or fail to recognize that a statute was amended or a case was overruled. Jurisdiction blending is related. You ask about Texas law and receive an answer blending Texas principles with California law and general common-law language. The answer may sound coherent while quietly mixing legal regimes that should never be mixed.

Use a checklist when you review AI output. Look specifically for hallucinations, omissions, anchoring, inconsistency, overconfidence, temporal confusion, and jurisdiction blending. Reading "for general quality" is not enough.

2.5 Bias and Uneven Performance Across Topics

The model learned from what it read. What it read was not a neutral cross-section of all legal knowledge. Training data overrepresents some jurisdictions, some practice areas, some perspectives, and some types of legal writing. This unevenness affects output quality in ways that are not always obvious. The model performs better on topics that were well represented in training and worse on topics that were sparse or absent.

Federal law and the law of large states like California, New York, and Texas are heavily represented in publicly available legal materials. State-specific law from smaller jurisdictions is less abundant. The practical effect is that the model handles federal questions and large-state questions more reliably than questions about Wyoming water rights or Rhode Island zoning law. When you ask about an underrepresented jurisdiction, the model might default to general principles, borrow from better-represented jurisdictions without flagging the substitution, or simply produce less accurate output. The confident tone remains the same regardless of the underlying data quality.

Practice area coverage is similarly uneven. Corporate transactions, commercial litigation, and intellectual property are well documented in public sources. Family law, immigration, public benefits, and criminal defense have different documentation patterns and may be less heavily represented in the training data. The model can still produce text about these areas, but the depth and accuracy may not match what you would get for a mainstream corporate question.

> Practice Point A useful rule of thumb is this: if a question would be easy for a generalist lawyer to research from public sources, the model is more likely to perform well. If the question depends on niche, local, or underrepresented materials, verification must be more rigorous.

Bias also appears in substantive framing. Legal materials reflect the perspectives of their authors, and some perspectives appear more frequently than others. Plaintiffs' briefs and defendants' briefs may not appear in equal proportions. Judicial opinions reflect the reasoning styles of different courts. The model absorbs all of this, producing output that reflects a weighted average of the perspectives it encountered. That average is not neutral, even if it is not intentionally skewed.

For transactional work, this might shape how the model frames risk or which party's interests it centers when drafting. For litigation or policy analysis, it may shape which arguments feel "natural" and which feel marginal. The model is not deciding what is fair. It is reproducing what is statistically common in the materials it saw.

The practical consequence is simple: the model's confidence does not tell you whether the topic was well represented or whether the framing is balanced. You must adjust your reliance based on the expected coverage of the topic and the stakes of the matter.

2.6 Reliability Ladder: Low-Risk to High-Risk Uses

The previous sections explained why AI output varies in reliability. This section provides a framework for using that understanding in practice. Not all tasks carry the same risk, and not all AI capabilities perform at the same level. Matching tasks to appropriate AI involvement requires a mental model that accounts for both dimensions. The reliability ladder provides that model.

Think of AI reliability as a spectrum rather than a binary. At one end are tasks where AI performs consistently well and errors are easy to catch. At the other end are tasks where AI performs unpredictably and errors can be catastrophic. Most legal work falls somewhere in between. Your job is to locate each task on this spectrum and calibrate your supervision accordingly.

Rung One: Internal extraction and organization. You are pulling information from documents for your own use, organizing materials, or creating internal work product that will not leave your desk without further processing. Errors here are inconvenient but often catchable in the normal course of work. *Example: extracting key dates from a deposition transcript for your own case timeline.*

Rung Two: Internal generation and transformation. You are using AI to produce first drafts or transformed content that remains internal and will receive significant human revision before anyone relies on it. *Example: converting your handwritten client interview notes into a structured intake memo you will review and revise.*

Rung Three: External routine communications. The output will reach someone outside your organization, but the content is routine and the stakes are modest. *Example: drafting a cover letter transmitting discovery responses to opposing counsel.*

Rung Four: External substantive work product. The output will reach clients, opposing counsel, or others who may rely on it, and it contains substantive legal content. *Example: a first draft of a demand letter in a breach-of-contract matter, intended for opposing counsel after your full review.*

Rung Five: External high-stakes filings and advice. The output involves court filings with major consequences, strategic opinions on which clients will make significant decisions, or advice in matters where errors could create serious harm. *Example: a motion for summary judgment in a seven-*

figure commercial dispute, or a formal opinion letter on which the client will base a major business decision.

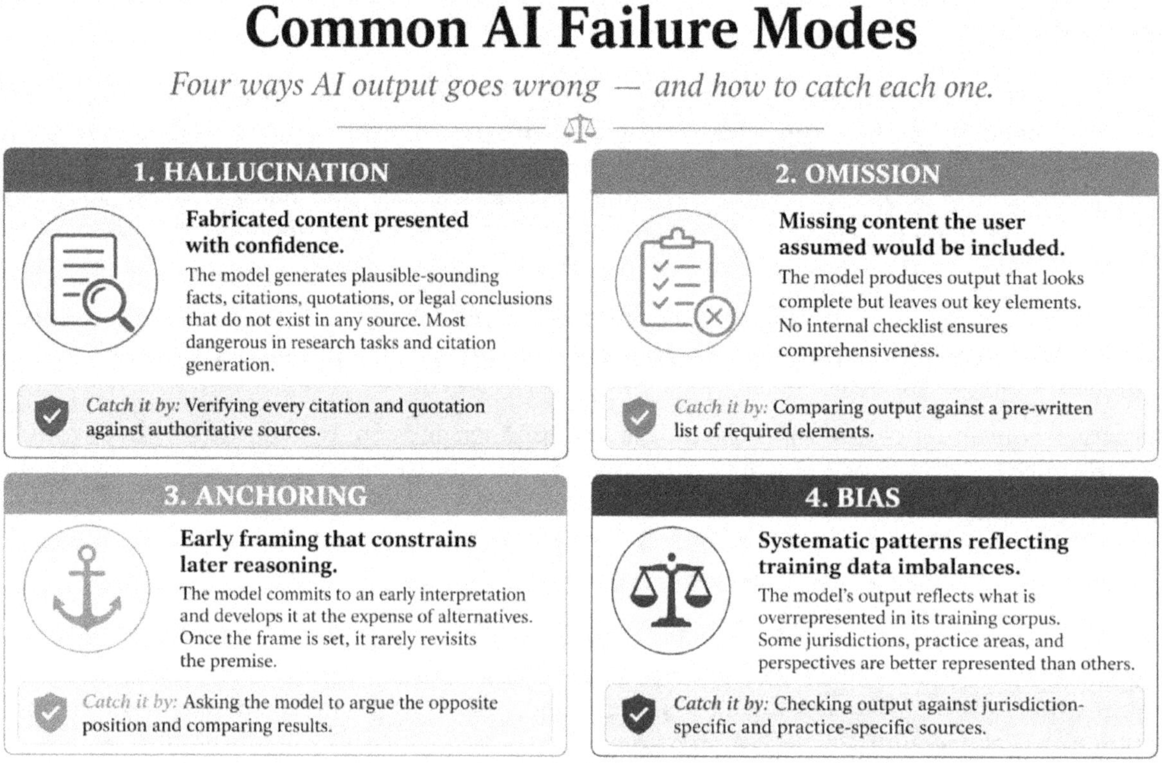

Figure 2.2 — Common AI Failure Modes

2.7 Output Triage: How to Review AI Work Competently

Verification is not optional, but neither is efficiency. If reviewing AI output takes as long as producing the work yourself, you have not gained anything. The goal is a triage process that catches errors reliably while preserving the time savings that made AI attractive in the first place. This requires structured review that allocates attention based on risk rather than reviewing everything with equal intensity.

The triage process has four sequential steps, and the failure modes from Section 2.4 run through all of them as a lens you apply at every stage. You are not checking for failure modes once as a separate step. You are carrying the list in your head as you move through structure, facts, reasoning, and tone — asking at each stage whether the specific failures most likely to surface there have surfaced.

Step 1: Check structure before substance. Before reading the output closely, scan its organization. Does it address what you asked for? Does it have the components you expected? Is anything obviously missing or obviously extraneous? Structural problems often signal deeper issues. If you asked for an analysis of three claims and the output only discusses two, you have an omission — the first failure mode to look for at this step.

Step 2: Verify hard facts and citations. This is the highest-yield verification step and the one most likely to catch catastrophic errors. Every case citation must be confirmed to exist, and every cited holding must be confirmed to say what the output claims it says. Every statutory reference must be checked against the actual text. Every factual assertion attributed to the record must be traced to its source. Hallucinations and temporal confusion are the failure modes to watch for here. This step is non-negotiable for any output going to a client, opposing counsel, or court.

Step 3: Assess reasoning and consistency. With hard facts confirmed, turn to the analytical content. Does the reasoning actually support the conclusions? Are there logical gaps where the analysis jumps from premise to conclusion without adequate connection? Does the output contradict itself? Inconsistent reasoning, anchoring, and jurisdiction blending are the failure modes most likely to surface at this step. This review requires slower reading than fact verification because you are evaluating coherence rather than simply checking sources.

Step 4: Evaluate tone and audience fit. Does the output sound like something you would actually send to this recipient? Is the level of formality right? Are explanations clear enough for the intended audience? Overconfidence is the failure mode to watch for here — output that sounds certain about things that should be hedged. Tone problems will not get you sanctioned, but they can still undermine your credibility or confuse your reader.

Verification fatigue degrades later sections. The middle and end of a document need the same attention as the beginning. If you find yourself skimming, stop and return later. Tired review is ineffective review.

Once triage is complete, decide whether to revise, regenerate, or abandon. If the output is basically sound, revise it. If it has structural flaws or repeated issues traceable to your prompt, regenerate with better instructions. If it is so far off that fixing it will take more time than starting over manually, abandon the AI-assisted approach for that task.

The point of triage is not just to catch errors. It is to preserve efficiency while catching the errors that matter most.

2.8 Mini-Lab: Spot the Failure Modes in a Sample Output

Pattern recognition is the core skill of AI output review. You cannot catch what you do not know how to look for. This Mini-Lab trains the eye by walking through a worked example — a realistic piece of AI-generated legal analysis that contains five embedded failure modes from Section 2.4. Your job is to find them. Then, in Part 2, you apply the same analysis to an AI output you generate yourself on one of your own matters.

Objective: Identify five failure modes in a worked sample AI output, then apply the same diagnostic process to an AI output generated from your own practice.

Time Required: 25 to 35 minutes

What You Will Need:

- The list of failure modes from Section 2.4: hallucination, omission, anchoring, inconsistent reasoning, overconfidence, temporal confusion, jurisdiction blending
- Access to an AI tool you can use for a quick output on a question from your practice
- A pen, highlighter, or note-taking tool

Part 1: Find the Failures in the Worked Sample (10 to 15 minutes)

The scenario: Sarah, a commercial litigation associate, asked a general-purpose AI tool for a preliminary analysis of her client's potential claims in a breach-of-contract matter. Her client, a mid-sized manufacturer, contracted with a supplier for specialized components. The supplier failed to deliver on schedule, forcing the client to source replacements at higher cost. Sarah wanted a fast overview of the strongest theories to pursue. Here is what the AI produced.

Sample AI Output:

Your client has a strong breach-of-contract claim against the supplier. Under the Uniform Commercial Code, a seller's failure to deliver conforming goods within the agreed timeframe constitutes a breach for which the buyer may recover expectation damages, including cover damages under UCC § 2-712. The controlling authority in this area is Reliant Industries v. Western Components, *487 F.3d 1129 (2d Cir. 2011), which held that cover damages are recoverable whenever the buyer acts in good faith and obtains substitute goods without unreasonable delay. The measure of damages is the difference between the cover price and the contract price, plus any incidental or consequential damages. This is well-established law and should form the core of the complaint.*

California courts apply the same framework under Cal. Comm. Code § 2712, and New York follows suit. Your client's decision to source replacement components at a higher price is exactly the kind of commercially reasonable cover that these authorities protect. The damages calculation should be straightforward: the difference between what your client paid for the replacement goods and what they would have paid under the original contract, together with any documented costs incurred due to the delay.

The supplier will likely argue that the delay was excusable due to supply chain disruptions, but this defense is unlikely to succeed absent a force majeure clause that specifically covers the type of disruption. Absent such a clause, the failure to deliver remains a breach. Your strongest path forward is to file a complaint seeking cover damages and consequential damages, and to resist any attempt by the supplier to invoke a general hardship defense.

Now review the sample. Five of the seven failure modes from Section 2.4 are embedded in this output. Read slowly. Mark what looks suspicious. Do not proceed to the answer box below until you have made your best attempt.

Answer Box — What's in the Sample

The five failure modes embedded in this sample are:

1. **Hallucination.** The case *Reliant Industries v. Western Components*, 487 F.3d 1129 (2d Cir. 2011), does not exist. The citation has the right structure, but no such case has been decided. A real Second Circuit decision at that volume and page number addresses a different matter entirely.
2. **Anchoring.** The output fixates on cover damages under UCC § 2-712 and develops that theory thoroughly, but never considers alternative or additional theories that might be stronger for this client. A careful analysis would also evaluate whether the supplier's conduct supports a claim for breach of the implied covenant of good faith and fair dealing.
3. **Omission.** The output entirely omits a potentially important theory: if the supplier knew about the client's downstream contracts and still failed to deliver, there may be a tortious interference claim against the supplier's own competing customers (if the supplier diverted inventory), or a promissory estoppel claim if the client relied on specific delivery assurances. The analysis covers one theory well and leaves others entirely invisible.
4. **Overconfidence.** The output states that the force majeure defense is "unlikely to succeed absent a force majeure clause that specifically covers the type of disruption." This is presented as settled, but post-2020 supply chain litigation has produced inconsistent

results on exactly this question, with some courts finding implicit excuse doctrines applicable and others rejecting them. The question is contested, not closed.

5. **Jurisdiction blending.** The output says "California courts apply the same framework under Cal. Comm. Code § 2712, and New York follows suit." This conflates three distinct legal regimes (UCC as adopted by the particular state, California's specific statutory variations, and New York's interpretive case law) and presents them as interchangeable. The matter is in one jurisdiction, and the output should identify that jurisdiction and stay within it, not blend principles across states.

Two failure modes from Section 2.4 are NOT in this sample: temporal confusion and inconsistent reasoning. A real AI output rarely contains every possible failure mode. Part of your job as a reviewer is knowing which failures are likely in which contexts.

Part 2: Apply the Same Analysis to Your Own AI Output (15 to 20 minutes)

Now take what you just practiced and apply it to your own work. Pick a legal question you currently need to answer — a research question, a clause analysis, a strategy assessment, anything that involves legal judgment. Enter it into an AI tool you have access to and ask for a preliminary analysis. The tool does not matter; any general-purpose AI will work for this exercise.

When the output comes back, do not read it for substance first. Read it for failure modes.

Walk through the seven categories from Section 2.4 in order:

1. **Hallucination.** Does every cited case, statute, or authority actually exist? Check the citations.
2. **Omission.** Are there theories, defenses, or issues you know are relevant that the output does not mention?
3. **Anchoring.** Did the output lock onto one interpretation and develop it at the expense of alternatives?
4. **Inconsistent reasoning.** Do the conclusions actually follow from the analysis? Does anything contradict itself?
5. **Overconfidence.** Are any contested propositions presented as settled?
6. **Temporal confusion.** Does the output reference any law that has been amended, superseded, or overruled since the training data cutoff?
7. **Jurisdiction blending.** Does the output mix principles from multiple jurisdictions without identifying which applies?

Mark every failure you find. For each one, write one sentence about why it would be a problem if you had taken the output at face value.

Deliverable

At the end of this Mini-Lab, you should have:

1. **Your answers on the worked sample** — five failure modes identified, with a short note on how you recognized each one.
2. **A marked-up copy of your own AI output** — every failure mode identified, with notes on practical consequence.

3. **A short pattern recognition log** — two or three sentences on which failure modes were easiest and hardest for you to spot, and what that tells you about where your review attention needs to go in future work.

Keep the log. Over time, the failure modes that are hardest to spot in your own review are the ones most likely to reach your work product. Knowing your own blind spots is how you close them.

Bonus Challenge

Select one failure mode that did NOT appear in your own output — temporal confusion or inconsistent reasoning are good candidates if the sample above did not surface them — and deliberately try to generate an example of it by prompting the AI in a way that is likely to trigger that specific failure. For temporal confusion, ask about a legal rule you know was recently amended. For inconsistent reasoning, ask a complex multi-part question and watch whether the parts of the answer actually cohere.

Producing failures on demand is how you build intuition for when they will appear uninvited.

Key Takeaways

1. Large language models predict plausible text, not accurate text. The model generates responses by pattern-matching against training data, not by retrieving verified information or reasoning logically.
2. Context windows limit what the model can process and remember. Long documents may be truncated or processed unevenly, and extended conversations lose earlier context.
3. Training is not retrieval, which is why citations are unreliable when produced by general-purpose AI. Every citation requires independent verification.
4. Several recurring failure modes appear in AI output: hallucinations, omissions, anchoring, inconsistency, overconfidence, temporal confusion, and jurisdiction blending.
5. Bias and coverage gaps make performance uneven across jurisdictions, practice areas, and perspectives.
6. The reliability ladder helps match AI tasks to supervision levels.
7. Triage lets you review AI output efficiently by allocating attention where the risk is greatest.

Chapter Checkpoint

Before moving on, make sure you can:

☐ Explain why an AI can produce a confident, well-formatted citation to a case that does not exist

☐ Identify situations in your practice where context window limitations might cause problems

☐ Review AI-generated analysis and actively check for specific failure modes

☐ Place an AI-assisted task on the reliability ladder and describe the review it requires

What's Next

You now understand what generative AI does, why it fails in predictable ways, and how to calibrate your review based on task type and risk level. The missing piece is how to communicate with these tools effectively. Chapter 3 addresses prompt engineering for legal work: how to structure your instructions, provide context, specify outputs, and build verification into the prompting process itself. Good prompts do not eliminate the need for verification, but they dramatically improve the quality of what you are verifying. The triage process from this chapter becomes faster and more productive when it operates on output generated from well-constructed prompts.

Key Terms

Large language model (LLM) — A type of AI trained on massive text datasets to predict what words should follow a given input. It generates new text by repeatedly predicting the most likely next word from statistical patterns, not by retrieving stored facts or applying logical reasoning.

Context window — The maximum amount of text that an AI model can process in a single interaction.

Hallucination — Output that is fabricated rather than grounded in fact.

Omission — A failure mode in which important information that should appear in AI output is missing, even though the output otherwise looks complete. Omissions are especially dangerous because the reader cannot see what was left out.

Anchoring — The model's tendency to commit to an early interpretation and develop it at the expense of alternatives.

Inconsistent reasoning — A failure mode in which the model's conclusions do not follow from its analysis, or in which different parts of the output contradict each other. It occurs because the model generates text sequentially without maintaining a global check for logical coherence.

Overconfidence — A failure mode in which the model presents uncertain or contested propositions in the same confident register as well-established rules. The tone of the output does not track the underlying certainty of the content.

Temporal confusion — A failure mode in which the model conflates current law with superseded law, or fails to recognize that a statute was amended or a case was overruled. The model's training data has a cutoff, and it does not reliably know what it does not know.

Jurisdiction blending — A failure mode in which the model's output mixes legal principles from different jurisdictions without flagging the substitution. The answer may sound coherent while quietly blending legal regimes that should never be mixed.

Training bias — Systematic patterns in AI output reflecting imbalances in training data.

Reliability ladder — A framework for matching AI tasks to appropriate supervision levels.

Triage (in verification) — Allocating review effort based on consequence.

Chapter 3: Prompt Engineering for Legal Work

The difference between useful AI output and useless AI output is rarely the tool. It's the instruction. Two lawyers using the same model on the same task will get dramatically different results based on how they frame their requests. One gets a generic, hedge-filled response that requires complete rewriting. The other gets a focused first draft that needs only targeted revision. The gap between them is prompt engineering, and it's the most leveraged skill you can develop for AI-assisted legal work.

Chapters 1 and 2 focused on what happens after you receive AI output: how to assess it, verify it, and decide whether to use it. This chapter focuses on what happens before. The quality of your prompt determines the quality of your starting material. Better prompts produce output that's closer to usable, which means less verification burden and faster time to finished work product. Prompting well doesn't eliminate the need for supervision, but it makes supervision more productive.

What This Chapter Covers

- A reusable prompt structure that works across legal task types
- How to set roles, audiences, and tone to calibrate output for your specific needs
- Techniques for packaging context so the model has what it needs without confidentiality exposure
- How to specify output format, structure, and length to get deliverable-ready drafts
- The prompt patterns that experienced legal AI users rely on repeatedly
- How to iterate and debug when your first prompt doesn't produce what you need
- Methods for building verification into the prompting process itself
- Source-bounded prompting that constrains the model to authorities you provide

Chapter Summary

Prompt engineering is the practice of structuring AI instructions to produce better output. For lawyers, this means learning to communicate with precision about role, context, task, format, and constraints. The Prompt Blueprint introduced in this chapter provides a reusable template: define who the AI should be, give it the information it needs, specify exactly what to produce, describe the desired format, and establish guardrails for what it should avoid. Each element of the blueprint addresses a common failure mode. Role instructions prevent generic output. Context packaging prevents hallucinated facts. Output specifications prevent structural mismatches. Guardrails prevent overreach. Beyond the blueprint, this chapter covers iteration strategies for when prompts don't work on the first try, verification loops that make the AI check its own work, and source-bounded prompting that constrains output to authorities you control. The goal is not to become a prompt technician. The goal is to communicate clearly enough that the AI produces output worth supervising.

3.1 The Prompt Blueprint (Reusable Template)

Most prompts fail for predictable reasons. They're vague about what they want. They omit context the model needs. They don't specify format, leaving the model to guess. They fail to exclude unwanted content, so the output includes material that requires deletion. These failures aren't about the model's limitations. They're about communication gaps between you and the tool.

The Prompt Blueprint addresses these gaps with a five-element structure that works across legal task types, as illustrated in Figure 3.1. Think of it as a checklist for prompt construction. Not every prompt needs all five elements at full intensity, but considering each element before you submit prevents the most common failure modes.

The five elements are Role, Context, Task, Format, and Guardrails. The blueprint applies across the task types introduced in Chapter 1 — generate, transform, extract, classify, and critique — though different elements carry more weight for different task types. Generation tasks rely heavily on role and guardrails. Extraction tasks rely heavily on context and format. Critique tasks rely heavily on role and task specification. As you learn each element, notice which task types it most affects.

Role tells the model who it should be when responding. Without role instructions, the model defaults to a generic helpful assistant, producing output calibrated for a general audience with no particular expertise or perspective. With role instructions, the model adjusts its vocabulary, depth, assumptions, and framing to match the specified persona. "You are a senior litigation associate at an AmLaw 100 firm" produces different output than "You are a paralegal summarizing documents for attorney review," even for identical underlying tasks.

Role instructions work because the model learned from text written by people in different roles. It has patterns for how senior associates write, how partners communicate, how judges reason, how opposing counsel argues. By specifying a role, you activate the relevant patterns and suppress the irrelevant ones. The output becomes more targeted because the model is drawing from a narrower, more appropriate slice of its training.

Context provides the information the model needs to produce accurate, relevant output. This includes facts about your matter, the jurisdiction you're working in, the procedural posture, the relevant documents, and any constraints that shape the analysis. Without adequate context, the model fills gaps with assumptions that may not match your situation. With adequate context, the model can tailor its output to the actual circumstances rather than generic possibilities.

Context packaging requires judgment about what to include and how to include it. Too little context produces generic output. Too much context can exceed token limits, dilute focus, or create confidentiality exposure. The goal is sufficiency, not completeness. Include what the model needs for this specific task, presented in a form the model can use effectively. Section 3.3 addresses context packaging techniques in detail.

Task specifies exactly what you want the model to produce. This is the core of your request, but it's often the element lawyers state too vaguely. "Analyze this contract" is a task instruction, but it doesn't tell the model what kind of analysis, for what purpose, or at what depth. "Identify all provisions that allocate risk between the parties, explain how each provision allocates risk, and flag any provisions that are unusually favorable to one side" is a task instruction that produces actionable output.

Effective task instructions use specific verbs: identify, draft, compare, summarize, extract, revise, critique, explain, list, outline. Vague verbs like "review" or "look at" invite the model to guess what you actually want. Specific verbs constrain the output to the type of work product you need. The task element should make clear what a successful response looks like, so the model knows when it has met your request.

Format tells the model how to structure its output. Do you want a memo with headings, or flowing prose? A numbered list, or a narrative explanation? A table comparing options, or a recommendation with supporting reasoning? How long should the response be? Should it include citations in a particular format? Should it flag items for follow-up with specific markers?

The Prompt Blueprint

Five elements of an effective legal prompt

1. ROLE

Who the AI should be

Example: "You are an experienced transactional attorney reviewing a software licensing agreement for a mid-sized healthcare company."

2. CONTEXT

What the AI needs to know

Facts. Record. Constraints. Definitions. Governing law. Client posture. The information the model cannot guess on its own.

3. TASK

What the AI should produce

The specific deliverable. Scope. Depth. What "done" looks like. Clarity here prevents output that is adjacent to what you wanted.

4. FORMAT

How the output should be structured

Length. Organization. Headings. Tone. Numbered or prose. Without format instructions, the model guesses — often badly.

5. GUARDRAILS

What the AI should not do

Confidentiality limits. Scope exclusions. What to avoid. Missing guardrails produce predictable failure modes.

Every effective prompt addresses all five elements. Missing elements produce predictable failure modes.

Figure 3.1 — The Prompt Blueprint

> **Practice Point** When format isn't specified, the model defaults to whatever structure seems most natural given the task and the patterns in its training data. That default may not match what you need. A five-minute investment in specifying format can save thirty minutes of reformatting after you receive the output.

Format instructions also manage length. If you need a one-paragraph summary, say so. If you need comprehensive analysis regardless of length, say that instead. The model has no way to know your length expectations unless you state them. Requests for "brief" or "concise" output are too vague; different people interpret these words differently, and so does the model. Specific length guidance works better: "approximately 500 words," "no more than two pages," "one paragraph per issue."

Guardrails specify what the model should not do. This element is often omitted, but it addresses failure modes that the other elements don't reach. Guardrails might prohibit the model from citing cases (when you'll supply the authorities yourself), from making assumptions about facts not provided (when you need only analysis of stated facts), from giving legal advice (when you want information without recommendations), or from discussing certain topics (when confidentiality or relevance concerns apply).

Guardrails are particularly important for generation tasks, where the model has the most latitude. Without guardrails, the model may include content you didn't ask for, make assumptions you didn't authorize, or stray into territory that creates problems. Explicit prohibition of unwanted behaviors is more effective than hoping the model will exercise appropriate restraint on its own.

The blueprint works as an integrated system. Role shapes how the model interprets context. Context informs what's possible for the task. Task determines what format makes sense. Format affects whether guardrails are needed. When all five elements align, the model has clear direction and produces focused output. When elements conflict or gaps exist, the model guesses, and its guesses may not match your needs.

PROMPT IN ACTION: The Blueprint Applied

Task: Analyze a force majeure clause from the client's perspective

ROLE: You are a senior transactional associate reviewing a vendor agreement for a technology company client.

CONTEXT: Our client is the customer under this agreement. The vendor provides cloud hosting services critical to our client's operations. The agreement is governed by Delaware law. I've pasted the force majeure clause below.

TASK: Analyze this force majeure clause and identify: (1) what events qualify as force majeure, (2) what obligations are excused during a force majeure event, (3) what notice or procedural requirements apply, and (4) any provisions that are unfavorable to our client as the customer.

FORMAT: Provide your analysis in four numbered sections corresponding to the four items above. For item 4, include specific recommended revisions to address each unfavorable provision. Total length should be approximately 600–800 words.

GUARDRAILS: Do not analyze any provisions other than the force majeure clause. Do not assume facts about the parties' negotiations or intentions. Flag any ambiguities rather than resolving them with assumptions.

[Paste force majeure clause here]

Why It Works: Every blueprint element is present. Role establishes perspective. Context provides party position, governing law, and the specific clause. Task breaks the analysis into concrete components. Format specifies structure and length. Guardrails prevent scope creep and unwarranted assumptions.

Not every prompt needs this level of detail. For simple tasks with low stakes, a shorter prompt may suffice. But when output quality matters, the blueprint pays dividends. The time you spend constructing the prompt reduces the time you spend fixing the output. As you develop fluency, blueprint thinking becomes automatic, and you'll construct well-structured prompts without consciously working through each element.

The blueprint is also a diagnostic tool. When output disappoints, review your prompt against the blueprint. Which element was weak or missing? If the output was too generic, the role instruction may have been inadequate. If the output contained incorrect assumptions, the context may have been insufficient. If the output was disorganized, the format may have been unspecified. If the output included unwanted content, guardrails were needed. The blueprint helps you identify what to fix for the next iteration.

> **Key Point** The Prompt Blueprint has five elements: Role (who the AI should be), Context (information it needs), Task (what to produce), Format (how to structure it), and Guardrails (what to avoid). Using all five elements consistently produces better output than ad hoc prompting. When output disappoints, diagnose against the blueprint to identify what to improve.

Appendix A contains a copy-paste version of the blueprint template along with element-by-element guidance for quick reference. As you work through the remaining sections of this chapter, each section will develop one or more blueprint elements in greater depth, with examples tailored to legal work.

The next section examines the Role element in detail: how to set roles, audiences, and tone to calibrate output for different purposes and recipients.

3.2 Roles, Audiences, and Tone Controls

The first element of the Prompt Blueprint tells the model who to be. This might seem like a formality, but role instructions fundamentally shape output quality. They determine what expertise the model assumes, what vocabulary it uses, what level of detail it provides, and what perspective it brings to the analysis. Getting the role right is often the difference between output you can use and output you have to rewrite.

Role instructions work by activating specific patterns in the model's training. The model learned from millions of documents written by people in different professional roles. It absorbed how senior partners communicate differently from junior associates, how litigators frame arguments differently from transactional lawyers, how in-house counsel approach risk differently from outside counsel. When you specify a role, you're telling the model which patterns to draw from. The output reflects the conventions, assumptions, and communication style associated with that role.

Consider the difference between these two prompts for the same task:

Without role: "Analyze whether the non-compete clause is enforceable."

With role: "You are a senior employment litigation partner advising a corporate client who wants to enforce a non-compete against a departing executive. Analyze whether the non-compete clause is enforceable."

The first prompt might produce a generic legal encyclopedia entry covering enforceability factors in the abstract. The second prompt produces analysis oriented toward a specific client objective, with attention to the practical considerations that matter for enforcement litigation. The underlying legal principles are the same, but the framing, emphasis, and actionability differ significantly.

Effective roles for legal prompts share certain characteristics. They're specific enough to activate useful patterns without being so narrow that the model has limited training examples to draw from. "Senior litigation associate" works better than either "lawyer" (too broad) or "third-year associate at a 50-attorney regional firm specializing in ERISA litigation" (so narrow that the model may not have enough matching patterns).

Roles can specify experience level, practice area, institutional context, and task orientation. A prompt might use "junior associate conducting initial research," which produces thorough, exploratory output that surfaces multiple angles rather than jumping to conclusions. Alternatively, "partner preparing for a client meeting" produces more concise, conclusory output focused on practical recommendations. The experience level signals how much hand-holding versus confidence the output should reflect.

Match the role to your actual need, not to flattery. Asking the model to be a "world-renowned expert" doesn't make the output more accurate. It may make the output more confident, which is dangerous when the underlying analysis is uncertain. Use roles that produce appropriate calibration between confidence and nuance.

Audience instructions complement role instructions by specifying who will receive the output. The same analysis should be communicated differently depending on whether the reader is a sophisticated general counsel, a business executive with no legal background, a federal judge, or opposing counsel. Audience instructions tell the model what to assume about the reader's knowledge, what terminology requires explanation, and what level of formality is appropriate.

Audience affects more than vocabulary. It affects structure, emphasis, and what counts as a complete explanation. Analysis for a partner who will review and sign a filing can assume familiarity with procedural posture and focus on substantive arguments. Analysis for a client who needs to make a business decision should translate legal conclusions into practical implications. Analysis for a court should be formally structured, precisely cited, and carefully hedged where the law is uncertain.

Role and audience can work together or create productive tension. "You are a senior associate drafting a memorandum for a partner" produces internal work product calibrated for an expert reader. "You are a senior associate drafting an explanation for the client" produces the same substantive analysis reframed for a different reader. The role stays constant while the audience changes, and the output adjusts accordingly.

Tone controls fine-tune how the output sounds beyond what role and audience establish. Tone instructions address register, confidence level, and stylistic qualities that affect how the reader experiences the content. They're useful when you need output that departs from what the role and audience would naturally produce, or when you want to emphasize particular qualities.

Common tone instructions for legal work include:

Formal and precise, which produces careful language suitable for court filings and official documents. This tone avoids contractions, colloquialisms, and casual phrasing while emphasizing technical accuracy.

Confident but measured, which produces authoritative analysis that acknowledges uncertainty where it exists. This tone avoids both excessive hedging and overstatement, striking the balance appropriate for professional advice.

Direct and actionable, which produces recommendations-focused output that leads with conclusions and supporting reasoning rather than building slowly to a point. This tone works well for busy readers who need to know what to do.

Accessible and explanatory, which produces output that prioritizes clarity over technical precision, explaining concepts rather than assuming familiarity. This tone works well for client communications and materials for non-lawyer audiences.

Adversarial and persuasive, which produces output framed to advance a position rather than present balanced analysis. This tone works for brief drafting and argument preparation but requires guardrails to prevent overreach.

Be cautious with adversarial tone instructions. "Argue as persuasively as possible" without constraints can produce output that overstates holdings, mischaracterizes facts, or makes claims you would never sign. Pair persuasive tone instructions with guardrails requiring accuracy and flagging of weak points.

Combining role, audience, and tone in a single prompt creates layered control over output characteristics. A complete specification might read: "You are a senior litigation associate (role) preparing a summary of case developments for the general counsel of our corporate client (audience). Use a direct, actionable tone that leads with key takeaways and implications for the ongoing litigation (tone)." Each layer adds specificity, and the combination produces output more precisely calibrated than any single instruction would achieve.

When role, audience, and tone conflict, the model will attempt to reconcile them, sometimes in unexpected ways. If you specify a formal role but a casual tone, the output may be inconsistent or awkward. If you specify an expert audience but request extensive explanation, the output may feel condescending. Alignment across these instructions produces more coherent output than mixing signals that pull in different directions.

PROMPT IN ACTION: Role, Audience, and Tone Combined

Task: Summarize deposition testimony for a client update

The Prompt: ROLE: You are a senior litigation associate who has been handling discovery in a breach of contract case for the past eight months.

AUDIENCE: The client is the CEO of a mid-sized manufacturing company. She is intelligent and engaged but not a lawyer. She has limited time and wants to understand the key developments without wading through legal jargon.

TONE: Direct and accessible. Lead with the most important takeaways. Translate legal significance into business implications. Avoid unnecessary hedging while being honest about uncertainties.

TASK: Summarize the key testimony from the deposition of the defendant's operations manager, which I've pasted below. Focus on testimony relevant to our claims that the defendant failed to meet delivery deadlines and provided defective components.

FORMAT: Begin with a 2-3 sentence executive summary of the most important points. Then provide a section on delivery deadline testimony and a section on defective components testimony. Total length approximately 400-500 words.

[Paste deposition excerpts here]

Why It Works: The role establishes expertise and case familiarity. The audience specifies the reader's characteristics and constraints. The tone instruction bridges the gap between the associate's natural writing style and what this particular reader needs. The combination produces a summary calibrated for exactly this communication.

One advanced technique is using role instructions to generate adversarial perspectives. Asking the model to respond as opposing counsel helps you identify weaknesses in your arguments. Asking it to respond as a skeptical judge helps you pressure-test your legal theories. Asking it to respond as the other party in a negotiation helps you anticipate counteroffers and objections. These adversarial roles produce output you would never send to anyone but that sharpens your own thinking and preparation.

Keep a library of role, audience, and tone combinations that work for your practice. Once you've dialed in a combination that produces good output for client updates, deposition summaries, or research memos, save it for reuse. Consistency in prompting produces consistency in output quality.

Role, audience, and tone instructions have limits. They calibrate style and perspective, not accuracy. A confident role doesn't make fabricated citations real. An expert role doesn't give the model knowledge it lacks. These instructions shape how the model presents information, not whether that information is correct. Verification remains essential regardless of how well-calibrated the presentation appears.

Role instructions activate patterns associated with specific professional personas. Audience instructions calibrate output for the intended reader. Tone instructions fine-tune register, confidence, and style. Used together, these three controls produce output tailored to your specific communication needs. They shape presentation but don't affect underlying accuracy, so verification remains necessary.

The next section addresses Context, the second blueprint element: how to package facts, documents, and constraints so the model has what it needs without confidentiality exposure.

3.3 Context Packaging: Facts, Record, Constraints, Definitions

The model knows nothing about your matter until you tell it. Every fact it states, every assumption it makes, every conclusion it reaches derives from the intersection of its training data and the context you provide. When context is thin, the model fills gaps from training patterns, which means it substitutes generic assumptions for your specific circumstances. When context is rich and well-structured, the model can produce output tailored to your actual situation. The difference between useful and useless output often comes down to how you package context.

Context packaging involves decisions about what to include, how to present it, and what to protect. Each decision affects output quality, and the decisions interact. Including too little produces generic output. Including too much can exceed token limits, dilute focus, or create confidentiality exposure. Presenting information poorly can confuse the model about what matters. Protecting information through redaction or abstraction affects what the model can do with what remains. Effective context packaging navigates these tensions to give the model what it needs while managing what you risk.

What context the model needs depends on the task. A request to draft a standard contract clause might need only the clause type and governing law. A request to analyze whether a client has a viable claim needs the relevant facts, the potential legal theories, the jurisdiction, and possibly the procedural posture. A request to revise a brief needs the current draft, the arguments being made, and what's wrong with the current version. Before constructing your prompt, identify what information is actually necessary for the model to complete the task at the quality level you need.

The principle is sufficiency, not comprehensiveness. You don't need to provide everything you know about a matter. You need to provide what the model needs for this specific task. A prompt asking the model to extract key dates from a contract doesn't need the full procedural history of the litigation.

A prompt asking for strategic analysis of settlement options does. Match the context to the task rather than defaulting to either minimal or maximal disclosure.

Fact presentation should be clear, organized, and explicit about what is established versus uncertain. The model cannot distinguish between facts you know with confidence and facts you're still investigating unless you signal the difference. If you present tentative information as established, the model will treat it as established and build analysis on that foundation. If you flag uncertainties, the model can acknowledge them and adjust its confidence appropriately.

Structure helps. Rather than providing facts as a narrative paragraph, consider organizing them under headings: Parties, Key Dates, Relevant Agreements, Disputed Facts, Undisputed Facts. This structure makes it easier for the model to locate and use specific information. It also forces you to think clearly about what you actually know, which sometimes reveals gaps you hadn't noticed.

When facts are complex, consider providing them incrementally. Give the model the basic setup, ask for initial analysis, then add complications in a follow-up prompt. This approach prevents the model from being overwhelmed by complexity and allows you to see how it handles each layer. It also produces a clearer record of your interaction if you need to review the process later.

Before writing your prompt, list the facts the model needs to know for this task. Then review the list and cut anything that isn't actually necessary. Leaner fact sets produce more focused output. If you're unsure whether a fact matters, leave it out initially and add it in a follow-up if the output suggests the model needed it.

Document handling presents particular challenges. Legal work often involves long documents, and the model's context window limits how much text it can process at once. You have several options depending on document length and task requirements.

For short documents that fit comfortably within context limits, you can paste the full text into your prompt. Label the document clearly so the model knows what it's looking at: "Below is the Asset Purchase Agreement dated March 15, 2024, between Acme Corp and Beta Industries."

For longer documents, consider providing relevant excerpts rather than the full text. Identify the sections that matter for your task and paste only those, with clear labels indicating what each excerpt is and where it comes from. The model can analyze Section 8.2 (Indemnification) without seeing the entire fifty-page agreement, as long as you've identified dependencies on other sections like definitions or related provisions.

For multiple documents, provide them with clear separation and labeling. The model can lose track of which document is which if you paste several texts together without demarcation. Use formatting to distinguish documents: headers, horizontal rules, or explicit labels like "[BEGIN DOCUMENT: Employment Agreement]" and "[END DOCUMENT]."

When documents are too long even for excerpting, consider summarizing the irrelevant portions yourself and providing full text only for the portions requiring detailed analysis. You might write: "The Agreement contains standard representations and warranties in Sections 3 and 4 (not included here). The relevant provisions for this analysis are the termination clause (Section 9, pasted below) and the dispute resolution clause (Section 12, pasted below)."

> **Risk Note** When you excerpt documents, you risk omitting provisions that interact with the excerpted text. Definitions, cross-references, and generally applicable terms may affect the meaning of the specific clause you're analyzing. Before relying on analysis of excerpted text, verify that no omitted provisions change the analysis.

Jurisdiction and procedural posture are context elements that lawyers often underspecify, leading to jurisdiction blending and procedurally inappropriate analysis. State the governing law explicitly: "This contract is governed by Texas law" or "This claim would be brought in federal court in the Eastern District of Pennsylvania, applying Pennsylvania substantive law." If the procedural posture matters, state it: "We are at the motion to dismiss stage; plaintiff has not yet had discovery" or "This is an appeal from summary judgment; we're limited to the record below."

The model doesn't automatically infer jurisdiction from other contextual clues, at least not reliably. It might notice that the parties are in California and the contract mentions San Francisco, but it won't necessarily apply California law unless you instruct it to. Explicit jurisdiction instructions prevent the blending problem where the model mixes rules from multiple states into an incoherent composite.

Constraints and limitations that affect the analysis should be stated explicitly. If there's a short statute of limitations that affects claim viability, say so. If the client has a limited budget that affects what litigation strategies are realistic, include that. If there's a business relationship that makes aggressive tactics undesirable, mention it. These constraints shape what analysis is useful, and the model can't account for constraints it doesn't know about.

The same applies to internal constraints on the task itself. If you only want analysis of certain issues and not others, say so. If you want the model to assume certain facts for purposes of analysis without vouching for their accuracy, make that explicit. If you want the model to flag issues for further investigation rather than reaching conclusions, state that preference.

PROMPT IN ACTION: Structured Context Packaging

Task: Analyze potential claims arising from a failed business acquisition

ROLE: You are a senior litigation associate evaluating potential claims for a client who was the buyer in a failed acquisition.

CONTEXT:

Parties:

- Our client: Summit Manufacturing, Inc. (Buyer)
- Opposing party: Greenfield Industries, LLC (Seller)

Key Dates:

- Letter of Intent signed: January 15, 2024
- Asset Purchase Agreement executed: March 1, 2024
- Scheduled closing: April 15, 2024
- Seller refused to close: April 14, 2024

Governing Law: The APA specifies Delaware law.

Key Facts:

- Seller refused to close, citing alleged breach of Buyer's confidentiality obligations
- Buyer denies any breach and contends Seller's refusal is pretextual
- Buyer incurred 40,000 in due diligence costs and financing fees
- Buyer claims the acquisition would have generated M annual profit

Disputed Issues:

- Whether Buyer breached confidentiality provisions

- Whether Seller's refusal to close was in good faith

Relevant APA Provisions: [Pasted below]

TASK: Identify and analyze the potential claims Buyer may have against Seller. For each claim, assess the elements, likely defenses, and available damages.

FORMAT: Organize by claim type. For each claim, provide separate paragraphs on elements, defenses, and damages. Approximately 800–1000 words total.

GUARDRAILS: Do not analyze potential counterclaims Seller might assert. Do not assume facts beyond what I've provided. Flag any additional facts that would be important to know.

[Paste relevant APA provisions here]

Why It Works: Context is organized under clear headings. Key facts are distinguished from disputed issues. Governing law is explicit. The relevant document portions are identified. The model has what it needs without extraneous information.

Definitions and terminology matter more than you might expect. If your matter involves terms of art or defined terms from an agreement, the model may or may not use them consistently unless you specify. Providing a short definitions section in your context, especially for terms that have specific meanings in the documents you're analyzing, prevents the model from drifting into general usage.

This is particularly important for contract analysis, where defined terms carry precise meanings that differ from ordinary language. "Material Adverse Effect," "Knowledge," "Permitted Liens," and similar terms have definitions that control interpretation. If you want analysis that respects these definitions, include them in your context rather than assuming the model will infer them from the document excerpts.

Confidentiality considerations constrain context packaging in ways that affect output quality. Chapter 5 addresses confidentiality in depth. For now, recognize the core tension: more detailed context generally produces better output, but more detailed context also means more sensitive information entering the AI system. You must balance these considerations based on the sensitivity of the matter, the protections offered by your AI tool, and your client's expectations.

Techniques for managing confidentiality while preserving utility include anonymizing party names (using "Company A" and "Company B"), redacting specific financial figures while preserving relative relationships ("the purchase price exceeded $10 million" rather than the exact figure), and describing factual patterns abstractly rather than with identifying details. Each technique has costs. Anonymization can make output feel generic. Redaction can remove information the model needs. Abstraction can lose nuance. Choose techniques based on what the task actually requires and what information actually creates risk.

Context packaging determines whether the model has what it needs to produce useful output. Organize facts clearly, handle documents with attention to length and dependencies, specify jurisdiction and procedural posture explicitly, state constraints that affect the analysis, and define terms that carry special meanings. Balance thoroughness against focus and confidentiality against utility.

The next section addresses Format and Output Specifications: how to tell the model what structure, length, and deliverable characteristics you need so the output requires minimal reformatting.

3.4 Output Specs: Structure and Deliverable-First Prompting

When you don't specify format, the model guesses. Sometimes it guesses well, producing output structured roughly the way you would have structured it. Often it guesses poorly, producing output that requires substantial reformatting before you can use it. The time you spend restructuring AI output is time you could have saved by specifying structure in your prompt. Format instructions are low effort and high yield, yet lawyers routinely omit them.

Format encompasses several dimensions: overall structure, section organization, length, style conventions, and any special formatting requirements like citation format or document markers. Each dimension can be specified independently, and the more precisely you specify, the closer the output comes to what you actually need. Think of format instructions as a blueprint for the deliverable, not just guidance about presentation.

Deliverable-first prompting inverts the typical approach to prompt construction. Instead of starting with what you want the model to do, start with what you want to receive. Envision the final work product in detail: What does it look like? How is it organized? How long is it? What sections does it contain? What makes it useful for your purpose? Then work backward to construct a prompt that produces that deliverable.

This inversion matters because it forces clarity about your actual needs. Lawyers often prompt for "analysis" or "a memo" without having a clear picture of what they want. The vagueness transfers to the output. When you start by specifying the deliverable precisely, you discover what you actually need, and you communicate that need to the model in terms it can act on.

Structure specifications tell the model how to organize content. The simplest structure specification is an ordered list of sections: "Organize your response with these sections: Background, Legal Standard, Analysis, Conclusion." More detailed specifications describe what each section should contain: "The Background section should summarize the relevant facts in 2-3 paragraphs. The Legal Standard section should state the elements of a breach of fiduciary duty claim under Delaware law. The Analysis section should apply each element to the facts. The Conclusion section should state the overall assessment and identify any additional facts that would strengthen or weaken the claim."

For familiar work products, you can specify structure by reference: "Format this as a standard legal memorandum with Issue, Brief Answer, Facts, Discussion, and Conclusion sections." The model has seen enough legal memos to recognize this structure and produce output that follows it. For less standard deliverables, explicit section-by-section specification works better than hoping the model infers what you want.

Hierarchical structure can also be specified. If you need a memo with multiple issues, each containing multiple sub-issues, describe that hierarchy: "Organize the Discussion section by claim. For each claim, include sub-sections on Elements, Application to Facts, and Likely Defenses." The model follows hierarchical instructions reasonably well, producing output with the nested structure you specified.

Before writing your format specification, sketch the outline you want the output to follow. Write down the section headings and, for each section, a brief note about what it should contain and roughly how long it should be. Then translate that sketch into format instructions. The sketch takes two minutes and often saves twenty.

Length specifications prevent two common problems: output that's too long and requires cutting, and output that's too short and requires expansion. Both problems waste time. Explicit length guidance puts you in control of how much content the model produces.

Length can be specified in several ways. Word counts are precise: "approximately 800 words" or "between 500 and 700 words." Page references are useful for longer documents: "approximately 3-4 pages single-spaced." Paragraph counts work for shorter outputs: "respond in 2-3 paragraphs." Section-specific lengths provide granular control: "the Background section should be approximately 200 words; the Analysis section should be approximately 500 words."

Relative length instructions work when you don't have a specific target: "keep this concise, no more than one page" or "provide comprehensive analysis; length is less important than thoroughness." These instructions are less precise but still better than no length guidance at all.

When length specifications conflict with thoroughness, the model will attempt to balance them, usually by becoming more concise. If you specify 500 words but the task genuinely requires 1,000 words of analysis to cover adequately, the model will compress, often by omitting detail or skipping issues. When you notice output that seems thin relative to the task complexity, check whether your length specification was too restrictive.

Specifying very short length for complex tasks produces superficial output. The model follows your length instruction by cutting content, not by becoming more efficient. If analysis seems incomplete, try removing the length constraint or increasing the word count to see if the model produces more thorough work.

Style specifications address how the content is written, distinct from how it's organized. Style instructions might address formality ("use formal legal prose appropriate for a court filing"), citation practices ("include full Bluebook citations for all cases"), voice ("use active voice throughout"), or paragraph structure ("use topic sentences that state conclusions, followed by supporting analysis").

For legal work, citation format deserves particular attention. If you need citations in a specific format, say so. The model has patterns for Bluebook, ALWD, jurisdiction-specific formats, and informal citation styles. Without instruction, it may default to something inconsistent or inappropriate for your jurisdiction. If you don't want citations at all because you'll add them yourself, specify that too: "do not include case citations; I will add citations during revision."

Style specifications interact with role and tone instructions from the previous section. A formal role plus a formal style specification reinforces formality. Conflicting instructions produce inconsistent output. Align your style specifications with the role and tone you've already established.

Document markers and special formatting help when you need output ready for a specific use. If you want issues flagged for follow-up, specify a marker: "flag any issues requiring additional research with [RESEARCH NEEDED]." If you want recommendations distinguished from analysis, specify how: "present recommendations in bold text" or "begin each recommendation with 'Recommendation:' on its own line." If you need the output in a format compatible with another system, describe that format.

These markers are particularly useful for workflow integration. When AI output feeds into a larger process, markers help you or others identify what needs attention. A document with clear [VERIFY] tags on factual assertions is easier to review than one where you must identify verification needs yourself. A contract analysis with [NEGOTIATE] markers on problematic provisions is immediately actionable.

Why It Works: The format specification describes exactly what each section should contain and approximately how long it should be. Special markers are defined for action items and flags. Style guidance is specific to the document type. The partner receiving this output gets something immediately useful rather than something requiring restructuring.

Iterative format refinement improves your specifications over time. When output structure doesn't match your needs, note what was wrong and adjust your specification for next time. If the model consistently produces sections in a different order than you specified, your order instruction may need emphasis. If sections consistently run long or short relative to your specifications, adjust the word counts. If the model ignores certain formatting instructions, try rephrasing them more prominently.

PROMPT IN ACTION: Detailed Format Specification

Task: Prepare a summary of case developments for partner review

ROLE: You are a litigation associate preparing a case status summary for the supervising partner.

CONTEXT: [Key case developments from the past month, pasted here]

TASK: Summarize the significant developments in this matter over the past 30 days for partner review.

FORMAT:

Overall length: 600–800 words

Structure the summary as follows:

1. **EXECUTIVE SUMMARY** *(2–3 sentences)* State the most important development and any immediate action items.
2. **DISCOVERY UPDATE** *(1–2 paragraphs)* Summarize discovery activity, including documents produced, depositions taken or scheduled, and any disputes.
3. **MOTION PRACTICE** *(1–2 paragraphs, if applicable)* Summarize any motions filed, pending, or decided. State deadlines.
4. **OPPOSING COUNSEL COMMUNICATIONS** *(1 paragraph, if applicable)* Note any significant communications regarding settlement, scheduling, or case management.
5. **UPCOMING DEADLINES AND ACTION ITEMS** *(bulleted list)* List all deadlines in the next 60 days with dates. Mark any items requiring partner attention with [PARTNER ACTION].

Use formal but efficient prose. Omit background the partner already knows. Focus on what changed, what's next, and what needs decisions.

GUARDRAILS: Do not include strategic recommendations unless explicitly supported by developments. Flag any concerns with [FLAG FOR DISCUSSION].

Building a library of tested format specifications accelerates your work. Once you've refined a specification for client update memos that produces good results, save it for reuse. Once you've dialed in the structure for research summaries or contract analyses, you don't need to reinvent it each time. The Mini-Lab at the end of this chapter will have you build initial versions of these reusable specifications.

Create template prompts for your recurring work products. A template for client status updates, one for research summaries, one for contract clause analyses. Each template locks in the format specification and key structural elements, leaving blanks only for the matter-specific context. Templates enforce consistency and save construction time.

When to leave format open is worth considering. Not every task benefits from detailed format specification. For brainstorming tasks where you want the model to explore freely, rigid structure can constrain useful output. For very short responses where format hardly matters, detailed specification is wasted effort. For tasks where you genuinely don't know what structure makes sense, letting the model choose can reveal options you hadn't considered.

The default should still be to specify format when output quality matters. But recognize that format specification is a tool for getting usable output, not a ritual to perform on every prompt. Match the precision of your specification to the precision you need in the output.

Format specifications tell the model how to structure and present its output. Deliverable-first prompting starts with a clear picture of what you want to receive and works backward to construct the prompt. Specify structure, length, style, and special markers to produce output that requires minimal reformatting. Build and reuse template specifications for recurring work products.

The next section catalogues the prompt patterns that experienced legal AI users rely on repeatedly: role reversal, structured comparison, issue spotting, iterative refinement, and others.

3.5 Prompt Patterns Lawyers Actually Use

Experienced AI users don't start from scratch with every prompt. They rely on patterns that work repeatedly across different matters and task types. These patterns are reusable structures that solve common problems: how to find weaknesses in your own arguments, how to compare alternatives systematically, how to ensure comprehensive issue coverage, how to build complex analysis from simpler components. Learning these patterns accelerates your prompting because you're not inventing technique each time. You're selecting from a toolkit of proven approaches.

The patterns in this section emerged from observing how lawyers who've integrated AI into their practices actually use it. Each pattern addresses a specific need, and each can be combined with the Prompt Blueprint elements from previous sections. Think of the blueprint as the foundation and these patterns as specialized techniques built on that foundation.

Pattern 1: Adversarial Role Reversal

Task: Challenge your own motion for summary judgment from opposing counsel's perspective

> **ROLE:** You are opposing counsel in this litigation. Your goal is to defeat our client's motion for summary judgment.
>
> **CONTEXT:** [Summary of our motion and key arguments]
>
> **TASK:** Identify the three strongest arguments against our motion. For each argument, explain the legal basis and how you would present it to the court. Be aggressive but realistic—focus on arguments that would actually concern a judge.
>
> **FORMAT:** Present each counterargument in a separate section with a descriptive heading.

The adversarial output isn't something you'd send to anyone. Its value is internal: it shows you what you're up against. Review the counterarguments and assess whether your motion adequately addresses them. If not, strengthen your motion before filing. If the model identifies weaknesses you hadn't considered, you've gained insight cheaply.

> Practice Point Run adversarial role reversal before finalizing any significant work product. If the model playing opposing counsel can poke holes in your argument, actual opposing counsel probably can too. Better to discover weaknesses on your own screen than in a courtroom.

Variations include asking the model to respond as a skeptical judge ("what questions would concern you about this motion?"), as a hostile witness ("how would you explain your way out of this document?"), or as the other party's business team ("why would you reject this contract term?"). Each variation surfaces different concerns and helps you prepare for different audiences.

Pattern 2: Structured Comparison

Task: Compare two indemnification clauses from the buyer's perspective

> **TASK:** Compare these two indemnification clauses from the buyer's perspective.
>
> **ITEMS TO COMPARE:**
>
> - [Clause A from Seller's draft]
> - [Clause B from our proposed revision]
>
> **COMPARISON DIMENSIONS:**
>
> 1. Scope of covered claims
> 2. Limitations and carve-outs
> 3. Procedural requirements for indemnification
> 4. Interaction with liability cap
>
> **FORMAT:** Create a table with rows for each dimension and columns for Clause A, Clause B, and Practical Significance. After the table, provide a 2-paragraph recommendation on which clause better serves our client and what further revisions to pursue.

The comparison dimensions are key. Without them, the model may compare on dimensions you don't care about while ignoring dimensions you do. Specifying dimensions produces analysis aligned with your actual decision criteria. For complex comparisons, the dimension list itself becomes a useful forcing function, requiring you to articulate what actually matters before asking for analysis.

This pattern works for case comparison ("compare how these two circuit courts applied the doctrine"), strategy comparison ("compare settlement versus trial given these assumptions"), and vendor evaluation ("compare these two AI tools on security, accuracy, and cost"). The structure transfers across domains.

Pattern 3: Comprehensive Issue Spotting

Task: Identify all potential legal issues arising from a new client matter

> **ROLE:** You are a senior associate conducting initial review of a potential new matter.
>
> **CONTEXT:** [Client's description of the situation]
>
> **TASK:** Identify all potential legal issues arising from these facts. Cast a wide net—include issues that might apply even if you're not certain they're viable. For each issue, note the relevant area of law and what additional facts would help assess viability.
>
> **FORMAT:** Organize issues by category (contract, tort, statutory, regulatory, etc.). Within each category, list issues in order of likely significance. After the issue list, identify any factual gaps that would be priorities for investigation.
>
> **GUARDRAILS:** Err on the side of inclusion. It is better to flag an issue that turns out to be inapplicable than to miss one that matters.

The instruction to "err on the side of inclusion" addresses the omission problem directly. The model defaults to producing a reasonable number of items; explicit instruction to be comprehensive pushes against that default. The request to identify factual gaps acknowledges that initial issue spotting often reveals what you don't know yet.

Follow-up prompts can drill into specific issues identified in the initial sweep: "From the list above, focus on the potential breach of fiduciary duty claim. Analyze the elements in detail based on the facts provided." This decomposition moves from comprehensive breadth to focused depth.

Issue spotting output is a starting point for your own analysis, not a substitute for it. The model may miss issues outside its training coverage or issues that depend on subtle factual combinations it doesn't recognize. Use the output as a checklist to consider, then apply your own judgment to identify what's missing.

Pattern 4: Decomposition and Chaining

Task: Draft a comprehensive research memo through sequential prompt chaining

Rather than asking for the complete memo in one prompt, decompose:

Prompt 1: "Identify the legal questions this issue raises and the likely sources of authority for each."

Prompt 2: "For the first legal question, outline the competing approaches courts have taken." (Feed the identified questions from Prompt 1 as context.)

Prompt 3: "Analyze how each approach would apply to our facts, identifying which approach favors our client." (Feed the approaches from Prompt 2 as context.)

Prompt 4: "Draft the Discussion section of a memo synthesizing this analysis." (Feed outputs from Prompts 2 and 3 as context.)

Pattern 5: Constraint Narrowing

Task: Research personal jurisdiction requirements through iterative narrowing

Start broad and iteratively add constraints based on results:

Initial prompt: "What are the general requirements for establishing personal jurisdiction over a foreign defendant?"

Second prompt: "Narrow this to the requirements in the Ninth Circuit for a defendant whose only contact with the forum is an interactive website accessible to forum residents."

Third prompt: "Further narrow to address how the Ninth Circuit has applied these requirements when the website involves commercial transactions but the defendant has no physical presence in the forum."

Pattern 6: Gap Analysis

Task: Review a software development agreement for missing standard provisions

TASK: Review this software development agreement and identify any standard provisions that are missing or inadequately addressed.

CONTEXT: [Agreement text]

STANDARD PROVISIONS TO CHECK:

- Intellectual property ownership and assignment
- Confidentiality obligations
- Representations and warranties
- Limitation of liability
- Indemnification
- Term and termination
- Data protection and security
- Dispute resolution
- Change order procedures
- Acceptance testing

FORMAT: For each standard provision, state whether it is Present, Partially Addressed, or Missing. For items marked Partially Addressed or Missing, explain what is lacking and provide recommended language to address the gap.

Pattern 7: Steelmanning

Task: Construct the strongest argument that an arbitration clause is unenforceable

TASK: Present the strongest argument that the arbitration clause in this contract is unenforceable under California law.

CONTEXT: [Clause text and relevant facts]

INSTRUCTIONS: Assume you are trying to persuade a skeptical judge who generally favors arbitration. What is the most compelling legal theory? What facts support it most effectively? What is the best response to the likely counterargument that the clause is standard and freely negotiated?

FORMAT: Present the argument as you would in a brief, with a clear statement of the legal theory, supporting authorities, and application to these facts. Approximately 400-500 words.

Combining Patterns

These patterns aren't mutually exclusive. A single workflow might use comprehensive issue spotting to identify what matters, decomposition to analyze each issue separately, structured comparison to evaluate alternatives, and adversarial role reversal to stress-test your conclusions. The patterns are modular tools that combine based on task requirements.

As you gain experience, you'll develop intuition for which patterns suit which tasks. Research tasks often benefit from constraint narrowing. Contract review tasks often benefit from gap analysis. Litigation tasks often benefit from adversarial perspectives. Transactional tasks often benefit from structured comparison. Match the pattern to the problem, and the output improves.

Seven reusable prompt patterns address common legal AI needs: adversarial role reversal finds weaknesses, structured comparison illuminates tradeoffs, comprehensive issue spotting ensures breadth, decomposition and chaining manage complexity, constraint narrowing enables iterative exploration, gap analysis identifies omissions, and steelmanning constructs the best opposing case. Learn these patterns and select among them based on task requirements.

The next section addresses what to do when your first prompt doesn't produce what you need: iteration and debugging strategies for improving AI output across multiple attempts.

3.6 Iteration and Debugging

Your first prompt will rarely produce exactly what you need. This isn't failure; it's the normal workflow. Even experienced AI users iterate, refining their prompts based on what the initial output reveals about gaps in their instructions. The skill isn't writing perfect prompts on the first try. The skill is diagnosing what went wrong and fixing it efficiently.

Iteration is distinct from the output triage process in Chapter 2. Triage is what you do with output you have decided to keep — catching problems before reliance. Iteration is what you do with output you have decided to replace — fixing the prompt to produce something better next time. The two processes are complementary: triage validates individual outputs, iteration improves your prompt writing over time.

Iteration works because each output teaches you something about how the model interpreted your instructions. Figure 3.2 shows the iteration and debugging loop as a cycle: generate, diagnose, adjust, regenerate. When the output misses the mark, the specific way it misses reveals what your prompt lacked. Generic output means your role or context was underspecified. Structural mismatch means your format instructions were unclear. Irrelevant content means your guardrails were missing. Thin analysis means your task description didn't convey the depth you needed. The output is feedback, and debugging is the process of translating that feedback into prompt improvements.

Diagnosing output problems starts with comparing what you received against what you wanted. Identify the specific gap. Resist the temptation to characterize the output as simply "bad" or "not useful." Precision about what's wrong points toward what to fix. Did the output miss the point entirely, or did it address the right topic with wrong execution? Is the problem substance or presentation? Are there good elements mixed with problematic ones, or is the whole output off track?

The Prompt Blueprint provides a diagnostic framework. When output disappoints, work through each element:

Role: Did the model adopt the perspective you intended? Does the output sound like it came from the persona you specified? If the tone or framing seems wrong, your role instruction may need strengthening or clarification.

Context: Did the model use the information you provided? Are there signs it made assumptions beyond your facts or ignored facts you included? If the analysis doesn't track your specific situation, your context packaging may need restructuring.

Task: Did the model do what you asked, or something adjacent? Is the scope right? If you got analysis when you wanted recommendations, or a summary when you wanted a draft, your task instruction may have been ambiguous.

Format: Does the structure match what you specified? Is the length appropriate? If you're spending time reformatting, your format specifications need detail.

Guardrails: Did the model include content you wanted excluded? Did it make assumptions you prohibited? If unwanted content appears, you need guardrails or stronger guardrails.

When output disappoints, don't immediately rewrite the entire prompt. Diagnose first. Identify which blueprint element failed, then adjust only that element. Targeted fixes teach you more than wholesale rewrites and build your understanding of how each element affects output.

Common problems and their fixes recur across prompt types. Learning to recognize these patterns accelerates your debugging.

Problem: Output is too generic, lacks specificity to your situation.

Likely cause*: Insufficient context or weak role instruction.* Fix: Add more relevant facts, specify jurisdiction explicitly, strengthen the role with more detail about perspective and purpose.

Problem: Output addresses wrong topic or misses the point.

Likely cause*: Ambiguous task instruction or context that sent the model in a wrong direction.* Fix: Rewrite the task instruction with more precise verbs and clearer specification of what you want. Check whether your context might have implied a different question than you intended.

Problem: Output is disorganized or structured differently than you need.

Likely cause*: Missing or vague format instructions.* Fix: Specify structure explicitly, including section order, headings, and what each section should contain.

Problem: Output is too long or too short.

Likely cause*: Missing length specification, or length specification that conflicts with task complexity.* Fix: Add explicit length guidance. If the task is complex and you need brevity, acknowledge the tension and tell the model what to prioritize.

Problem: Output includes content you didn't want.

Likely cause*: Missing guardrails.* Fix: Add explicit instructions about what to exclude. Use phrasing like "do not include" or "do not discuss."

Problem: Output is superficial, doesn't go deep enough.

Likely cause*: Task instruction didn't convey depth expectations, or context didn't provide enough material for deep analysis.* Fix: Specify that you want thorough analysis. Add detail to your context that enables depth. Consider decomposing into subtasks that each receive focused attention.

Problem: Output makes assumptions about facts you didn't provide.

Likely cause*: Insufficient context combined with missing guardrails against assumption.* Fix: Provide the facts you want the model to use. Add guardrail: "Do not assume facts beyond what I have provided. If additional facts would be relevant, identify them as questions rather than assumptions."

If the same problem recurs across multiple prompt revisions, you may be trying to fix the wrong element. Step back and reconsider your diagnosis. Sometimes what looks like a context problem is actually a task problem, or what looks like a format problem is actually a role problem.

Refinement versus restart presents a choice when iteration is needed. Refinement adjusts the existing prompt incrementally, preserving what worked while fixing what didn't. Restart abandons the current prompt and approaches the task from a different angle. Both have their place.

Refinement works when the output is partially successful. If the model understood your basic intent and produced useful content that needs adjustment, refine. Add the missing context, strengthen the weak instruction, specify the format more precisely. Refinement is efficient because you're building on progress rather than starting over.

Restart works when the output suggests fundamental misunderstanding. If the model went in a completely wrong direction, incremental adjustment may not redirect it effectively. A fresh prompt with reconceived structure may produce better results faster than trying to patch a flawed approach. Restart is also appropriate when you realize your initial conception of the task was wrong. The output showed you something about what you actually need, and that insight warrants a new approach rather than patches to the old one.

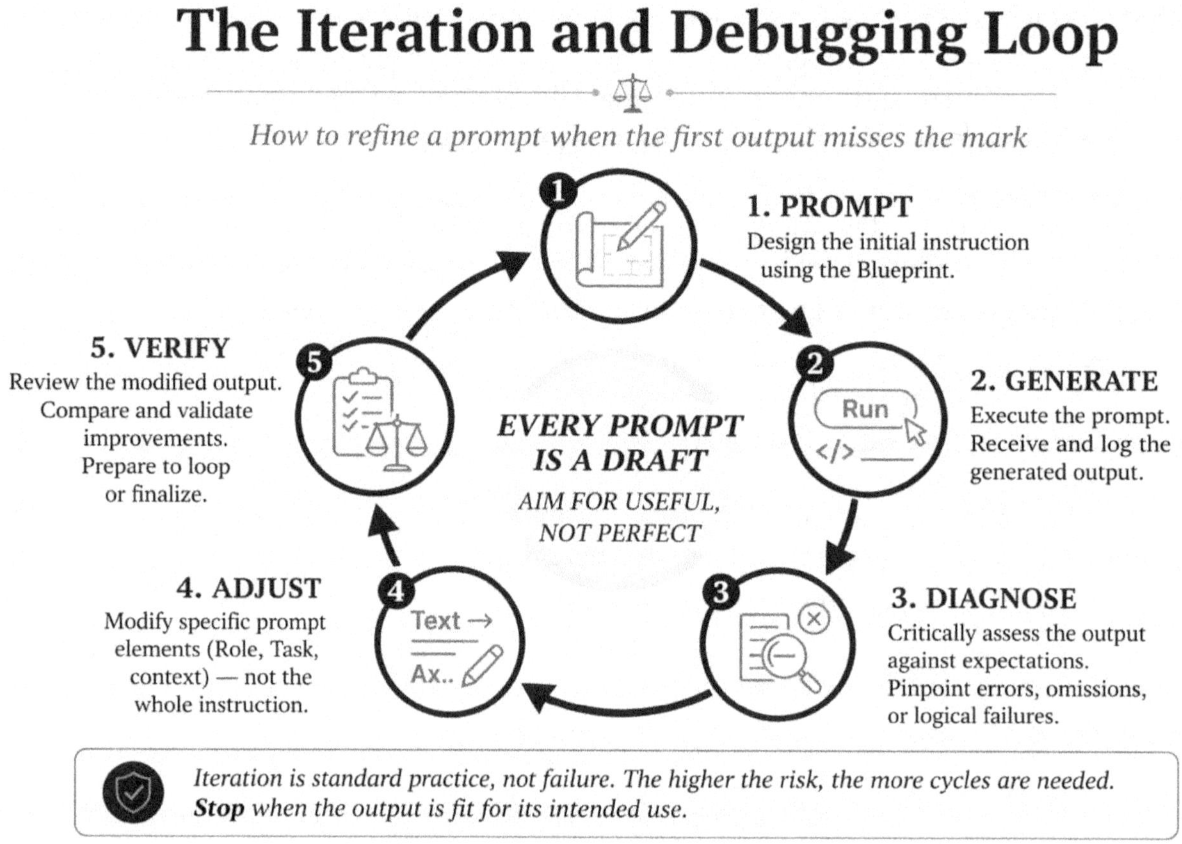

Figure 3.2 — The Iteration and Debugging Loop

The decision between refinement and restart is judgment, not formula. As a rough guide, if more than half the output is usable, refine. If less than a quarter is usable, consider restarting. In between, let the nature of the problems guide you. Structural problems often warrant restart; content problems often warrant refinement.

Iterative refinement strategies make debugging more systematic. One approach is to fix one element at a time, testing after each change to see whether the fix worked. This isolation helps you learn which adjustments have which effects. If you change three things at once and the output improves, you don't know which change mattered.

Another approach is progressive elaboration. Start with a minimal prompt and add elements incrementally until the output meets your needs. This approach works well when you're uncertain what the model requires. Each iteration adds one element, and you stop when the output reaches acceptable quality. What you learn transfers to future prompts for similar tasks.

A third approach is A/B comparison. When you're uncertain whether a particular instruction will help, create two versions of the prompt, one with and one without the instruction, and compare the outputs. This comparison reveals whether the instruction makes a meaningful difference. If it doesn't, you've learned that element isn't crucial for this task type.

PROMPT IN ACTION: Iterative Debugging

Task: Improve a vague contract analysis prompt through systematic diagnosis and revision

ORIGINAL PROMPT:

"Analyze this contract for issues."

OUTPUT:

Generic list of contract issues unrelated to client's specific concerns, no prioritization, wrong jurisdiction assumed.

DIAGNOSIS:

Role missing: no perspective established.

Context missing: no party position, jurisdiction, or specific concerns.

Task too vague: what kind of issues? for what purpose?

Format missing: no structure specified.

Guardrails missing: no scope limitations.

REVISED PROMPT:

ROLE: You are a transactional associate reviewing a software licensing agreement for our client, the licensee.

CONTEXT: Our client is a mid-sized healthcare company licensing practice management software. The agreement is governed by California law. Client's primary concerns are data security, service levels, and exit rights. The vendor has more leverage because switching costs are high.

TASK: Identify provisions in this agreement that are unfavorable to our client or create risk, focusing on the three areas of primary concern. For each issue identified, explain the risk and suggest specific language improvements.

FORMAT: Organize by concern area (data security, service levels, exit rights). Within each area, address issues in order of significance. Approximately 800 words total.

GUARDRAILS: Focus only on the three areas of primary concern. Do not address standard provisions that are not problematic.

[Contract text]

OUTPUT:

Targeted analysis of data security provisions with specific concerns about breach notification timing, service level analysis identifying missing uptime guarantees, exit rights analysis noting problematic termination-for-convenience limitation. Each issue includes recommended revision language.

Revised output: Targeted analysis of data security provisions with specific concerns about breach notification timing, service level analysis identifying missing uptime guarantees, exit rights analysis noting problematic termination-for-convenience limitation. Each issue includes recommended revision language.

Knowing when to stop iterating prevents diminishing returns. Not every prompt can be perfected. At some point, the cost of further iteration exceeds the value of marginal improvement. Signs you should stop include: the output is good enough for your purpose even if not ideal, further refinements produce negligible improvement, and you've already spent more time iterating than you would have spent doing the task manually.

The "good enough" threshold depends on how you'll use the output. A brainstorming prompt that generates ideas for your consideration can tolerate imperfection. A prompt generating text you'll substantially revise can tolerate imperfection. A prompt generating near-final work product requires higher quality and may justify more iteration. Match your iteration investment to the output's eventual use.

Track your iteration patterns. If you consistently need three or four rounds of refinement for a particular task type, your initial prompts for that task type are probably underspecified. Capture what you learn during iteration and incorporate it into your starting prompt next time.

Building skill through deliberate iteration makes debugging a learning process rather than just a problem-solving exercise. Each iteration cycle teaches you something about how the model responds to different instructions. Over time, you internalize these patterns and write better first drafts of prompts.

Keep notes on what works. When you discover that a particular phrasing produces better results, record it. When you learn that certain task types require certain blueprint elements at higher intensity, remember that. When you find combinations that work well together, save them as templates. This accumulated knowledge becomes your personal prompt engineering expertise.

The goal is to reduce iteration over time, not to eliminate it. Even experts iterate. But the number of iterations needed for acceptable output should decrease as your skill increases. If you're still taking five rounds to get usable output after months of practice, examine whether you're learning from your iterations or just repeating the same mistakes.

Iteration is normal, not failure. Diagnose output problems against the Prompt Blueprint to identify which element needs adjustment. Fix one element at a time to learn what works. Choose between refinement and restart based on how much of the output is salvageable. Know when to stop iterating and accept good enough. Capture what you learn to improve future first-draft prompts.

The next section addresses verification loops: techniques for building self-checking into your prompts so the model catches some of its own errors before you review the output..

3.7 Verification Loops: Cross-Checks and Red Flags

Chapter 2 covered how to verify AI output after you receive it. This section addresses something different: building verification into the prompting process itself so the model catches some of its own errors before you begin review. Verification loops don't replace human review, but they can surface problems earlier and reduce the burden on your post-output triage.

The basic insight is that you can ask the model to check its work as part of producing that work. The same capabilities that generate content can be directed toward critiquing content, including content the model just generated. This self-checking is imperfect. The model has systematic blind spots, and it may miss errors that a human reviewer would catch. But it catches enough to be useful, particularly for errors that are easy to verify once someone thinks to look.

Embedded verification instructions add checking steps to your primary prompt. Instead of asking only for output, you ask for output plus verification of that output. The verification becomes part of the task rather than a separate follow-up.

A simple version adds a verification requirement at the end of your task instruction:

TASK: Draft a motion to compel discovery responses based on the facts I've provided. After drafting, review your draft and identify any factual assertions that are not directly supported by the facts I've provided. Flag these assertions with [VERIFY] so I can check them.

This instruction tells the model to generate the motion and then audit it for unsupported facts. The model rereads its own output, compares assertions against the provided context, and flags discrepancies. You still need to review the flagged items and check for unflagged problems, but the model has done a first pass that focuses your attention.

More elaborate verification instructions can check multiple dimensions:

After completing the analysis, review your work for the following:

1. Are all legal citations to actual cases? Flag any citation you are not certain exists with [CITATION CHECK].
2. Does each conclusion follow logically from the preceding analysis? Flag any conclusion that might not follow with [LOGIC CHECK].
3. Have you addressed all elements of the claim, or are any missing? Note any gaps.
4. Are there counterarguments or weaknesses you did not address? List them at the end.

The model works through each verification dimension and surfaces concerns. This approach is particularly useful for generation tasks where multiple failure modes are possible. Rather than hoping you catch everything in review, you've instructed the model to look for specific problems.

Match your verification instructions to the failure modes most likely for the task type. Citation-heavy research memos warrant citation verification. Fact-intensive briefs warrant fact verification. Analytical work warrants logic verification. Target the checks to where errors are most likely and most consequential.

Two-pass verification separates generation from checking into distinct steps. The first pass produces output. The second pass critiques that output. This separation can produce more thorough verification because the model approaches the critique as a fresh task rather than an afterthought.

First prompt: Generate the work product without verification instructions.

Second prompt: "Review the following [draft/analysis/memo] and identify any errors, weaknesses, gaps, or concerns. Be thorough and critical. It's better to flag a potential problem that turns out to be fine than to miss an actual problem."

Then paste the output from the first prompt and let the model critique it. The critique prompt activates different patterns than the generation prompt. The model shifts from constructive mode to critical mode, often surfacing problems it wouldn't have flagged if asked to self-check in the same breath as generating.

Two-pass verification takes more time than embedded verification but often catches more. Use it when accuracy matters enough to justify the extra step, particularly for Rung Four and Five tasks on the reliability ladder from Chapter 2.

Structured critique prompts guide the verification pass with specific dimensions to evaluate. Rather than asking for general critique, you specify what to look for:

> Review the contract analysis above and evaluate it on these dimensions:
>
> **COMPLETENESS:** Are all material provisions addressed? What's missing?

> **ACCURACY:** Are the descriptions of clause contents accurate to the actual language? Flag any mischaracterizations.
>
> **BALANCE:** Does the analysis fairly represent risks to both parties, or does it skew toward one side?
>
> **ACTIONABILITY:** Are the recommendations specific enough to act on? Where would you need more detail?
>
> **ASSUMPTIONS:** What assumptions does the analysis make? Are they stated explicitly or hidden?
>
> For each dimension, provide a brief assessment and note specific concerns if any.

The structured critique produces organized feedback across the dimensions you care about. Different tasks warrant different critique dimensions. A litigation document might be evaluated for persuasiveness, citation support, and procedural accuracy. A client communication might be evaluated for clarity, completeness, and tone. Choose dimensions that match what matters for the specific work product.

Red flag detection asks the model specifically to identify problems rather than to assess quality generally. This framing produces more critical output because you've explicitly invited criticism:

> Your task is to find problems with this draft brief. Assume problems exist and look for them. Specifically:
>
> - Identify any statements that might be factually inaccurate or unsupported by the record
> - Identify any legal propositions that might be misstated or overclaimed
> - Identify any arguments that opposing counsel could effectively attack
> - Identify any requirements for this type of motion that might not be satisfied
>
> Do not tell me what's good about the brief. Focus only on problems and concerns.

The instruction to assume problems exist and to omit positive feedback pushes the model toward critical output. Without this framing, the model often defaults to balanced assessment that softens criticism with praise. When you want problems surfaced, ask for problems specifically.

> **Risk Note** Verification loops catch some errors but not all. The model has the same knowledge limitations during verification that it has during generation. It cannot verify that a citation exists by looking it up; it can only assess whether the citation looks plausible. It cannot check facts against reality; it can only check internal consistency. Human verification remains essential.

Verification for specific failure modes targets the problems identified in Chapter 2. For each failure mode, you can construct verification instructions that look specifically for that type of error.

For hallucination risk: *"Review each factual assertion in your response. For any assertion not directly derived from the information I provided, either remove it or flag it with [UNVERIFIED]."*

For omission risk: *"After completing your analysis, review it against this checklist of issues that should be addressed: [checklist]. Note any items from the checklist that your analysis does not cover."*

For anchoring risk: *"Before finalizing, consider whether there are alternative approaches or interpretations you did not explore. Briefly note any significant alternatives and why you did or didn't address them."*

For inconsistency risk: *"Review your conclusions and confirm that each follows logically from your analysis. Flag any conclusion that might not be fully supported with [LOGIC CHECK]."*

For overconfidence risk: *"Review any statements that express certainty about legal outcomes. If reasonable lawyers could disagree, soften the statement or note the uncertainty."*

These targeted instructions don't guarantee the model will catch the relevant errors, but they increase the probability by directing attention to specific problem types.

PROMPT IN ACTION: Comprehensive Verification Loop

Task: Research memo with embedded verification

ROLE: You are a litigation associate preparing a research memo for partner review.

CONTEXT: [Facts and legal question provided here]

TASK: Analyze whether our client has a viable claim for tortious interference with contract under New York law based on the facts provided.

FORMAT: Standard memo format with Issue, Brief Answer, Facts, Discussion, and Conclusion. Approximately 1000–1200 words.

VERIFICATION REQUIREMENTS: After drafting the memo, append a Verification Notes section that addresses:

1. CITATIONS: List each case cited and note your confidence level (HIGH if you're certain it exists and says what you claim; MEDIUM if you believe it exists but aren't certain of the exact holding; LOW if you're uncertain whether it exists). I will verify all citations regardless, but this helps me prioritize.
2. ELEMENTS: Confirm that your analysis addresses each element of the claim. If any element is not fully analyzed, note why.
3. COUNTERARGUMENTS: List the strongest counterarguments to our position that you did not fully address in the memo.
4. FACTUAL GAPS: Identify any additional facts that would strengthen or clarify the analysis.

The Verification Notes are for my use and will not appear in the final memo.

Why It Works: The verification requirements are specific and structured. The confidence-level system for citations provides useful triage information. The explicit request for unaddressed counterarguments and factual gaps surfaces omissions. The verification section is clearly separated from the substantive memo.

When verification loops help most depends on task characteristics. Verification loops add the most value when errors are consequential, when the model can plausibly detect the error type, and when the verification doesn't require external information the model lacks.

Verification loops help with:

- Internal consistency checking (does the conclusion follow from the analysis?)
- Completeness checking against a provided checklist (are all required elements addressed?)
- Identifying unstated assumptions (what did the analysis assume without saying?)
- Surfacing alternative approaches (what other angles weren't explored?)
- Flagging uncertainty (where is confidence potentially unwarranted?)

Verification loops help less with:

- Citation accuracy (the model cannot verify citations exist)
- Factual accuracy against external reality (the model cannot check facts against the world)
- Current law verification (the model may not know about recent changes)
- Jurisdiction-specific accuracy (the model may not recognize jurisdiction errors)

For the limitations, human verification is the only reliable check. Verification loops are a supplement, not a substitute.

Use verification loops to reduce your burden, not to replace your review. Think of them as a research assistant who does a first pass before you do your own review. The assistant catches obvious problems and surfaces concerns, but you still read the work yourself before relying on it.

Verification loops and iteration interact in useful ways. When the model's self-verification flags problems, you can address them in a follow-up prompt:

> *"Your verification notes identified that the analysis doesn't fully address the requirement of existing contract. Please expand the Discussion section to analyze this element more thoroughly."*

The flagged concern becomes the target for iteration. This approach is more efficient than general requests for improvement because you're directing the model to specific identified gaps rather than asking it to make the output generically better.

Over time, patterns in what verification loops catch can inform your prompting. If the model consistently flags incomplete analysis of a particular element, your initial prompts may need to emphasize that element more explicitly. If the model consistently surfaces the same type of counterargument, consider addressing it proactively in future prompts. Verification feedback loops back into prompt improvement.

Verification loops build self-checking into the prompting process. Embedded verification adds checking steps to the primary prompt. Two-pass verification separates generation from critique for more thorough review. Target verification to specific failure modes relevant to the task type. Remember that verification loops supplement but don't replace human review: the model cannot verify citations, check external facts, or catch errors outside its knowledge.

The next section addresses source-bounded prompting: techniques for constraining the model to work only with authorities you provide, reducing hallucination risk by eliminating the model's incentive to generate authorities from training patterns

3.8 Source-Bounded Prompting

The hallucination problem discussed in Chapter 2 arises because the model generates citations from patterns rather than retrieving them from verified sources. When you ask for legal analysis, the model predicts what a helpful response would look like, including case citations that fit the pattern even if they don't correspond to real cases. Source-bounded prompting addresses this problem by removing the model's need to generate authorities. You provide the authorities; the model works with what you provide and nothing else.

The technique is simple in concept: supply the relevant cases, statutes, or other sources in your prompt, and instruct the model to use only those sources in its analysis. The model becomes an analytical tool rather than a research tool. It synthesizes, compares, applies, and explains the authorities you've given it, but it doesn't venture beyond them. Since the authorities are real (you verified them), the outputs that rely exclusively on those authorities avoid the fabrication problem.

Source-bounded prompting requires you to do the research. You find the relevant authorities through traditional legal research, using Westlaw, Lexis, or other tools designed to retrieve actual sources. Then you use AI to work with those authorities: summarizing holdings, comparing approaches across cases, applying rules to facts, drafting analysis sections that synthesize multiple sources. The division of labor plays to each tool's strengths. Research platforms retrieve; AI analyzes.

The basic structure provides sources and prohibits venturing beyond them:

CONTEXT: I have identified the following authorities relevant to whether a forum selection clause is enforceable under New York law. Use ONLY these authorities in your analysis. Do not cite any other cases or sources.

AUTHORITIES PROVIDED:

[Case 1: Full text or detailed summary with citation]

[Case 2: Full text or detailed summary with citation]

[Statute: Relevant statutory text with citation]

TASK: Analyze whether the forum selection clause in our client's contract is likely enforceable based on these authorities. Apply the legal standards from the provided cases to the facts I've described.

GUARDRAILS: Cite only the authorities I have provided. If you believe additional authorities would be relevant, note what type of authority would help, but do not fabricate citations. If the provided authorities are insufficient to answer the question, say so.

The explicit instruction to use only provided authorities is essential. Without it, the model may supplement with additional citations from its training, reintroducing hallucination risk. The guardrail must be clear and prominent.

What to provide depends on your task and how much context window space you can allocate. For a focused analysis involving a few key cases, you can provide full case text or extensive excerpts. For broader analysis involving many authorities, you'll need to provide summaries or the most relevant portions of each source.

Full text provides the richest basis for analysis. The model can identify nuances, distinguish holdings from dicta, and recognize how different parts of an opinion interact. But full text consumes context window space rapidly. A single appellate opinion might run thousands of tokens. If you're working with multiple authorities, full text for all of them may exceed your limits.

Detailed summaries preserve key information while managing length. Your summary might include: the citation, the facts relevant to your analysis, the court's holding on the relevant issue, the reasoning supporting that holding, and any important limitations or qualifications. The model works with your summary, so the quality of your summary determines the quality of the analysis.

Selective excerpts provide the exact language you want the model to work with, such as the specific paragraphs stating the legal standard or applying the test. This approach is precise but risks missing context that affects interpretation. If you excerpt a holding without the factual predicates, the model may misapply it.

When providing case summaries rather than full text, include more context than you think necessary. The model benefits from understanding why a court reached its holding, not just what the holding was. A summary that captures the reasoning supports better analysis than one that states only the result.

Combining bounded and unbounded elements accommodates tasks that need both constrained analysis and broader input. You might bound the authorities while allowing the model to generate other content:

> **AUTHORITIES:** Use only the following cases for legal analysis. Do not cite any other cases.
>
> *[Provided cases]*
>
> **OTHER CONTENT:** You may draft argument structure, transitional language, and factual summaries without restriction. Only legal citations must come from the provided authorities.
>
> **TASK:** Draft the Argument section of a brief arguing that the defendant's motion to dismiss should be denied, using only the provided authorities for legal support.

This hybrid approach lets the model contribute drafting and organizational value while keeping citation risk contained. The output may still contain errors in the unbounded elements, but the specific risk of fabricated authorities is managed.

Verification remains necessary even with source-bounded prompting. The model can still make errors when working with provided authorities. It might misstate what a case holds, misapply a rule to facts, or draw conclusions that don't follow from the sources. Source-bounding eliminates fabrication of authorities but doesn't eliminate analytical errors.

Your verification for source-bounded output focuses on accuracy of use rather than existence of sources. Check that the model correctly characterizes each authority's holding. Check that applications of rules to facts are sound. Check that the analysis doesn't slip in references to unprovided sources despite your instructions. The model generally follows bounding instructions, but occasional violations occur, especially when the task seems to call for an authority you didn't provide.

> **Risk Note** The model may resist bounding when it "knows" a relevant authority exists. If the model includes a citation you did not provide despite your instructions, it is violating the constraint. Either remove the citation or verify it through research, but do not assume it is accurate just because the model included it confidently.

When source-bounded prompting fits depends on your workflow and the task at hand. The technique works well in several scenarios.

After completing traditional research, when you have the authorities and need to work with them. The analysis, drafting, and synthesis tasks that follow research are ideal for source-bounded prompting. You've done the retrieval; now you want help with the processing.

When working within a closed universe of authorities. Contract interpretation often involves a defined set of documents: the agreement, relevant amendments, and perhaps some defined extrinsic evidence. Statutory analysis may focus on a specific statute and its implementing regulations. Source-bounded prompting keeps the analysis tightly connected to the actual source materials.

When accuracy is paramount and you cannot tolerate fabrication risk. For court filings and formal advice, source-bounded prompting ensures that every cited authority is real because you provided every cited authority. This protection may justify the extra work of conducting research separately.

When training junior lawyers or building your own understanding. Working through authorities yourself before asking AI to analyze them keeps you engaged with the material. The AI output becomes a complement to your own analysis rather than a replacement for it.

Source-bounded prompting pairs naturally with traditional legal research. Use research platforms to find and verify authorities. Use AI to synthesize, compare, and apply those authorities. Each tool does what it does best, and the combination produces better results than either alone.

When source-bounded prompting doesn't fit includes scenarios where you want broader exploration or where the bounding constraint interferes with your goals.

Brainstorming and issue-spotting tasks may benefit from the model's broader knowledge, even if that knowledge requires verification. If you're trying to identify what legal theories might apply to a new situation, constraining to pre-identified authorities limits the exploration.

Efficiency-critical tasks where separate research adds too much friction may warrant accepting some hallucination risk in exchange for speed. For a quick internal analysis that you'll verify before relying on, the overhead of source-bounded prompting may not be justified.

Learning and orientation tasks where you're trying to understand an unfamiliar area may benefit from the model's ability to provide general frameworks, even if specific citations need verification. You can use unbounded prompting to orient yourself, then shift to source-bounded prompting for the formal work product.

Hybrid workflows integrate source-bounded and unbounded prompting at different stages. A research project might proceed through phases:

Phase 1 (unbounded): "What are the key legal issues raised by these facts, and what authorities would typically govern each issue?" The model provides orientation and identifies what to research.

Phase 2 (traditional research): You conduct research using Westlaw, Lexis, or other tools, finding and verifying authorities for each identified issue.

Phase 3 (source-bounded): "Using only the following authorities, analyze whether our client has a viable claim. Do not cite any sources I have not provided." The model produces analysis grounded exclusively in verified sources.

Phase 4 (verification): You review the analysis for accuracy of characterization and application, catching analytical errors even though fabrication risk has been eliminated.

This hybrid workflow uses each tool appropriately. The model helps you think broadly in Phase 1 and work precisely in Phase 3. Traditional research provides the verified foundation. Verification catches what remains.

PROMPT IN ACTION: Source-Bounded Analysis

Task: Analyze enforceability of non-compete using only provided authorities

ROLE: You are a senior employment associate analyzing whether a non-compete agreement is enforceable.

CONTEXT: Our client is an employer seeking to enforce a non-compete against a former sales executive who left to join a competitor. The agreement restricts the employee from working for any competitor within 50 miles for 18 months. The employee had access to customer lists and pricing information. The employee worked for our client for 4 years. This is governed by Georgia law.

AUTHORITIES PROVIDED:

> **[1] O.C.G.A. § 13-8-53 (Georgia statute governing enforceability of restrictive covenants in employment):**
>
> *[Paste the relevant statutory text verbatim from your research source.]*
>
> **[2] Paramount Tax & Accounting, LLC v. H&R Block Eastern Enters., Inc., 683 S.E.2d 141 (Ga. Ct. App. 2009):**
>
> *[Paste or summarize the court's analysis of how the non-compete's territorial limitation failed the reasonableness test. The court found the noncompetition clause unenforceable because its territorial restriction was so overbroad as to be almost nonexistent — the clause would have prevented the employee from working for any employer that had an office or advertised within ten miles of the relevant district, regardless of where the employee herself would be working.]*

TASK: Analyze whether our client's non-compete is likely enforceable under Georgia law. Apply the statutory requirements from the provided O.C.G.A. § 13-8-53 text and the reasonableness analysis from Paramount Tax to our specific facts. Address the statutory compliance, the geographic scope, the temporal scope, and the protectable interest. Identify which aspects of the clause are most vulnerable under the Paramount Tax reasoning.

FORMAT: Organize by issue (statutory compliance, geographic scope, temporal scope, protectable interest). For each issue, state the applicable rule from the provided authorities, apply it to our facts, and provide a conclusion. Note that Paramount Tax is a cautionary case — the court invalidated the non-compete there — so your analysis should explicitly assess whether our client's clause avoids the overbreadth problems the court identified. Approximately 800–1000 words.

GUARDRAILS: Use ONLY the authorities I have provided. Do not cite any other cases, statutes, or sources. If you believe additional authority would be helpful, note what type of authority at the end of your analysis, but do not fabricate a citation. If my provided authorities are insufficient to answer any aspect of the question, state that explicitly.

Why It Works: The authorities are explicitly identified and provided in a form the model can work with. The task asks the model to apply the standards from the provided sources to the specific facts, using a known cautionary case as the reasonableness anchor. The guardrails prohibit fabrication and provide an escape valve when the provided authorities are insufficient. Notice that a single strong case can anchor a source-bounded analysis effectively — breadth of authorities matters less than depth of engagement with the authorities you provide.

Building your authority libraries over time makes source-bounded prompting more efficient. As you work in particular practice areas, you develop collections of key authorities for recurring issues. Standard cases on contract formation, typical holdings on summary judgment standards, established frameworks for particular regulatory questions. These authority libraries become reusable context that you can drop into prompts without conducting fresh research each time.

Organize authority libraries by issue rather than by matter. The key cases on personal jurisdiction in your jurisdiction serve across many matters. The standard framework for analyzing restrictive covenants applies whenever that issue arises. Building issue-based libraries creates reusable infrastructure that makes source-bounded prompting a sustainable practice rather than one-off extra work.

Source-bounded prompting eliminates citation fabrication by constraining the model to authorities you provide. Conduct research through traditional tools, then use AI to analyze, synthesize, and apply the verified authorities. This technique requires more upfront work but produces outputs where every cited source actually exists. Combine bounded and unbounded approaches across workflow stages to get the benefits of each.

The final section of this chapter provides a hands-on exercise: building a prompt pack for a matter in your own practice, integrating the blueprint, patterns, and techniques covered throughout the chapter.

3.9 Mini-Lab: Build a Prompt Pack for One Matter

Objective: Create a collection of tested, reusable prompts tailored to a specific matter or practice area from your own work. This prompt pack becomes a practical resource you can use immediately and refine over time.

Time Required: 60 to 90 minutes

What You'll Need:

- A current matter or a representative matter type from your practice
- The Prompt Blueprint framework (Role, Context, Task, Format, Guardrails)
- The list of prompt patterns from Section 3.5
- Access to an AI tool to test your prompts
- A document to capture your prompt pack for future reuse

Instructions

1. Select Your Matter and Identify Recurring Tasks: Choose a matter you're currently working on or a matter type you handle regularly. List the tasks you perform repeatedly on this type of matter. Aim for six to ten tasks.
2. Prioritize Three Tasks for Prompt Development: From your list, select three tasks where AI assistance would provide the most value. Consider frequency, time consumption, and AI suitability.
3. Build Your First Prompt Using the Blueprint: Take your highest-priority task and construct a prompt using all five blueprint elements. Work through each element systematically. Assemble the five elements into a complete prompt.
4. Test and Refine Your First Prompt: Run your prompt through an AI tool with actual or representative matter content. Review the output against what you would consider good work product for this task. Diagnose any gaps using the framework from Section 3.6.

5. Build Prompts for Your Second and Third Tasks: Repeat Steps 3 and 4 for your other two priority tasks. For at least one of your three prompts, incorporate one of the prompt patterns from Section 3.5.
6. Add Verification Elements: Review each of your three prompts and add appropriate verification instructions from Section 3.7.
7. Assemble Your Prompt Pack: Organize your three tested prompts into a prompt pack document. Before adding any prompt to the pack, cross-check it against the Personal AI Rules of Use you drafted in Chapter 1. Every prompt in your pack should satisfy the commitments you made there about what information goes in, what comes out, how you verify, and how you document. A prompt that violates your own Rules of Use does not belong in your pack.

Deliverable

By the end of this Mini-Lab, you should have:

1. A list of six to ten AI-eligible tasks for your selected matter type
2. Three fully developed prompts using the Prompt Blueprint structure
3. Testing notes documenting what refinements you made and why
4. Verification instructions incorporated into each prompt
5. An organized prompt pack document ready for reuse

Bonus Challenge

Extend your prompt pack with two additional elements:

First, create a meta-prompt for this matter type: a prompt that helps you generate other prompts. The meta-prompt might read: "I'm working on a [matter type] and need to [describe task]. Help me construct a prompt using the following framework: Role, Context, Task, Format, Guardrails. Ask me clarifying questions about the task before generating the prompt."

Second, create an iteration prompt for this matter type: a prompt you use when initial output needs improvement. The iteration prompt might read: "The output above doesn't meet my needs. The specific problem is [placeholder]. Revise the output to address this problem while preserving what worked. Explain what changes you made."

Store your prompt pack somewhere you'll actually access it: a note-taking app, a document folder you use daily, or a practice management system. Prompts that live in obscure files don't get used. Prompts that are one click away become part of your workflow. Make reuse easy.

Extending Your Library Over Time

The three prompts you built today are a starting point. A mature prompt library for a practice area might include dozens of tested prompts covering every recurring task. Building that library happens incrementally: each time you face a new task, you construct a prompt, test it, refine it, and add the successful version to your collection.

Schedule periodic reviews of your prompt library. Prompts that worked six months ago may need updating as AI tools evolve, as your practice changes, or as you learn better techniques. Remove prompts you never use. Refine prompts that produce inconsistent results. Add prompts for tasks you've been doing manually that could benefit from AI assistance.

The investment in building and maintaining your prompt library pays returns every time you reuse a prompt instead of constructing one from scratch. The time you spend in this Mini-Lab is time you'll save many times over on future matters.

Key Takeaways

1. The Prompt Blueprint provides a reusable structure for effective prompts. Five elements address the most common failure modes: Role (who the AI should be), Context (information it needs), Task (what to produce), Format (how to structure it), and Guardrails (what to avoid). Using all five consistently produces better output than ad hoc prompting.
2. Role, audience, and tone instructions calibrate output for your specific needs. Role activates patterns associated with professional personas. Audience adjusts vocabulary and emphasis for the intended reader. Tone fine-tunes register and confidence level. These controls shape presentation but don't affect underlying accuracy.
3. Context packaging determines whether the model has what it needs. Organize facts clearly, handle documents with attention to length and dependencies, specify jurisdiction explicitly, and state constraints that affect the analysis. Balance thoroughness against focus and confidentiality against utility.
4. Deliverable-first prompting produces output closer to what you actually need. Start with a detailed picture of the desired work product, then work backward to construct the prompt. Specify structure, length, style, and special markers to minimize post-output reformatting.
5. Seven prompt patterns solve recurring problems. Adversarial role reversal finds weaknesses. Structured comparison illuminates tradeoffs. Comprehensive issue spotting ensures breadth. Decomposition and chaining manage complexity. Constraint narrowing enables iterative exploration. Gap analysis identifies omissions. Steelmanning constructs the best opposing case.
6. Iteration is normal workflow, not failure. Diagnose output problems against the blueprint to identify which element needs adjustment. Choose between refinement and restart based on how much is salvageable. Capture what you learn to improve future prompts.
7. Verification loops and source-bounded prompting reduce error risk. Embedded verification and two-pass verification catch some errors before your review. Source-bounded prompting eliminates citation fabrication by constraining the model to authorities you provide. Neither technique replaces human verification, but both reduce the burden.

Chapter Checkpoint

Before moving on, make sure you can:

☐ Construct a prompt using all five blueprint elements and explain how each element addresses a specific failure mode

☐ Select the appropriate prompt pattern for a given task and explain why that pattern fits better than alternatives

☐ Diagnose a disappointing output by identifying which blueprint element failed and articulate a specific fix

☐ Design a source-bounded prompt that eliminates citation fabrication risk while still producing useful analytical output

What's Next

You now have techniques for communicating effectively with AI tools. But technique alone doesn't make AI use responsible. Part II of this book addresses the professional responsibility dimensions that constrain and guide how lawyers can use these tools. Chapter 4 examines the ethical obligations that apply when AI enters your workflow: competence requirements, supervision duties, candor obligations, confidentiality rules, and the documentation practices that demonstrate compliance. The prompting skills you've developed become professionally sustainable only when they operate within these boundaries. Understanding the rules isn't separate from using AI well; it's part of using AI well.

Key Terms

Prompt Blueprint — A five-element structure for constructing effective AI prompts: Role (who the AI should be), Context (information it needs), Task (what to produce), Format (how to structure it), and Guardrails (what to avoid). Each element addresses a common failure mode.

Audience instructions — Specifications about who will receive the AI output, allowing the model to calibrate vocabulary, formality, assumed knowledge, and emphasis for the intended reader.

Context packaging — The process of selecting, structuring, and presenting information in a prompt so the model has what it needs to produce useful output. Effective packaging balances sufficiency against token limits, focus, and confidentiality concerns.

Deliverable-first prompting — An approach that begins with a detailed specification of the desired output, then works backward to provide the role, context, and task instructions needed to produce it.

Prompt chaining — Using the output of one prompt as input for a subsequent prompt, building complex analysis through a sequence of simpler steps. Each step can be verified before proceeding, reducing compounding errors.

Two-pass verification — Separating content generation from content critique into distinct prompts. The first prompt produces output; the second reviews that output for errors and weaknesses.

Source-bounded prompting — Constraining the model to work only with authorities you provide, prohibiting it from generating additional citations. This eliminates hallucination risk for the bounded portion of the analysis.

Part II: Risk and Professional Responsibility

Chapter 4: Professional Responsibility in the AI Era

The rules of professional conduct were written before generative AI existed, but they were not written for a world without technological change. Lawyers have always adopted new tools, and the ethics rules have always applied to how those tools are used. AI is not an exception. The duties of competence, supervision, candor, and confidentiality apply to AI-assisted work just as they apply to work done any other way. What changes is not whether the rules apply but how they apply to this particular technology.

Part I of this book focused on using AI effectively: understanding what the tools do, recognizing their limitations, and communicating with them through well-constructed prompts. This chapter reframes those same capabilities through the lens of professional obligation. The techniques you learned are not just good practice; several of them are ethically required. Verification is not optional caution. It is a competence obligation. Supervision is not perfectionism. It is what the rules demand. The gap between "AI helped me draft this" and "I take professional responsibility for this" must be bridged by specific practices, and those practices have names in the ethics rules.

What This Chapter Covers

- Why technological competence is now part of baseline professional competence
- How supervision duties extend to AI tools and to staff who use them
- What candor and accuracy obligations mean for AI-assisted court filings
- How confidentiality rules constrain what information enters AI systems
- What clients should know about AI use and how billing practices must adapt
- Why documentation matters and what audit-ready practice looks like
- How courts are responding with AI-specific rules and disclosure requirements

Chapter Summary

The Model Rules of Professional Conduct establish obligations that apply directly to AI use, even though the rules never mention AI by name. Competence requires understanding AI well enough to use it appropriately and to recognize when it produces unreliable output. Supervision requires treating AI output with the same scrutiny you would apply to work from a junior lawyer or staff member, and it extends to supervising others in your organization who use AI. Candor prohibits submitting AI-generated content to courts without verification that would catch fabricated citations or misstated law. Confidentiality limits what client information can enter AI systems based on the data protections those systems provide. AI disclosure obligations are emerging rapidly through court orders, bar opinions, and firm policies, with significant variation by jurisdiction; billing duties require honest representation of the time and work involved. Documentation practices create the record that demonstrates compliance when questions arise. Courts are increasingly adopting rules that require AI disclosure or impose specific verification obligations. This chapter maps these obligations to concrete practices, transforming abstract ethical duties into specific habits that make AI use professionally sustainable.

4.1 Competence and Technology

Model Rule 1.1 requires lawyers to provide competent representation, which includes "the legal knowledge, skill, thoroughness and preparation reasonably necessary for the representation." For most of the rule's history, competence focused on substantive legal knowledge and analytical ability. Technology was a tool lawyers might use, but fluency with technology was not itself a competence requirement.

That changed in 2012 when the ABA amended Comment 8 to Rule 1.1. The comment now states that competence includes keeping abreast of "the benefits and risks associated with relevant technology." This amendment did not create a new obligation so much as make explicit what was already implicit: lawyers who use technology without understanding it cannot provide competent representation. The comment applies to all relevant technology, but its application to AI is particularly significant because AI's risks are substantial and not always obvious.

Understanding "benefits and risks" means more than knowing that AI exists or that it can draft documents. It means understanding how the technology works well enough to recognize its failure modes, to identify tasks where it is reliable versus unreliable, and to implement verification processes appropriate to the risks. A lawyer who uses AI to draft a brief without knowing that AI can hallucinate citations is not competently using the technology, even if the lawyer is otherwise an excellent brief writer.

> **Authority Note** Comment 8 to Model Rule 1.1 states that, to maintain the requisite knowledge and skill, a lawyer should keep abreast of changes in the law and its practice, including the benefits and risks associated with relevant technology.

The competence obligation has two dimensions in the AI context. The first is competence to use AI: understanding the technology well enough to deploy it appropriately. The second is competence despite AI: ensuring that AI assistance does not degrade the quality of your representation below what competent practice requires.

Competence to use AI means knowing what you are doing when you prompt an AI tool. It means understanding that the model predicts plausible text rather than retrieving verified information. It means recognizing that citations require independent verification. It means appreciating how context windows work and why long documents may be processed unevenly. It means knowing the difference between task types and how reliability varies across them. The foundational knowledge from Chapters 1 and 2 is not just helpful background. It is what competence requires before you integrate AI into client work.

Competence to use AI also means knowing what you do not know. If you lack the expertise to evaluate whether AI output on a specialized topic is correct, you either need to develop that expertise, consult someone who has it, or refrain from using AI for that task. The model's confident tone does not substitute for your own ability to assess the work. A lawyer who submits AI output on a topic they do not understand is not practicing law competently. They are gambling with a client's interests.

> **Risk Note** "The AI made an error" is not a defense to an incompetence claim. The client hired you, not the AI. If you submitted work product containing errors that competent review would have caught, the failure is yours.

Several state bar ethics opinions have addressed technological competence in the AI context. While the specific guidance varies, common themes emerge. Lawyers must understand the technology before using it on client matters. Lawyers must implement appropriate safeguards based on the technology's known limitations. Lawyers must verify AI output rather than accepting it uncritically. Lawyers must maintain the ability to perform the underlying legal work even when AI assists with production.

AI and the Model Rules

Six professional responsibility obligations AI implicates directly

Rule 1.1 Competence	**The duty to maintain legal knowledge and skill includes understanding the benefits and risks of technology you use.** AI competence means knowing how the tools work, where they fail, and when verification is required.
Rule 1.6 Confidentiality	**Information relating to the representation is protected.** Submitting client information to AI tools is a disclosure. Enterprise deployment, data classification, and careful abstraction are how you comply.
Rule 3.3 Candor to the Tribunal	**You cannot make false statements of fact or law to a court.** AI-generated citations and quotations must be verified against authoritative sources before filing. Fabricated citations violate Rule 3.3 even if the lawyer did not know they were fabricated.
Rule 5.1 / 5.3 Supervision	**Managerial and supervisory lawyers must ensure subordinate lawyers and non-lawyer staff follow the rules.** When AI is a "non-lawyer assistant," supervision includes reviewing AI output before it becomes work product.
Rule 1.5 Reasonable Fees	**Fees must be reasonable.** AI can reduce time spent on tasks, which affects hourly billing. Fee disclosures may need to reflect the use of AI-assisted work where courts or clients require it.
Rule 1.4 Communication	**The duty to keep clients reasonably informed may include disclosing the use of AI tools in their matters**, particularly when AI is generating substantial output.

Figure 4.1 — AI and the Model Rules

In July 2024, the ABA Standing Committee on Ethics and Professional Responsibility issued Formal Opinion 512, "Generative Artificial Intelligence Tools" — the ABA's first formal ethics opinion directly addressing generative AI in legal practice. Opinion 512 applies across the full range of professional obligations discussed in this chapter: competence (Rule 1.1), confidentiality (Rule 1.6), communication with clients (Rule 1.4), meritorious claims and candor toward the tribunal (Rules 3.1 and 3.3), supervisory responsibilities (Rules 5.1 and 5.3), and reasonable fees (Rule 1.5). The opinion does not create new obligations. It confirms that existing rules apply to generative AI use and explains how they apply in practice.

On competence, Opinion 512 reinforces that lawyers must understand the capacity and limitations of generative AI tools and must periodically update that understanding as the technology evolves. On confidentiality, it addresses the risks of inputting client information into AI systems and the duty to understand how those systems process, store, and retain data. On communication, it provides guidance on when and to what extent lawyers must inform clients about AI use in their representation. On candor, it reaffirms the obligation to verify AI-generated content before submitting it to courts. On supervision, it extends existing supervisory duties to cover the work of nonlawyer assistants using AI and the lawyer's responsibility to ensure AI-assisted work product meets professional standards. On fees, it addresses whether lawyers may bill for time spent using AI and how efficiency gains from AI should be reflected in billing practices.

Opinion 512 is one of the leading national reference points for the topics covered in the remaining sections of this chapter. Where this chapter explains how specific rules apply to AI-assisted legal work, Opinion 512 provides the ABA's formal position confirming and elaborating those applications.

The last point deserves emphasis. Competence requires that you could do the work without AI, even if you choose to use AI for efficiency. If you cannot evaluate whether a research memo is correct because you do not know the relevant law, AI does not make you competent. It makes you dangerous. AI amplifies existing legal expertise. It does not create expertise where none exists.

The competence obligation is ongoing, not one-time. AI capabilities evolve rapidly. What the tools could do a year ago differs from what they can do today, and both differ from what they will do next year. A lawyer who understood AI risks in 2023 but has not updated that understanding may no longer be competently informed. Comment 8's requirement to keep abreast implies continuous learning, not a single educational session followed by static practice.

This does not mean you need to track every AI development or become a technical expert. It means maintaining working knowledge sufficient for your actual AI use. If you use AI for contract review, you should understand current capabilities and limitations for contract review tasks. If you use AI for legal research, you should understand the hallucination risks and verification requirements that apply to research. Match your knowledge to your use.

> **Key Point** Competence under Rule 1.1 requires understanding AI's benefits and risks before using it on client matters. This includes knowing how the technology works, recognizing its failure modes, implementing appropriate verification, and maintaining the legal expertise to evaluate AI output. "The AI made an error" is not a defense. If you submitted it, you own it.

4.2 Supervision: Staff, Vendors, and Nonhuman Assistance

Model Rules 5.1 and 5.3 establish supervision obligations for lawyers with managerial or supervisory authority. Rule 5.1 addresses supervision of other lawyers. Rule 5.3 addresses supervision of nonlawyer assistants. Neither rule mentions AI, but both apply to AI-assisted work in ways that matter for everyday practice.

The supervision framework rests on a simple principle: lawyers are responsible for ensuring that work performed under their authority complies with professional obligations. When a partner assigns research to an associate, the partner has supervisory duties regarding that research. When an attorney delegates document review to a paralegal, the attorney must ensure the paralegal's work meets appropriate standards. The person doing the work may be competent and trustworthy, but the supervising lawyer cannot simply assume compliance. Reasonable measures to ensure proper performance are required.

AI fits awkwardly into this framework because it is neither a lawyer nor a nonlawyer assistant in the traditional sense. It is a tool. But the supervision principles still apply for two reasons. First, someone must take responsibility for AI output, and that someone is the lawyer who uses the output. Second, when lawyers and staff in an organization use AI, the supervisory lawyers must ensure that AI use across the organization complies with professional standards.

Supervising AI output directly means treating AI-generated content with appropriate scrutiny before incorporating it into your work product. The level of scrutiny should match the scrutiny you would apply to work from a junior professional whose reliability you have not yet established. This framing has ethical as well as practical dimensions.

You would not sign a brief drafted by a new associate without reading it. You would not file a motion prepared by an unfamiliar contract attorney without reviewing the analysis. You would not send a client letter written by an intern without checking it for accuracy and tone. The same principle applies to AI output. The source of the draft does not relieve you of responsibility for its contents. If anything, AI's known failure modes make scrutiny more important, not less.

> **Practice Point** When reviewing AI output, imagine you are reviewing work from a junior lawyer you have never worked with before. That junior lawyer is smart and writes well but has been known to make confident mistakes. Check AI output the same way you would check that work.

What constitutes adequate supervision depends on the task and the stakes. For a low-risk internal summary, a quick review for obvious errors may suffice. For a court filing, adequate supervision requires verification of every citation, confirmation of every legal proposition, and assessment of whether the analysis actually supports the conclusions. Higher-stakes work requires more intensive supervision.

Supervising others who use AI extends your responsibility to the lawyers and staff in your organization. If you are a partner, senior associate, or manager with supervisory authority, you must ensure that those under your supervision use AI in ways that comply with professional obligations. This means you need to know whether and how they are using AI, and you need to have reasonable measures in place to ensure their AI use meets standards.

Rule 5.1 makes partners and supervising lawyers responsible for ensuring that lawyers they supervise conform to ethics rules. If a junior associate uses AI to draft a brief and includes hallucinated citations, the supervising partner who failed to implement reasonable supervision measures may share responsibility. The partner does not need to personally verify every citation the associate produces, but the partner needs systems that make it reasonably likely such errors will be caught.

Rule 5.3 extends similar principles to nonlawyer assistants. Paralegals, legal assistants, and other staff who use AI must be supervised to ensure their AI-assisted work complies with professional standards. This includes ensuring they understand the limitations of AI tools, that they verify output appropriately, and that they do not input confidential information into systems that lack adequate protection.

> **Risk Note** Telling subordinates they can use AI without providing guidance on how to use it responsibly is a supervision failure. If you authorize AI use, you must also establish standards for that use and mechanisms to ensure compliance. Permission without parameters is an abdication of supervisory duty.

Reasonable supervision measures in the AI context generally include training, policies, review processes, and monitoring. Training ensures people understand the technology and the organization's expectations. Policies establish what tools are approved, what information can be entered, and what verification is required. Review processes catch errors before they cause harm. Monitoring provides feedback about how tools are actually being used.

Vendor relationships add another supervision layer. When you use AI tools provided by outside vendors, you rely on those vendors to deliver technology that functions as represented. That means conducting due diligence before adoption and maintaining appropriate oversight during use. You should understand the vendor's data handling practices, security measures, and the accuracy characteristics of their tools. You should have contractual terms that address confidentiality, data use, and liability.

Supervisory responsibility is nondelegable in a fundamental sense. You can delegate tasks to AI, to junior lawyers, and to staff. You cannot delegate your professional responsibility for the work product that results. The supervision requirement exists precisely because delegation creates distance between the supervising lawyer and the final product. That distance must be bridged by processes that ensure quality and compliance.

> **Practice Point** When estimating time for AI-assisted work performed by you or by someone you supervise, budget verification time explicitly. If the verification burden is not planned for, it will either be skipped or done poorly.

4.3 Candor and Accuracy: The Court-Facing Standard

Model Rule 3.3 prohibits lawyers from making false statements of fact or law to a tribunal. The rule imposes an affirmative duty: lawyers must not knowingly make false statements, and they must correct false statements they have made. When you sign a brief or motion, you represent that its contents are accurate to the best of your knowledge and belief. AI assistance does not change this obligation. It intensifies it.

The candor obligation intersects with AI use most directly in the hallucination problem. When an AI tool generates a citation to a case that does not exist, and a lawyer includes that citation in a court filing without verification, the lawyer has made a false statement to the tribunal. The falsity may be unknowing if the lawyer genuinely believed the case existed, but that belief was unreasonable given what competent practice now requires. At this stage in the game, lawyers are expected to know that AI can fabricate citations. Relying on AI citations without verification is not reasonable. It is a failure of the inquiry the rules require.

Rule 11 of the Federal Rules of Civil Procedure reinforces this standard in federal court. By presenting a filing to the court, an attorney certifies that to the best of their knowledge, information, and belief, formed after an inquiry reasonable under the circumstances, the legal contentions are warranted by existing law or by a nonfrivolous argument for extending, modifying, or reversing existing law. A filing based on fabricated citations fails this standard. The inquiry was not reasonable because it omitted the basic step of verifying that the cited authorities exist.

The consequences of violating candor obligations with AI-assisted filings have already materialized. In *Mata v. Avianca*, lawyers submitted a brief containing multiple fabricated case citations generated by ChatGPT. The court imposed sanctions, and the reputational damage was severe. Similar incidents have emerged in other jurisdictions. In each instance, the pattern is the same: lawyers used AI to generate legal content, included that content in court filings without adequate verification, and faced consequences when the fabrications were discovered.

Rule 11 requires that legal contentions in filings be warranted by existing law or by a nonfrivolous argument for changing the law. AI-generated citations do not satisfy that standard unless the lawyer has verified that the authorities exist and support the propositions for which they are cited.

What verification requires in the court-filing context is more demanding than in other contexts. Every citation must be confirmed to exist. Every case must be verified as real, with the correct citation format, court, and date. Every holding attributed to a case must be confirmed against the actual opinion. Every quotation must be checked against the source text. These steps were always good practice. They are now essential practice when AI is involved in the drafting process.

Verification must use authoritative sources. Westlaw, Lexis, court websites, and official reporters are authoritative. The AI tool itself is not an authoritative source. Asking the AI whether a citation is real does not satisfy your verification obligation, because the AI may confidently affirm a fabricated citation just as it confidently generated it.

Beyond citations, the candor obligation extends to the accuracy of legal propositions stated in your filing. AI can misstate legal standards, conflate rules from different jurisdictions, present minority positions as majority rules, or describe holdings in ways that do not match what cases actually say. These errors are harder to catch than outright fabrication because the cited case exists. Verification must include reading the authorities you cite, not just confirming they exist.

Factual accuracy in court filings carries similar obligations. If AI drafts a statement of facts, those facts must be grounded in the record. AI may embellish, infer, or simply invent factual details that sound plausible but are not supported. Every factual assertion in a court filing should trace to a specific record citation, and you should confirm that the cited record actually contains what you claim it contains.

Courts are increasingly rejecting "the AI did it" as a defense or mitigation. When you sign a filing, you take responsibility for its contents. Ignorance of AI's tendency to fabricate is no longer an excuse.

The certification standard provides a useful frame for thinking about AI-assisted filings. When you sign, you certify that you conducted a reasonable inquiry and that the contents are accurate to the best of your knowledge. A reasonable inquiry includes understanding that AI can fabricate and misstate, verifying citations through authoritative sources, reading the authorities you cite to confirm they support the propositions for which you cite them, tracing factual assertions to the record, and reviewing the logic of the filing as a whole.

The burden of this inquiry is part of the cost of using AI. If the time required for verification exceeds the time saved by AI drafting, you have not gained efficiency on that task. This is particularly likely for complex filings where accuracy matters most. The calculus may still favor AI assistance for generating a starting draft, but you must budget verification time realistically. AI does not reduce the need for careful lawyering. It shifts when the careful lawyering occurs.

Opposing counsel and courts are watching more carefully than before. The publicity around AI-related sanctions has made lawyers and judges alert to the possibility of fabricated citations. Some judges now require AI disclosures precisely because they want to know when additional scrutiny is warranted. Being caught with a fabricated citation is more likely now than it would have been a few years ago, not because AI errors are more common, but because people are looking for them.

Candor to the tribunal requires that every statement of fact and law in your court filings be accurate. AI assistance does not change this obligation. Every citation must be confirmed through authoritative sources, every holding must be verified against the actual opinion, and every fact must trace to the record.

4.4 Confidentiality Overview

Model Rule 1.6 prohibits lawyers from revealing information relating to the representation of a client unless the client gives informed consent, the disclosure is impliedly authorized to carry out the representation, or an exception applies. This duty is broad. It covers not just privileged communications but all information relating to the representation, whatever its source. It applies regardless of whether disclosure would harm the client. The duty exists from the moment representation begins and continues indefinitely after it ends.

When you enter client information into an AI system, you are potentially revealing that information to the system's operator, to the infrastructure providers that host the system, and, depending on the system's data practices, to an unknown range of other parties and purposes. Whether this constitutes a prohibited disclosure under Rule 1.6 depends on factors including the nature of the information, the system's data handling practices, and whether appropriate safeguards are in place.

The confidentiality analysis for AI use differs from the candor analysis in the previous section. Candor concerns what comes out of the AI process: the accuracy of work product submitted to courts. Confidentiality concerns what goes in: the client information you provide to enable the AI to do useful work. Both risks must be managed, but they require different strategies.

The core tension is that useful AI output often requires detailed input. To get a tailored contract analysis, you need to provide the contract. To get a relevant case assessment, you need to provide the facts. To get useful drafting assistance, you need to explain what you are drafting and why. Vague prompts that protect confidentiality produce generic outputs of limited value. Detailed prompts that produce useful outputs require disclosing information that may be confidential.

This tension cannot be eliminated, only managed. Management requires understanding what data protections different AI systems provide and matching your use to systems with protections appropriate for the sensitivity of the information involved. It also requires understanding what information is truly necessary for each task and providing only what is needed rather than more than needed.

Consumer versus enterprise AI presents different risk profiles. Consumer-facing AI products, the kind you access through a web browser without an enterprise agreement, typically offer weaker confidentiality protections. Their terms of service often permit the provider to use input data for model training, which means your client's information could influence outputs that other users see. Even when training use is disclaimed, data may be retained for other purposes, accessed by the provider's personnel, or stored in ways that create security risks.

Enterprise AI deployments, accessed through contracts that specify data handling terms, can offer stronger protections. These deployments may include contractual commitments against using customer data for training, data isolation that keeps your information separate from other customers, defined retention and deletion policies, and security measures appropriate for sensitive information. The protections vary significantly across vendors and contract terms, so evaluation of the specific arrangement is essential.

> **Risk Note** Using consumer AI tools for confidential client information without understanding and accepting the confidentiality implications creates a serious Rule 1.6 problem. If you would not email the information to a stranger, think carefully before entering it into a consumer AI system.

Informed consent can authorize disclosures that would otherwise violate Rule 1.6. If a client consents to your use of AI tools on their matter after being informed of the associated risks, the confidentiality barrier may be cleared. But informed consent requires actual information. The client must understand what AI tools you will use, what information will enter those tools, what happens to that information, and what risks are involved. A general engagement letter clause mentioning technology use probably does not constitute informed consent to specific AI data practices.

Whether to seek explicit AI consent for every matter or to rely on implied authorization for routine technology use is a judgment call that depends on the sensitivity of the matter, the client's sophistication, and the risk profile of the AI tools you use. The key point is that consent is available as a tool for managing confidentiality concerns, but it must be genuine informed consent to be effective.

Practical safeguards reduce confidentiality risk regardless of the consent analysis. These include using enterprise tools with appropriate data protections, anonymizing or redacting sensitive identifiers where possible, providing only the information necessary for the specific task, and understanding your tools' data practices before using them.

4.5 Billing, Disclosure, and Client Expectations

The economics of AI-assisted legal work create tensions that the profession is still learning to navigate. A task that once took four hours might now take one hour of prompting and verification. The work product may be equivalent or better. The client benefited from your legal judgment in both cases. But the time billed differs dramatically, and that difference raises questions about what you are charging for, what you should disclose, and how to maintain trust when the production process has fundamentally changed.

Model Rule 1.5 requires that fees be reasonable. Time is one factor among several, not the sole determinant. A fee can be reasonable even if the time spent was minimal, provided the other factors support it. Conversely, extensive time does not automatically make a fee reasonable if the time was spent inefficiently.

AI complicates the time factor because it compresses production time while potentially increasing the skill required to use the technology effectively. The lawyer who produces a quality brief in two hours using AI has done something valuable, arguably more valuable than the lawyer who produced the same brief in eight hours without AI. The client got the same result faster, freeing capacity for other matters. Charging less simply because less time elapsed may undervalue the expertise involved in achieving that efficiency.

Billing for AI-assisted work requires honesty about what you actually did. If you bill by the hour, you cannot bill for time you did not spend. Four hours of billed time must represent four hours of actual work. AI-assisted work that took one hour cannot be billed as four hours simply because it would have taken four hours without AI. That is not creative billing. It is fraud.

The honest approach when billing hourly is to bill the time actually spent: the time prompting, the time reviewing and verifying, the time revising and finalizing. If that time is one hour, you bill one hour. The client benefits from your efficiency, and you maintain integrity in your billing practices.

Value-based or flat-fee arrangements avoid some of this tension. When you charge a flat fee for a deliverable, the time you spend producing it is your concern, not the client's. If AI helps you produce a quality result in less time, you have increased your effective hourly rate without any billing ethics issue. The client agreed to pay a specified amount for a specified result. How you achieved that result is your business.

Disclosure to clients about AI use is distinct from billing but related. The Model Rules themselves do not explicitly require client disclosure of AI use, but the broader landscape is shifting rapidly. Court orders requiring AI disclosure in filings are appearing across federal and state courts. Bar associations are issuing guidance that frequently favors transparency. Firm policies and client engagement standards increasingly address AI disclosure directly. The question is no longer purely one of professional judgment — it is increasingly a function of the rules, orders, and policies governing a particular matter. If a client would want to know that AI is involved, failing to disclose deprives them of information relevant to their engagement.

> Practice Point Consider adding AI disclosure to your standard engagement process. A paragraph in your engagement letter or your intake materials explaining your AI practices normalizes the disclosure and ensures consistent communication.

The right level of disclosure depends on context. For institutional clients with their own AI policies, more detailed disclosure may be expected. For individual clients less familiar with AI, a simple explanation that you use AI tools with appropriate safeguards may be enough. The important thing is honesty and clarity. Whether disclosure is required by a court is a separate question — Chapter 13 addresses the growing landscape of court-imposed disclosure rules and the pre-filing check that practice point recommends.

Billing for AI-assisted work requires honesty about time actually spent. Bill the hours you worked, not the hours you would have worked without AI. Consider whether flat-fee or value-based pricing better aligns incentives when AI produces genuine efficiency gains.

4.6 Documentation: Audit-Ready Habits

Documentation serves two purposes that reinforce each other. The first is protection: when questions arise about your AI use, a clear record demonstrates that you acted responsibly. The second is improvement: documenting what you did helps you learn from experience and refine your practices over time.

The professional responsibility framework makes documentation valuable even when it is not explicitly required. If a court asks how you verified a filing, you want an answer more specific than "I reviewed it carefully." If a client questions whether confidential information was adequately protected, you want evidence of the safeguards you employed. If a malpractice claim alleges that AI-assisted work was deficient, you want records showing the supervision you exercised.

Think of documentation as creating a contemporaneous record of your professional judgment. At the time you used AI and verified its output, you made decisions about what tools to use, what information to provide, what verification steps to take, and what changes to make before relying on the work product. Documentation captures that judgment while it is fresh, before memory fades and before any dispute arises that might color your recollection.

For significant work product, documentation should capture at minimum what AI tool you used, what task you asked it to perform, what verification steps you completed, and what changes you made before finalizing the work. This core information establishes that you exercised supervision and did not simply accept AI output uncritically.

A simple documentation format for most purposes is a brief note in the matter file. The note might read: "Motion to compel: initial draft generated using [Tool] on [Date]. All citations verified against Westlaw. Argument structure revised to add proportionality analysis. Fact section rewritten to match record cites. Final version reflects attorney work product after verification and revision." That note takes thirty seconds to write and establishes the essential facts about your process.

Documentation can be discoverable. Records you create about your AI use might be subject to discovery in litigation, including malpractice litigation. Document facts about your process, not self-critical commentary that could be used against you.

Prompt preservation and output preservation raise additional questions. Some lawyers preserve their actual prompts and outputs for significant work product. Others preserve summaries or logs rather than the complete text. Either approach can work, but be consistent and think through discoverability and confidentiality implications before choosing a method.

The important point is that documentation should be proportionate to the stakes. Not every AI interaction requires a formal record. But for significant work product, and especially for court filings or major client advice, documentation is part of responsible practice.

4.7 Court Rules and Orders About AI

Courts have not waited for bar associations or legislatures to address AI in legal practice. Judges who saw *Mata v. Avianca* and similar incidents recognized that existing rules, while applicable, might not be sufficient to ensure lawyers take AI risks seriously. The response has been a wave of court-specific requirements addressing AI use in litigation.

The landscape is evolving rapidly. As of this writing, multiple federal and state courts have adopted some form of AI-related requirement, whether through local rules, standing orders, or administrative directives. Common elements include mandatory disclosure of AI use in certain filings, certification that AI-generated content has been verified, and affirmation that the attorney takes responsibility for the filing regardless of how it was produced.

Standing orders from individual judges are currently the most common form of AI regulation in federal courts. These orders apply to cases before a particular judge and reflect that judge's assessment of what safeguards are needed. Some standing orders are brief, requiring only a certification that AI output has been reviewed. Others are elaborate, specifying verification procedures, disclosure formats, and consequences for noncompliance.

> **Authority Note** Judge Brantley Starr of the Northern District of Texas issued one of the earliest AI-specific standing orders, requiring lawyers either to certify that no portion of a filing was drafted by generative AI or to certify that all AI-generated language was checked for accuracy using traditional legal sources.

Disclosure requirements are the most common element of court AI rules. These requirements typically ask lawyers to state whether AI was used in preparing a filing and, if so, what role AI played. The rationale is straightforward: judges want to know when filings may carry heightened risk of the errors AI is known to produce.

Certification requirements go beyond disclosure to require affirmative statements about verification. A typical certification may require the attorney to affirm that all AI-generated content has been reviewed for accuracy, that all citations have been verified, and that the attorney takes full responsibility for the filing. These certifications make explicit what Rule 11 already implies.

Lawyers should expect that AI disclosure and certification requirements will become more common, not less. Planning your practice around the assumption that such requirements will apply positions you better than treating each requirement as an anomaly.

The practical lesson is simple: before filing in any court, check the local rules, standing orders, and any case-specific orders for AI-related requirements. Build compliance into your filing workflow so that disclosure and certification are not afterthoughts.

4.8 Mini-Lab: Risk-Score an AI-Assisted Filing

Objective: Apply the professional responsibility framework from this chapter to assess the risks in a realistic AI-assisted filing scenario. Develop a systematic approach to identifying ethical issues before they become problems.

Time Required: 45 to 60 minutes

What You Will Need:

The professional responsibility concepts from this chapter: competence, supervision, candor, confidentiality, billing and disclosure, documentation, and court rules

The three-risk framework from Chapter 1: accuracy, confidentiality, accountability

A method for taking notes

A willingness to think critically about practices that might seem routine

Instructions

Step 1: Read the Scenario

You are a senior associate at a mid-sized litigation firm. A partner has asked you to review a motion for summary judgment that a junior associate prepared with AI assistance. The motion will be filed in federal court in the Eastern District of Pennsylvania. The case involves a contract dispute with approximately $2 million at stake. The filing deadline is tomorrow afternoon.

The junior associate tells you the following about how the motion was prepared:

She used a general-purpose AI chatbot, consumer version, accessed through her personal account, to generate an initial draft of the legal argument section. She provided the AI with the relevant contract provisions, a summary of the key facts, and the question she wanted addressed. She did not include party names but did include specific dollar figures and dates from the actual transaction.

The AI produced a twelve-page draft with citations to seven cases. The junior associate read through the draft, thought it looked good, and made some revisions to the introduction and conclusion. She added the statement of facts herself based on the record. She did not independently verify the case citations because she recognized two of the case names as leading cases in the area and assumed the others were similarly legitimate.

The firm has no formal AI use policy. The client's engagement letter is a standard form that does not mention AI. The partner who assigned the matter has not asked about AI use and does not know AI was involved. The junior associate did not document her AI use or verification process.

The assigned judge has a standing order requiring attorneys to certify whether generative AI was used in preparing any filing and, if so, to affirm that all AI-generated content was verified for accuracy.

Step 2: Identify the Professional Responsibility Issues

Work through each area of professional responsibility covered in this chapter and identify the specific issues present in this scenario.

Competence

Supervision

Candor

Confidentiality

Disclosure and Client Communication

Documentation

Court Rules

Step 3: Assign Risk Scores

For each area of professional responsibility, assign a risk score from 1 to 5.

1 = no significant risk

2 = minor risk

3 = moderate risk

4 = serious risk

5 = critical risk

Step 4: Develop a Remediation Plan

For each issue you identified, determine what must be done before the motion can be filed. Distinguish between actions that are absolutely required and actions that are strongly advisable.

Step 5: Create a Prospective Protocol

Based on what you learned from analyzing this scenario, draft a brief protocol that would prevent similar issues in future matters.

Deliverable

By the end of this Mini-Lab, you should have:

1. A list of professional responsibility issues organized by category
2. Risk scores for each category with brief explanations
3. A remediation plan distinguishing required actions from advisable actions
4. A prospective protocol capturing key safeguards for future AI-assisted filings

Bonus Challenge

Draft the conversation you would have with the junior associate after the filing is complete. The conversation should ensure she understands what went wrong, provide clear guidance on verification requirements going forward, and identify what support or training she needs.

Then draft the conversation you would have with the partner about implementing firm-wide AI guidance. Explain why a policy is needed, what it should address, and how to implement it without creating excessive burden.

Key Takeaways

1. Competence under Rule 1.1 now includes technological competence. Lawyers must understand AI well enough to use it responsibly and to supervise its output adequately.
2. Supervision duties extend to AI output and to lawyers and staff who use AI under your authority.
3. Candor to the tribunal requires full verification of AI-assisted court filings. Every citation, holding, quotation, and factual assertion must be checked.
4. Confidentiality under Rule 1.6 constrains what client information can enter AI systems and requires understanding the protections of the tools you use.
5. Billing and disclosure require honesty about what you actually did and transparency with clients about AI use when appropriate.
6. Documentation protects you, supports compliance, and enables improvement.
7. Courts are increasingly adopting AI-specific rules, orders, and certifications, and lawyers must check for and comply with those requirements.

Chapter Checkpoint

Before moving on, make sure you can:

- ☐ Explain how Comment 8 to Rule 1.1 applies to AI use and what technological competence requires in practice
- ☐ Describe your supervision obligations when reviewing AI-assisted work prepared by a junior lawyer or staff member
- ☐ Identify the verification steps required before filing an AI-assisted document with a court
- ☐ Assess whether a particular AI tool is appropriate for confidential client information based on its data handling practices
- ☐ Draft an AI disclosure paragraph for a client engagement letter
- ☐ Create a documentation protocol for AI-assisted work that balances thoroughness with practicality
- ☐ Check for and comply with AI-specific court requirements in any jurisdiction where you practice

What's Next

This chapter provided an overview of how professional responsibility obligations apply to AI use. Chapter 5 goes deeper on one of those obligations: confidentiality. The confidentiality analysis is more complex than the overview suggested, involving data classification frameworks, distinctions among deployment types, safe input workflows, and incident response protocols. If confidentiality risk is what keeps many lawyers awake at night when thinking about AI, Chapter 5 provides the detailed guidance needed to manage that risk systematically rather than through anxiety and avoidance.

Key Terms

Technological competence — The obligation under Comment 8 to Model Rule 1.1 to understand the benefits and risks of technology used in legal practice. For AI tools, this includes understanding how the technology works, its limitations and failure modes, and the verification processes necessary for responsible use.

ABA Formal Opinion 512 — The ABA Standing Committee on Ethics and Professional Responsibility's July 2024 opinion on generative AI in legal practice. Addresses competence, confidentiality, communication, candor, supervision, and fees under the Model Rules as applied to generative AI tools.

Supervision (under the Model Rules) — The obligation of lawyers with managerial or supervisory authority to ensure that conduct by those under their supervision conforms to professional rules. In the AI context, this extends to supervising AI output directly and to supervising lawyers and staff who use AI.

Candor to the tribunal (Model Rule 3.3) — The duty prohibiting lawyers from making false statements of fact or law to a court. In the AI context, this requires verification that all citations exist, that characterizations of holdings are accurate, and that factual assertions are supported by the record.

Confidentiality (Model Rule 1.6) — The duty not to reveal information relating to the representation of a client. In the AI context, this duty constrains what information lawyers can enter into AI systems based on those systems' data handling practices and the protections they provide.

Data handling practices — How an AI system processes, stores, retains, and uses information entered into it. Understanding these practices is essential for assessing whether use of a particular system is compatible with confidentiality obligations.

Reasonable fees (Model Rule 1.5) — Fees determined by multiple factors including time, difficulty, skill required, results obtained, and customary charges. AI use affects the time factor but may increase the skill required for effective deployment.

Audit-ready documentation — Records sufficient to demonstrate, if later questioned, that AI-assisted work was performed in compliance with professional obligations. The goal is a contemporaneous record without excessive administrative burden.

Standing order — A directive issued by an individual judge that applies to all cases on that judge's docket. AI-related standing orders typically require disclosure of AI use, certification of verification, or both.

Chapter 5: Confidentiality and AI: Protecting Client Information in Every Prompt

A lawyer enters a client's trade secret into a consumer AI chatbot to help draft a licensing agreement. The chatbot's terms of service allow the provider to use inputs for model training. Six months later, a competitor prompts the same AI about that industry and receives suggestions that incorporate language suspiciously similar to the client's proprietary formulation. This scenario isn't hypothetical paranoia. It's the logical endpoint of using AI tools without understanding how they handle the information you provide.

Chapter 4 introduced confidentiality as one of several professional responsibility obligations. This chapter goes deeper, treating confidentiality not as a single rule to follow but as a system to implement. The duty to protect client information predates AI by centuries, but AI creates confidentiality exposures that traditional practice never contemplated. Every prompt is a potential disclosure. Every AI tool has a data policy. Every failure to understand those policies creates risk that travels with you long after the conversation ends.

What This Chapter Covers

- How Rule 1.6 applies to AI inputs and why "relating to the representation" is broader than you think
- A data classification framework for determining what can enter which AI systems
- The critical distinctions among consumer, enterprise, and API deployments
- Safe input workflows that protect confidentiality without eliminating AI's usefulness
- Anonymization and abstraction techniques that preserve analytical value while removing identifying information
- What to do when confidentiality incidents occur and how to prevent them from recurring
- How to build confidentiality protections into your AI practice systematically

Chapter Summary

Model Rule 1.6 protects all information relating to the representation of a client, regardless of source. When you enter information into an AI system, you disclose it to that system's operator and potentially to its training processes, downstream users, or security vulnerabilities. Managing this risk requires understanding what information you have, how sensitive it is, and what protections each AI tool provides. This chapter introduces a four-tier data classification framework that maps information sensitivity to AI tool requirements. It explains the differences among consumer AI tools (which typically train on inputs), enterprise deployments (which offer contractual protections), and API access (which provides the most control). Safe input workflows operationalize these distinctions through practical techniques: anonymization that removes identifying details, abstraction that generalizes specific facts, and structural approaches that keep sensitive material out of AI systems entirely. When confidentiality incidents occur despite precautions, a response protocol minimizes harm and prevents recurrence. The goal is not to avoid AI use but to use AI in ways that honor your fundamental obligation to protect what clients share with you.

5.1 Rule 1.6 and the Scope of "Information Relating to the Representation"

Model Rule 1.6(a) states that a lawyer "shall not reveal information relating to the representation of a client unless the client gives informed consent." This language is deliberately broad. It doesn't say "confidential information" or "privileged information" or "secrets." It says "information relating to the representation." Understanding that breadth is the starting point for AI confidentiality analysis.

The scope of Rule 1.6 extends beyond what most lawyers intuitively consider confidential. It includes information the client told you directly. It includes information you learned from other sources during the representation. It includes information that's publicly available if your use of it would reveal something about the representation. It includes information about the fact of the representation itself—that you represent this client on this matter—unless that fact is already public.

Rule 1.6's protection extends beyond information communicated in confidence by the client to all information relating to the representation, whatever its source.

When you enter a prompt into an AI system, you're disclosing whatever information that prompt contains. The question is: to whom? The answer depends on the AI tool you're using, but the range of potential recipients is wider than many lawyers realize.

At minimum, your prompt goes to the AI provider—the company operating the model. Depending on the provider's terms of service and data practices, your input may be stored in logs, reviewed by employees for quality assurance, used to train future model versions, or retained indefinitely in systems you cannot access or audit. If the provider suffers a data breach, your inputs may be exposed to unauthorized third parties. If the provider is acquired or goes bankrupt, your data may transfer to entities you never anticipated.

This doesn't mean AI use automatically violates Rule 1.6. But it means every AI use is a disclosure that requires analysis. The analysis asks: What am I disclosing? To whom? With what protections? Does this disclosure fall within an exception to Rule 1.6, or does it require client consent?

Several ethics opinions have addressed the application of confidentiality rules to cloud computing and AI. While specific conclusions vary, common principles emerge. Lawyers must make reasonable efforts to prevent unauthorized disclosure. Reasonable efforts require understanding how a technology works and what risks it creates. If a technology involves disclosure to third parties, the lawyer must evaluate whether that disclosure is permissible and what safeguards exist.

The "reasonable efforts" standard from Rule 1.6(c) is crucial here. The rule doesn't require perfect security; it requires reasonable security. What's reasonable depends on the sensitivity of the information, the likelihood of disclosure, the available safeguards, and the cost of implementing them. Highly sensitive information requires stronger protections. Less sensitive information may tolerate more risk. But some level of analysis is always required.

> **Practice Point** Before you enter any client information into an AI tool, ask three questions: What information am I disclosing? What does this provider do with inputs? Is this disclosure reasonable given the sensitivity of the information? If you cannot answer all three questions, stop and find the answers before proceeding.

One common mistake is assuming that information already available publicly cannot be confidential. Rule 1.6 protects information relating to the representation, which can include public information. If your use of a public fact would reveal that you represent a client, that you're working on a particular matter, or that certain facts are relevant to your analysis, the public information becomes protected in that context. Entering a publicly known company name into an AI prompt might not be problematic standing alone, but combining it with details about a dispute reveals more than the individual facts suggest.

Another mistake is treating privilege and confidentiality as synonymous. Attorney-client privilege is a rule of evidence that protects communications from compelled disclosure in legal proceedings.

Confidentiality under Rule 1.6 is an ethical obligation that restricts voluntary disclosure in any context. Privilege is narrower and carries different exceptions. Confidentiality is broader and is the relevant framework for AI analysis. Information that isn't privileged can still be confidential. The AI analysis happens under Rule 1.6, not evidence law.

> **Key Point** Every prompt containing client information is a disclosure under Rule 1.6. The duty to protect confidentiality requires understanding what happens to that information after you enter it. "I didn't know" is not a defense. Understanding your tools is part of your professional obligation.

The practical implication is that AI confidentiality analysis requires due diligence on every tool you use. You need to know the provider's data practices before you enter client information. You need to read terms of service—or have someone read them for you—and understand what rights you're granting. You need to assess whether the tool's protections are adequate for the information you plan to enter. This due diligence isn't optional; it's what "reasonable efforts" requires in the AI context.

5.2 Data Classification: A Framework for Decision-Making

Not all information carries the same sensitivity. A client's publicly known business address requires less protection than their trade secret formula. A routine contract dispute involves less sensitivity than a whistleblower matter. Treating all information identically—either by protecting everything as if it were the crown jewels or by treating nothing as sensitive—produces either paralysis or exposure. A classification framework provides the middle path: protecting information in proportion to its sensitivity.

This section introduces a four-tier framework for classifying information that might enter AI systems, as illustrated in Figure 5.1. The tiers map to different AI deployment types and different safeguard requirements. The framework isn't the only possible approach, and you may need to adapt it for your practice. But it provides a starting point for making disciplined decisions about what goes where.

Tier 1: Public or Non-Sensitive Information

Tier 1 includes information that creates no confidentiality exposure if disclosed. This might include general legal questions not tied to any client, publicly available legal authorities, hypothetical scenarios without connection to actual matters, or your own work product when you own the rights and it contains nothing client-specific.

Tier 1 information can enter any AI system, including consumer tools with minimal privacy protections. The information isn't confidential, so disclosure doesn't implicate Rule 1.6.

Example: "Summarize the key holdings from the Supreme Court's personal jurisdiction cases from the past five years." This prompt contains no client information and can go anywhere.

Tier 2: De-Identified or Abstracted Information

Tier 2 includes information derived from client matters but processed to remove or obscure identifying details. The underlying facts come from a representation, but the prompt has been crafted so that no one reading it could connect it to a specific client, matter, or confidential fact.

Tier 2 information can enter enterprise AI tools that offer reasonable data protections but should generally stay out of consumer tools that train on inputs. The abstraction reduces sensitivity, but some residual risk remains if the de-identification is imperfect or if patterns could enable re-identification.

The Data Tier Framework

Classify the data first. Then select the deployment.

TIER	DESCRIPTION	EXAMPLE	DEPLOYMENT
TIER 1 — Public or Non-Sensitive	Publicly available information, general legal research, reference material. Nothing client-identifying.	• *Published case law citations.* • *Statutory research.* • *Secondary source identification.*	**Any AI tool** Consumer, enterprise, or API deployments all acceptable.
TIER 2 — De-Identified or Abstracted	Client information with identifiers removed, or abstracted to a general legal question. Scope narrowed to what the model needs.	• *'A client in an employment dispute with a non-compete clause...'*	**Enterprise AI preferred** Consumer AI acceptable with careful abstraction. Confirm data handling before submission.
TIER 3 — Identified but Controlled	Client-identifying information with governance controls in place. Contractual protections required.	• *Specific facts of a matter under NDA.* • *Internal memos about identified clients.*	**Enterprise AI with DPA** No consumer tools. Require contractual protections, retention controls, zero training on inputs.
TIER 4 — Highly Sensitive	Privileged communications, trade secrets, attorney work product, regulated (HIPAA, GLBA, state privacy laws). Maximum protection.	• *Privileged memos, M&A deal terms.* • *Protected health information under HIPAA.*	**On-premises only** Private deployment. Many matters require no AI at all. When in doubt, do not use AI.

The tier determines what tools are acceptable.
Never submit *Tier 3 or Tier 4 data to a tool approved only for lower tiers.*

Figure 5.1 — The Data Tier Framework

Example: Instead of "My client Acme Corp is being sued by their former CEO for wrongful termination after she reported financial irregularities," you might ask: "In a wrongful termination case where the plaintiff alleges retaliation for reporting financial misconduct, what are the key elements the employer would need to prove to establish a legitimate non-retaliatory reason for termination?" The facts are generalized, the parties are unnamed, and the prompt doesn't reveal that you're working on any particular matter.

Tier 3: Identified but Controlled Information

Tier 3 includes information that identifies clients or matters but doesn't involve the most sensitive categories. This might include client names and general matter descriptions, routine correspondence, standard contract provisions being negotiated, or factual scenarios where the specific identities matter for the analysis but the information isn't highly sensitive.

Tier 3 information requires enterprise AI deployments with contractual confidentiality protections, clear data handling policies, and appropriate security measures. It should not enter consumer AI tools regardless of convenience. The information is confidential under Rule 1.6, and adequate safeguards are required.

Example: Using an enterprise AI tool to draft a demand letter for "ABC Industries regarding the delayed shipment of component parts under Purchase Order 2024-4521." The client name and matter details are present, but the information isn't in the highest sensitivity category.

Tier 4: Highly Sensitive Information

Tier 4 includes information where disclosure could cause serious harm, where heightened protection obligations exist, or where the nature of the matter demands exceptional care. This includes trade secrets and proprietary business information, personal health or financial information protected by law, matters involving physical safety, whistleblower identities and allegations, criminal defense matters, and any information where the client has expressed particular sensitivity.

Tier 4 information requires the strongest available protections. This might mean API access with no data retention, on-premises AI deployments, or a determination that AI assistance isn't appropriate for this information at all. The heightened sensitivity demands heightened safeguards, and "enterprise" protections alone may not be sufficient.

If you're uncertain whether information belongs in Tier 3 or Tier 4, treat it as Tier 4. The cost of over-protection is inefficiency. The cost of under-protection is potential harm to clients and exposure to discipline. Err toward caution.

The framework requires judgment at every step. Classification isn't mechanical. Two matters involving the same general facts might warrant different tiers based on client sensitivity, regulatory exposure, or strategic considerations. The same information might move between tiers as a matter progresses—an allegation that starts as one theory among many might become central to the case and warrant heightened protection.

Document your classification decisions for significant matters. You don't need to write an essay, but a note in your file indicating how you classified the matter's information and why supports your compliance if questions arise later. It also forces you to actually make the decision rather than proceeding on autopilot.

When in doubt, ask the client. If you're uncertain whether information is sensitive enough for Tier 4 treatment, your client probably has a view. Sophisticated clients will appreciate that you're thinking carefully about data protection. Less sophisticated clients will learn something about the risks, which serves their interests too.

5.3 Consumer, Enterprise, and API: Understanding Deployment Types

The same underlying AI model can be accessed through different deployment types, and those deployment types carry dramatically different confidentiality implications. Understanding these distinctions is essential for matching your data classification to appropriate tools.

Consumer AI Tools

Consumer AI tools are the chatbots and applications marketed to individual users: the free tiers of popular AI assistants, browser-based interfaces, and mobile apps. They're accessible, convenient, and usually free or low-cost. They're also the most problematic for confidential information.

Consumer tools typically retain your inputs and outputs. They often use your conversations to improve their models, which means your prompts may influence what the model says to other users. Their terms of service usually grant the provider broad rights to your data. Their privacy protections are designed for consumer expectations, not professional obligations.

Some consumer tools now offer options to opt out of training data use or to delete conversation history. These features help but don't eliminate concerns. Opt-out may not be retroactive. Deletion may not be complete. The underlying data practices may still involve retention you can't control. And the security posture of consumer services may not meet the "reasonable efforts" standard for sensitive professional information.

> **Risk Note** The default assumption for consumer AI tools should be: your inputs may be retained, reviewed by provider employees, used for training, and potentially exposed in a breach. Unless you have confirmed otherwise through careful review of current terms and policies, treat consumer tools as unsuitable for any client-specific information.

Enterprise AI Deployments

Enterprise AI refers to AI services offered under business contracts rather than consumer terms of service. Enterprise arrangements typically include negotiated terms, confidentiality provisions, data handling commitments, and sometimes compliance certifications.

Common features of enterprise AI deployments include: no use of inputs for model training, data retention policies you can review and influence, contractual confidentiality obligations, security certifications such as SOC 2, deletion upon request or contract termination, and designated personnel for data protection inquiries.

Enterprise doesn't mean risk-free. The quality of protections varies by provider and by contract. "Enterprise" is a marketing term, not a guarantee. You still need to evaluate what protections each specific enterprise arrangement provides. But enterprise deployments are generally appropriate for Tier 3 information when the specific protections match the sensitivity.

Many law firms now have firm-wide enterprise AI agreements. If your firm has such an arrangement, understand what it covers. Which tools are included? What data handling terms apply? Are there usage restrictions you need to follow? The presence of an enterprise agreement doesn't eliminate your individual responsibility; it changes what protections you can rely on when you've confirmed the agreement applies to your use.

API Access

API (Application Programming Interface) access means connecting to an AI model programmatically rather than through a user interface. API access is typically used by developers building applications or by organizations integrating AI into their own systems.

API access usually provides the most control over data handling. Common features include: explicit policies on data retention (often with options for zero retention), no use of inputs for training, the ability to run requests through your own infrastructure, detailed logging and audit capabilities, and contractual terms negotiated for your specific use case.

API access is most appropriate for Tier 4 information when AI assistance is warranted at all. It allows organizations to build wrappers that add logging, filtering, or additional protections. It enables integration with secure environments. It provides the transparency and control that highest-sensitivity information requires.

The tradeoff is complexity. API access requires technical capability to implement. It lacks the polished interfaces of consumer tools. It's more expensive and harder to use casually. For most lawyers, API access will be provided through firm infrastructure rather than individual implementation. But understanding that it exists and what it offers helps you advocate for appropriate tools when you need them.

Ask your firm's technology team what AI tools are available, what deployment types they represent, and what data protections each provides. If the answer is "we haven't looked into that," you've identified a gap that needs attention. If the answer is detailed and thoughtful, you've found resources to help you use AI responsibly.

Mapping Tiers to Deployment Types

The four-tier classification framework maps to deployment types as follows:

- Tier 1 (public/non-sensitive): Any tool, including consumer AI
- Tier 2 (de-identified/abstracted): Enterprise AI with standard protections
- Tier 3 (identified but controlled): Enterprise AI with verified confidentiality terms
- Tier 4 (highly sensitive): API with enhanced controls, or no AI use

This mapping isn't absolute. Specific tools might offer stronger or weaker protections than their category suggests. Your firm might have negotiated enhanced terms for what would otherwise be a standard enterprise tool. The point is to have a framework for decision-making, not a mechanical rule that substitutes for judgment.

The deployment type determines what protections exist, not what category the tool falls into. "Enterprise" and "API" describe contractual arrangements, not inherent security properties. Always verify what protections actually apply before relying on categorical assumptions.

5.4 Safe Input Workflows: Techniques for Protecting Confidential Information

Understanding data classification and deployment types provides a framework. Safe input workflows provide the practice. This section covers specific techniques for using AI productively while maintaining confidentiality protections.

Anonymization

Anonymization replaces identifying information with placeholders while preserving the structure and relationships that matter for analysis. Effective anonymization removes the link between the prompt and any specific client or matter while retaining enough substance for the AI to provide useful output.

Basic anonymization replaces names with generic labels: "Company A," "Party 1," "the Plaintiff," "the Employer." It replaces specific locations with general descriptions: "a state in the northeast," "a major metropolitan area." It replaces dates with relative markers: "three months before the termination," "the following quarter." It replaces numerical values with ranges or categories: "a seven-figure amount," "more than one hundred employees."

Effective anonymization requires attention to indirect identifiers. A combination of industry, geography, company size, and transaction type might identify a matter even without naming the parties. If you describe "a Fortune 500 pharmaceutical company headquartered in New Jersey facing an FCPA investigation related to their Chinese subsidiary," you may have identified the client despite using no names. Anonymization must address these combinations, not just direct identifiers.

After anonymizing a prompt, read it through the eyes of someone trying to identify the matter. Could a journalist, opposing counsel, or competitor figure out who you're representing? If indirect identifiers create risk, generalize them further until identification becomes implausible.

One useful technique is creating an anonymization key that you keep separate from the prompt. You assign placeholders systematically—Party A is always the client, Party B is always the counterparty, Amount 1 is always the transaction value—and record the mapping in a document that never enters the AI system. This lets you translate AI output back to your actual facts while maintaining protection during the AI interaction.

Abstraction

Abstraction goes further than anonymization by generalizing specific facts into categories or hypotheticals. Instead of describing your actual case with identifying details removed, you describe a scenario that captures the analytical questions without corresponding to any specific facts.

Example of progression from specific to abstract:

- Original: "Acme Corp wants to know if their non-compete with John Smith is enforceable. Smith signed a two-year non-compete covering the tri-state area when he joined as VP of Sales. He just left to join competitor Omega Inc. as their Chief Revenue Officer. Acme is incorporated in Delaware, headquartered in New York, and Smith worked from the Connecticut office."
- Anonymized: "Company A wants to know if their non-compete with Employee 1 is enforceable. Employee 1 signed a two-year non-compete covering a three-state region when he joined as a senior sales executive. He just left to join a competitor in a similar senior role. The company is incorporated in State X, headquartered in State Y, and the employee worked from State Z."
- Abstracted: "What factors determine whether a geographic restriction in a non-compete agreement is reasonable under New York law? Specifically, how do courts analyze a multi-state restriction when the employee held a senior sales role with access to customer relationships across the restricted territory?"

The abstracted version asks the same analytical question without revealing anything about the specific parties, the specific geography, or even the fact that you're working on a live matter. The AI can provide useful analysis, and you can apply that analysis to your facts without having disclosed them.

Structural Separation

Some workflows avoid confidentiality risk by never combining AI with client information in the same step. The AI performs one task using only public or non-confidential inputs; the lawyer then separately applies that output to client facts.

For example, rather than asking AI to analyze your client's contract, you might ask AI to explain how courts generally interpret force majeure clauses in supply agreements, what factors they consider, and what common drafting pitfalls exist. You then separately read your client's clause against that general framework, applying the AI's educational output to your confidential facts through your own analysis.

This approach sacrifices some of AI's efficiency—the AI can't do the application work if it doesn't have the specific facts—but it eliminates confidentiality exposure entirely. For Tier 4 information or situations where even enterprise protections feel inadequate, structural separation provides a path forward.

Chunking and Limiting Disclosure

Sometimes a task requires confidential information, but not all of it. Chunking divides the task into pieces and provides only the information necessary for each piece.

Instead of uploading an entire contract and asking for comprehensive analysis, you might extract specific provisions and ask about each in isolation. The AI sees less information, reducing exposure if something goes wrong. The tradeoff is that the AI can't identify issues arising from interactions between provisions it hasn't seen.

Limiting disclosure also means providing minimum necessary detail. If you're asking about enforceability of a non-compete, does the AI need to know the executive's actual job title, or is "senior sales executive" sufficient? Does it need the exact dollar amount of the alleged damages, or is "significant" enough? Every additional detail is additional exposure. Include what's necessary; omit what's not.

Safe input workflows operate on a principle of minimum necessary disclosure. Anonymize identifying information. Abstract specific facts when possible. Separate AI analysis from confidential application. Limit what you share to what the task actually requires. Every omission reduces risk without necessarily reducing usefulness.

Pre-Submission Review

Before entering any prompt involving client matters, review what you're about to submit. Read it as if you were opposing counsel who intercepted it. Could they identify your client? Could they learn strategic information? Could they use it against your client's interests?

This review takes seconds but catches problems that post-hoc remediation cannot fix. Once information enters an AI system, you may not be able to retrieve it. The review before submission is your last line of defense.

> Practice Point Develop a habit of reading every prompt aloud before submitting it. This slows you down just enough to engage your judgment rather than your reflexes. It also makes the content of the prompt salient in a way that silent reading sometimes does not.

Consider implementing a brief pause before submission for any prompt involving even Tier 2 information. The pause forces conscious decision-making and prevents autopilot behavior from creating exposure.

5.5 Incident Response: When Confidentiality Is Compromised

Despite precautions, confidentiality incidents occur. An associate pastes a confidential document into a consumer AI tool without thinking. A prompt that seemed adequately anonymized turns out to contain an indirect identifier. A tool that was thought to be enterprise-grade turns out to have weaker protections than represented. Knowing how to respond when incidents happen limits harm and prevents recurrence.

Immediate Response

When you discover that confidential information has entered an AI system inappropriately, act quickly. The longer the information remains in the system, the greater the potential exposure.

First, stop the immediate risk. If the conversation is ongoing, end it. If the tool offers deletion options, use them immediately—though understand that deletion from the interface may not mean deletion from all storage.

Second, document what happened while it's fresh. What information was disclosed? Which tool received it? When did the disclosure occur? What data handling policies apply to that tool? This documentation supports both your own analysis and any required reporting.

Third, assess the scope of exposure. What's the worst-case scenario for this information? Could it appear in training data? Could it be accessed by provider employees? Could it be exposed in a breach? The answers guide how urgently you need to act and what steps are appropriate.

> **Risk Note** Do not assume that deleting a conversation removes all copies of your input. Many AI systems retain logs, training data, or backups that persist beyond user-accessible conversations. Deletion is harm reduction, not harm elimination. Your response should account for the possibility that the information persists despite your deletion.

Notification Considerations

Depending on the severity of the incident and the nature of the information, you may need to notify affected clients. Rule 1.4 requires keeping clients reasonably informed about significant developments affecting their interests. A confidentiality breach involving their information may qualify.

The notification analysis considers: How sensitive is the disclosed information? What's the likelihood it will actually cause harm? What would a reasonable client want to know? Are there regulatory notification requirements (such as data breach notification laws) that apply?

Not every incident requires client notification. If you anonymized information adequately but are concerned it might have been insufficient, the risk may be speculative enough that notification would alarm the client unnecessarily. But significant exposures of clearly sensitive information generally warrant disclosure. When in doubt, err toward transparency; clients are more likely to forgive a mistake you told them about than one they discovered independently.

Document your notification decision either way. If you decide to notify, document what you said and when. If you decide notification isn't required, document why. This contemporaneous record protects you if questions arise later.

Internal Reporting and Escalation

If you work in a firm or organization, follow internal incident reporting procedures. Most firms want to know about confidentiality incidents so they can assess firm-wide exposure, update guidance, and prevent recurrence. An incident that seems isolated might reveal a pattern. Your report contributes to the firm's ability to learn and adapt.

If your firm lacks incident reporting procedures for AI-related confidentiality concerns, that's a gap worth raising with firm leadership. The absence of a procedure doesn't eliminate your obligation to manage incidents responsibly; it just means you need to figure out the appropriate response without institutional support.

Treat AI confidentiality incidents the way you would treat any other breach of client confidence. The technology is new, but the obligations aren't. If a junior associate accidentally sent confidential documents to the wrong recipient, you wouldn't pretend it didn't happen. AI disclosures warrant the same seriousness.

Provider Engagement

For significant incidents, consider contacting the AI provider. Some providers have processes for handling data deletion requests, particularly for enterprise customers or significant breaches. They may be able to confirm what retention applies, expedite deletion, or provide documentation useful for your own response.

Provider responsiveness varies. Consumer tools may offer little support. Enterprise relationships may include incident response commitments. API providers may have technical mechanisms for addressing specific requests. Know what your provider offers before you need it.

Root Cause Analysis

After immediate response, step back and understand why the incident occurred. Did someone not understand the tool's data practices? Did the classification framework fail to capture this type of information? Did time pressure lead to shortcuts? Did the available tools make compliant behavior difficult?

Root cause analysis isn't about blame; it's about prevention. If the cause was individual error, training or reminders might help. If the cause was unclear guidance, improved policies might help. If the cause was inadequate tools, advocacy for better tools might help. Address the system that allowed the incident, not just the incident itself.

Incident response is damage control, not damage elimination. By the time you're responding, the information has already been disclosed. The goals are to limit further exposure, meet your professional obligations, and prevent recurrence. How you handle incidents reveals whether your confidentiality practice is genuinely systematic or just hopeful.

5.6 Building a Systematic Confidentiality Practice

The techniques in this chapter are more effective as a system than as ad hoc measures applied inconsistently. This section addresses how to build confidentiality protections into your AI practice so that compliance becomes habitual rather than effortful.

Default to Protection

Configure your AI environment to make protection the default. If your firm offers multiple AI tools, set the more secure option as your default. If tools offer data protection settings, enable them proactively rather than relying on case-by-case decisions. If your browser has bookmarks or shortcuts to AI tools, make sure they point to the appropriate enterprise versions.

Defaults matter because most decisions are made on autopilot. When you're focused on a legal problem and reach for an AI tool, you'll use whatever's most accessible. Make the accessible option the compliant option.

Establish Clear Policies

Document your rules for AI use and confidentiality in writing. This might be firm policy, practice group guidance, or your personal rules-of-use. The act of writing forces clarity. The document provides reference when you're uncertain. The existence of a policy supports your competence and reasonable-efforts arguments if questions arise.

A minimum viable policy addresses: which tools are approved for which data tiers, what anonymization is required for Tier 2 uses, what information is categorically off-limits for AI, how to handle incidents, and who to contact with questions.

If your firm lacks AI confidentiality policies, don't wait for firm-wide action. Develop your own practice-group or individual policies and follow them consistently. You can advocate for firm-wide policies while protecting yourself and your clients in the interim.

Train and Reinforce

If you supervise others, ensure they understand the confidentiality framework and know how to apply it. Training isn't a one-time event; it's ongoing reinforcement through supervision, feedback, and modeling.

When you see someone using AI appropriately, acknowledge it. When you see someone cutting corners, address it immediately and constructively. When new tools become available, proactively discuss how they fit into your framework. Keep confidentiality salient rather than letting it fade into background awareness.

Training yourself is equally important. Review your own practices periodically. Are you following your policies consistently? Have you drifted into shortcuts that seemed minor but aggregate into risk? Are there new tools or techniques that would let you use AI more safely or effectively?

Integrate with Matter Management

Make confidentiality classification part of matter intake. When you open a new matter, decide what data tier applies and document it. Review the classification when circumstances change—a routine matter might become highly sensitive if it attracts media attention or regulatory scrutiny.

Some practice management systems can tag matters with confidentiality levels and provide reminders when you're working in those matters. Even without technological support, a note in your file accomplishes the same purpose: conscious decision-making about how AI can appropriately support this work.

Audit and Improve

Periodically review your AI use patterns. What tools have you used? What types of information have you entered? Were your uses consistent with your policies? Did you encounter situations your policies didn't address?

Audits don't need to be elaborate. A quarterly review of your AI conversation history (if retained) or a mental inventory of how you've used AI in recent matters identifies patterns you might miss day-to-day. Audits surface both problems to address and successes to maintain.

Systematic practice beats episodic vigilance. If you rely on willpower to protect confidentiality in each interaction, you'll eventually slip. If you build protection into defaults, policies, training, and workflows, compliance becomes the path of least resistance. Design your system for the lawyer you are on a busy day, not the lawyer you aspire to be on a calm one.

5.7 Special Contexts: When Standard Protections Aren't Enough

Some practice contexts require heightened confidentiality protections that go beyond the standard framework. Recognizing these contexts and adjusting your approach protects clients whose matters carry exceptional sensitivity.

Criminal Defense

Criminal defense matters often involve liberty interests, stigma risks, and constitutional protections that elevate the stakes of confidentiality breaches. The attorney-client relationship in criminal matters exists partly to encourage full disclosure from clients who fear the consequences of candor. Anything that undermines that trust affects not just the individual client but the system's ability to function.

For criminal defense matters, consider treating all client information as Tier 4 regardless of surface sensitivity. The nature of the representation creates sensitivity even when specific facts seem unremarkable. Structural separation may be more appropriate than anonymization, keeping AI assistance in the realm of legal research and general analysis rather than matter-specific work.

Whistleblower and Retaliation Matters

Clients who report misconduct face serious risks if their identities or activities become known. A confidentiality breach might cost them their careers, their safety, or their willingness to proceed with legitimate complaints. The asymmetry of harm—your disclosure versus their destruction—counsels exceptional caution.

Anonymization for whistleblower matters must be particularly rigorous. The combination of company, industry, type of misconduct, and timing might identify the matter even without names. Consider whether AI assistance is appropriate at all for substantive analysis of these matters, reserving it for procedural questions or general legal research.

> Risk Note If a client's physical safety could be compromised by disclosure of your representation, standard confidentiality protections are insufficient. No AI tool, regardless of deployment type, offers the security appropriate for matters where lives may literally be at stake. Some matters simply should not involve AI.

Trade Secrets and Intellectual Property

Trade secrets lose their protected status if they're disclosed without adequate protection. Using AI to analyze a client's trade secret might be the disclosure that destroys its value. The legal harm from confidentiality breach is direct and potentially catastrophic.

When working with actual trade secrets, confirm that your deployment type provides contractual protections against use in training or retention beyond the immediate interaction. API access with zero retention is strongly preferable. For formulas, processes, or other information where the value is in the specifics, consider whether any AI use is appropriate.

Matters Under Protective Order

If you're operating under a protective order that restricts disclosure of certain information, AI input may violate that order. The analysis isn't just Rule 1.6; it's court-ordered confidentiality with potential contempt implications.

Review your protective orders before using AI on covered matters. If the order restricts disclosure to named individuals or within specific systems, AI use is probably prohibited for covered information. If the order is less specific, analyze whether AI input fits within permitted disclosures. When in doubt, seek clarification from the court or opposing counsel before proceeding.

Matters With Heightened Client Expectations

Some clients care more about confidentiality than others. A client who specifically requested information about your firm's data security practices before engaging you has expressed heightened expectations. A client operating in a heavily regulated industry may face their own compliance obligations affected by your data handling. A client who has experienced a breach may be particularly sensitive to data risks.

Identify clients with heightened expectations and adjust your practices accordingly. This might mean upgrading their matters to Tier 4 classification, providing additional disclosures about your AI practices, or agreeing to specific restrictions they request. Client expectations don't override your professional judgment, but they're a factor in what "reasonable efforts" means for that representation. When matter sensitivity exceeds what your available tools can safely support, the answer isn't to proceed with inadequate protection. The answer is to not use AI for that matter, or to obtain client consent after fully informed disclosure. Remember that convenience doesn't trump client protection.

Mini-Lab: Confidentiality Audit of Your Current AI Practice

Objective: This exercise helps you assess your current AI use against the confidentiality framework from this chapter and identify improvements.

Time Required: 30 to 45 minutes

What You'll Need:

- A list of your current AI tools
- Your recent AI usage history (mental or logged)
- The confidentiality framework from this chapter

Instructions

1. **Inventory Your AI Tools:** List every AI tool you've used for work purposes in the past three months. Include consumer tools, enterprise tools, and integrated AI. For each tool, note what you know about its data handling.
2. **Sample Your Recent Uses:** Select five recent AI interactions from across your tool inventory. For each one, document what tool you used, what information you entered, what data tier that information represents, and whether the tool's protections matched the sensitivity. Capture your sample in the worksheet below.
3. **Identify Gaps and Exposures:** Review your sample for confidentiality concerns. Did you enter client info into a consumer tool? Was anonymization sufficient? Note specific exposures and assess severity.
4. **Evaluate Your Workflow:** Assess your systematic practices. Do you have a consistent process for classifying information? Are defaults configured for protection? Do you have policies?
5. **Create an Improvement Plan:** Identify three specific changes to improve your confidentiality practice: one immediate change, one workflow change, and one systemic change.

Sample Worksheet

#	Tool Used	Information Entered (brief description)	Data Tier (1–4)	Protections Matched Sensitivity? (Y/N)	Notes
1					
2					
3					
4					
5					

A blank version of this worksheet is available on the companion site for download.

Deliverable

By the end of this Mini-Lab, you should have:

1. A complete inventory of your AI tools with data handling notes
2. A documented sample of five recent interactions with confidentiality analysis
3. A list of identified gaps and exposures with severity assessments
4. An honest evaluation of your systematic practices
5. A three-item improvement plan with implementation timelines

Bonus Challenge

Extend your audit to your team or practice group. Interview colleagues about their AI use patterns. Identify common gaps and propose practice-group guidance that addresses them.

Conduct this audit honestly. The goal isn't to produce a document that makes you look good; it's to identify real risks and address them. An audit that finds no problems is probably an audit conducted with blinders on. Everyone has room to improve, and acknowledging that is the first step toward improvement.

Key Takeaways

1. Rule 1.6 protects all information relating to the representation, which is broader than most lawyers assume. Every AI prompt containing client information is a disclosure requiring analysis. The duty to protect confidentiality requires understanding what happens to information after you enter it, including retention, training use, and potential exposure.
2. Data classification provides a framework for matching information sensitivity to AI tool capabilities. Four tiers—public, de-identified, identified-but-controlled, and highly sensitive—map to different deployment types and protection requirements. Classification should be a deliberate decision for each matter, not an afterthought.
3. Deployment types determine what protections exist. Consumer AI tools typically train on inputs and provide minimal protections. Enterprise deployments offer contractual safeguards and data handling commitments. API access provides the most control for the most sensitive information. Verify actual protections rather than relying on category labels.
4. Safe input workflows operationalize confidentiality protection. Anonymization removes identifying information. Abstraction generalizes facts into analytical questions. Structural separation keeps AI analysis and confidential information in separate steps. Pre-submission review catches problems before they become irreversible.
5. Incident response limits harm when confidentiality is compromised. Act quickly to delete where possible and document what happened. Consider notification obligations to clients. Report internally. Engage providers when appropriate. Conduct root cause analysis to prevent recurrence.
6. Systematic practice beats episodic vigilance. Build confidentiality protection into defaults, policies, training, and workflows. Design for the lawyer you are on a busy day, not the

lawyer you aspire to be on a calm one. Conduct periodic audits to identify drift and improvement opportunities.

7. Some contexts require heightened protection beyond standard frameworks. Criminal defense, whistleblower matters, trade secrets, matters under protective order, and matters with heightened client expectations may warrant elevated classification, stricter tool selection, or determination that AI isn't appropriate.

Chapter Checkpoint

Before moving on, make sure you can:

☐ Explain why information "relating to the representation" is broader than attorney-client privilege and identify examples of protected information that might not be obviously confidential

☐ Classify a given piece of information into the appropriate tier and articulate what AI tools are appropriate for that tier

☐ Describe the key differences among consumer, enterprise, and API deployment types and explain how those differences affect confidentiality analysis

☐ Apply anonymization and abstraction techniques to a client scenario, producing a prompt that preserves analytical value while eliminating identifying information

☐ Outline the steps you would take if you discovered confidential information had been entered into an inappropriate AI tool

☐ Evaluate your current AI practices against the systematic framework described in this chapter and identify specific improvements to implement

What's Next

This chapter addressed how to protect confidential information from disclosure through AI inputs. Chapter 6 turns to AI tools specifically designed for legal work: legal-specific AI platforms, AI-enhanced research tools, document automation systems, and the emerging landscape of AI applications built for lawyers. These tools offer capabilities that general-purpose AI doesn't, but they also present distinct considerations around reliability, integration, and professional responsibility. Understanding the legal AI tool landscape helps you choose the right tool for each task and use each tool effectively within the frameworks we've developed.

Key Terms

Information relating to the representation — The scope of Rule 1.6 protection. It includes all information learned during or connected to the representation, regardless of source, including publicly available information if its use would reveal something about the representation. This scope is broader than attorney-client privilege and broader than trade secrets.

De-identification and abstraction — De-identification removes or obscures identifying information, while abstraction goes further by transforming specific facts into general analytical questions. Both techniques reduce confidentiality risk, with abstraction usually providing stronger protection at the cost of more translation work.

Enterprise AI deployment — AI services provided under business contracts with negotiated terms, typically including confidentiality provisions, data handling commitments, and restrictions on using inputs for model training. Enterprise deployments offer stronger protections than consumer tools but still require due diligence on specific terms.

Chapter 6: AI Tools for Legal Work: Selection, Implementation, and Maintenance

The partner who brought you into the meeting wanted your opinion. The firm was evaluating three AI platforms, each promising to transform legal research and document drafting. The demonstrations had been impressive: queries answered instantly, contracts analyzed in seconds, research memos generated while you watched. The sales materials made each tool sound indispensable. The partner's question was simple: "Which one should we get?" The honest answer was more complicated: it depends on what problems you're trying to solve, what risks you're willing to accept, and whether anyone has actually verified that these tools do what the demos suggest.

Chapters 1 through 5 built frameworks for using AI responsibly: understanding how the technology works, prompting effectively, meeting professional obligations, and protecting confidentiality. This chapter addresses a different question: which tools should you use in the first place? The market for legal AI tools has exploded, offering everything from general-purpose assistants to specialized platforms for research, drafting, contract analysis, and discovery. Some tools deliver genuine value; others deliver impressive demos that don't survive contact with real work. Distinguishing between them requires evaluation skills that most lawyers never learned, because until recently, they didn't need them.

What This Chapter Covers

- The major categories of AI tools for legal work and what each category does well
- A framework for evaluating tools against criteria that actually matter: accuracy, security, cost, and governance fit
- How to match specific tools to specific tasks based on their strengths and limitations
- The diligence questions to ask vendors before adopting any tool
- Implementation pitfalls that undermine even well-selected tools, including shadow AI and inconsistent prompting
- How to maintain current knowledge as tools and the landscape evolve

Chapter Summary

AI tools for legal work fall into distinct categories, each with different strengths and appropriate uses. General AI assistants offer versatility but lack legal-specific grounding. Legal research platforms connect AI to verified authorities, addressing hallucination risks while still requiring verification of analysis. Drafting tools range from template-based systems to generation-based tools with different reliability profiles. Contract analysis tools excel at extraction and comparison but cannot replace contextual judgment. Discovery tools have the longest track record, accelerating human review without replacing it. Selecting tools requires evaluation against criteria that matter: accuracy for your intended uses, security that meets your confidentiality requirements, cost that accounts for total investment, and governance fit with your organizational capacity. Matching tools to tasks ensures deployment aligns with tool strengths. Vendor diligence provides answers that determine whether adoption is responsible. Implementation pitfalls, particularly shadow AI and inconsistent prompting, can undermine even excellent tools. Staying current is an ongoing obligation, not a one-time task. Tool cards document key information and provide frameworks for tracking changes. The goal is not to adopt the most tools or the newest tools but to deploy appropriate tools effectively within professional responsibility constraints.

6.1 The Legal AI Tool Landscape

The AI tools available for legal work have multiplied rapidly, but they cluster into recognizable categories. Understanding these categories helps you navigate vendor claims, identify what you actually need, and avoid both gaps and redundancy in your tool portfolio. No single tool does everything well; responsible practice often requires different tools matched to different tasks.

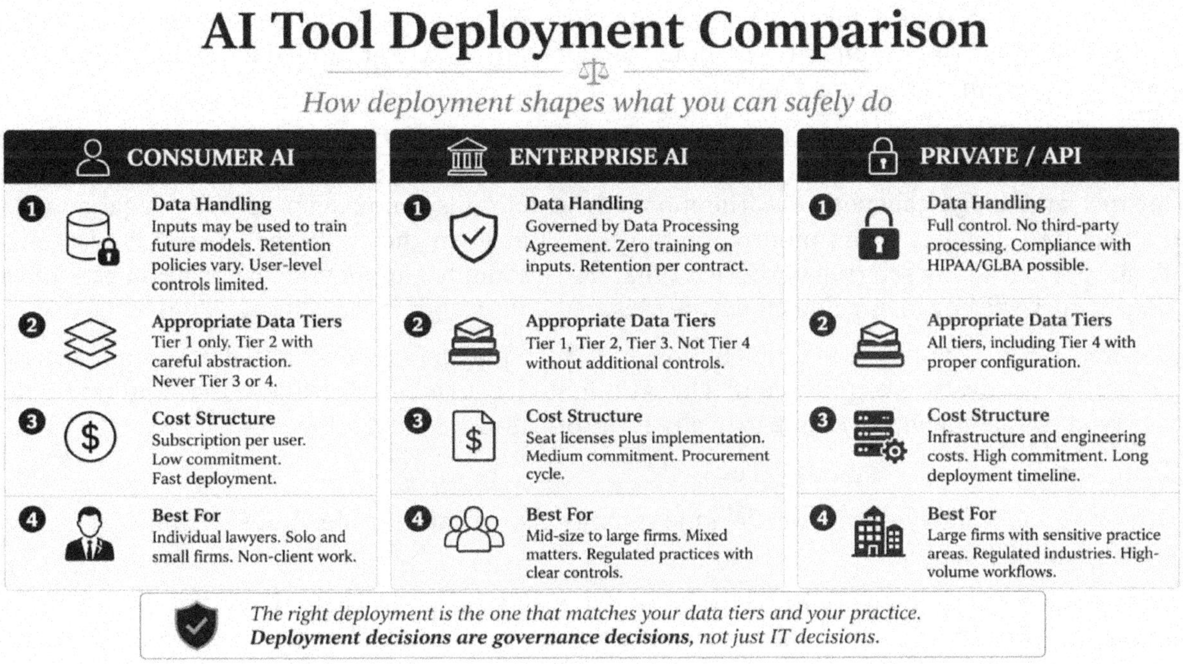

Figure 6.1 — AI Tool Deployment Comparison

General AI Assistants

General AI assistants are the large language models you access through consumer or enterprise interfaces: the chatbots, the productivity copilots, the AI features embedded in office software. They're trained on broad datasets spanning many domains and can engage with virtually any topic. Their versatility is their strength; their lack of legal-specific grounding is their limitation.

General assistants excel at tasks where broad language capability matters more than legal precision: brainstorming arguments, transforming content between formats, drafting initial versions of routine documents, explaining concepts in plain language, and summarizing long texts. They struggle with tasks requiring verified legal content: they hallucinate citations, blend jurisdictions, and state holdings that don't match actual cases. Chapter 2's reliability framework applies directly here.

For legal work, general assistants are useful but dangerous. Useful because they can accelerate many tasks that don't require legal-specific accuracy. Dangerous because their confident outputs look authoritative regardless of accuracy. The key is matching these tools to appropriate tasks and maintaining rigorous verification for anything that requires legal precision.

Legal Research Platforms

Legal research platforms represent the marriage of AI capabilities with traditional legal research infrastructure. These tools, offered by established legal publishers and newer entrants, connect language models to databases of verified legal authorities. When you ask a question, the AI can retrieve actual cases, statutes, and regulations rather than generating plausible-sounding fabrications.

This architecture addresses the hallucination problem that makes general assistants dangerous for legal research. The AI's outputs are grounded in real authorities that you can verify. Citations point to actual documents in the database. Holdings can be checked against the underlying opinions. The tool's usefulness comes not from generating content but from finding and synthesizing content that already exists in verified form. This architecture is sometimes called Retrieval-Augmented Generation (RAG) — the AI generates responses but its outputs are grounded in retrieved authorities rather than purely in its training data.

Legal research AI platforms vary in sophistication. Some offer basic natural language search that improves on traditional Boolean queries. Others provide AI-generated summaries, suggested authorities, and analytical synthesis. The most advanced tools can answer complex legal questions with structured analysis citing multiple authorities. But even the best legal research AI requires verification. The AI may retrieve authorities that don't actually support the proposition, summarize holdings incorrectly, or miss relevant authorities that a thorough researcher would find.

Legal research AI reduces the hallucination risk that plagues general assistants, but it doesn't eliminate the verification requirement. The AI might cite a real case while misstating its holding. Always read the actual authorities, especially for propositions central to your analysis.

Drafting and Document Assembly Tools

Drafting tools help lawyers create documents, but they span a wide range of approaches with different reliability profiles.

Template-based systems start with vetted templates and use AI to customize them based on your inputs. The foundation is human-created and reviewed; the AI handles population, variation selection, and adaptation to specific circumstances. These systems are more constrained than pure generation, which makes them more reliable for producing accurate documents, though less flexible for novel situations.

Generation-based tools produce documents from scratch based on your instructions. They can draft contracts, letters, memos, and briefs without starting from templates. The flexibility is appealing, but every word they produce requires the same scrutiny you'd apply to work from an unproven junior associate. The Chapter 1 framework applies fully: you are the supervisor, and supervision requires verifying accuracy across the entire document.

Hybrid approaches combine templates with generative capability: starting from vetted structures while allowing AI to draft content within those structures. These tools attempt to capture the reliability of templates with the flexibility of generation, with varying success.

The critical question for any drafting tool is: where does the content come from, and how much do you need to verify? Template-based content that's been vetted by experienced lawyers requires less verification than generated content that might contain errors anywhere. Your supervision burden scales with the tool's generative freedom.

Contract Analysis Tools

Contract analysis tools apply AI to the review and comparison of contracts. They can extract key terms from documents, compare provisions across agreements, identify deviations from standard language, flag missing clauses, and summarize obligations. These tools accelerate work that would otherwise require careful human reading of lengthy documents.

The strength of contract analysis tools is extraction and comparison: tasks where the AI processes existing text rather than generating new content. When the AI tells you that a contract contains a particular provision or that two agreements differ in specific ways, you can verify those claims by reading the provisions yourself. The AI has surfaced information for your review rather than created content you must evaluate for accuracy.

The limitation is contextual judgment. The AI can identify that a limitation of liability clause caps damages at contract value, but it cannot tell you whether that cap is acceptable for your client's risk profile in this transaction. It can flag that a termination provision differs from your standard form, but it cannot advise whether the difference matters or how to negotiate it. Extraction is reliable; evaluation requires human judgment.

Discovery and Document Review Tools

Discovery tools represent the most mature category of legal AI, with over a decade of development and substantial court acceptance. Technology-assisted review, predictive coding, and AI-powered document classification have become standard in large-scale litigation and investigations. Courts have repeatedly blessed these technologies when properly implemented.

Discovery AI accelerates human review rather than replacing it. The AI learns from examples to classify documents as relevant or not, privileged or not, responsive to particular issues or not. Human reviewers train the system, validate its classifications, and make final determinations on documents the AI flags as uncertain. The AI prioritizes what humans review rather than making unreviewable determinations.

The track record of discovery AI provides a model for responsible legal AI use generally: AI augments human capability, humans maintain oversight, validation processes confirm AI performance, and documentation demonstrates the reasonableness of the approach. Discovery workflows have developed precisely because courts required evidence that AI-assisted review was defensible. That rigor provides a template for other legal AI applications.

> **Authority Note** Courts have consistently approved technology-assisted review in discovery when parties demonstrate reasonable processes for training, validation, and quality control. The case law provides a framework for AI acceptance that requires human oversight, documented methodology, and defensible results. See, e.g., *Rio Tinto PLC v. Vale S.A.*, 306 F.R.D. 125 (S.D.N.Y. 2015).

Knowledge Management and Practice Operations Tools

A final category addresses firm operations rather than client work directly. AI tools can help with knowledge management by surfacing relevant precedents, identifying expertise within the firm, and connecting current matters to past work product. They can assist practice operations by analyzing matter data, predicting outcomes, pricing work, and identifying business development opportunities.

These tools raise different considerations than client-facing AI. Confidentiality concerns focus on information security within the firm rather than disclosure to external AI providers. Accuracy concerns affect business decisions rather than legal analysis. The evaluation framework still applies, but the specific criteria differ.

Different tool categories serve different purposes with different reliability profiles. General assistants offer versatility without legal grounding. Legal research platforms connect AI to verified authorities. Drafting tools range from template-based to fully generative. Contract analysis tools excel at extraction but not evaluation. Discovery tools have the longest track record and clearest court acceptance. Match tool category to task requirements rather than assuming any single tool does everything well.

6.2 Evaluating Tools: The Four-Criterion Framework

Selecting AI tools requires evaluation criteria that cut through marketing claims to identify what actually matters. Four criteria provide a framework for assessment: accuracy, security, cost, and governance fit. Each criterion addresses a different dimension of responsible adoption. Trade-offs among criteria are inevitable; the framework structures evaluation while judgment determines decisions.

Criterion 1: Accuracy

Accuracy is the threshold question. If a tool doesn't produce reliable outputs for your intended uses, nothing else matters. But accuracy assessment is harder than it sounds, because vendor demonstrations and published metrics rarely reflect performance on your specific tasks.

The accuracy question is always contextual: accurate for what? A tool might perform well on contract extraction but poorly on research synthesis. It might handle common fact patterns reliably but fail on unusual ones. It might work well in federal practice but struggle with state-specific questions. You need accuracy for your practice, your matters, and your tasks.

Evaluating accuracy requires testing beyond vendor demonstrations. Demonstrations showcase the tool's strengths on problems the vendor has optimized for. They don't reveal performance on the edge cases, unusual situations, and specific requirements that constitute your actual work. Request trial access and test on real problems from your practice. Compare outputs to work you've done manually. Note where the tool excels and where it fails.

> **Practice Point** Before adopting any AI tool, test it on at least five real tasks from your practice, including some difficult ones. Do not rely solely on vendor demonstrations or published accuracy statistics. Your evaluation should reveal how the tool performs on work you actually do, not work the vendor selected to showcase.

Published accuracy metrics, where they exist, require interpretation. What did the vendor measure? On what dataset? Using what definition of accuracy? A tool claiming 95% accuracy might mean 95% of documents were classified correctly in a particular study, which tells you little about whether it will correctly answer your research questions or draft your contracts without errors. Metrics provide data points, not conclusions.

Ask vendors about known limitations. Where does the tool perform less well? What task types should you avoid? What failure modes have users reported? Vendors who can't articulate limitations either haven't evaluated their tool thoroughly or aren't being candid. Either answer is a concern.

Criterion 2: Security

Security implements the Chapter 5 confidentiality framework at the tool selection stage. Before adopting any tool, you need to understand its data handling practices well enough to determine what information can appropriately enter it.

The security questions from Chapter 5 apply directly: What happens to information you enter? Is it retained, and for how long? Is it used for training? Who can access it? What certifications or audits validate the vendor's security claims? Does the tool involve subprocessors who receive your data? What happens to your data if the vendor is acquired or ceases operations?

Security evaluation should produce a clear determination: what data tier is this tool appropriate for? If the tool's practices don't support Tier 3 information, you need to know that before anyone enters client names or matter details. If it doesn't support Tier 4, certain matters must stay out entirely. The evaluation should inform usage guidelines, not just an adoption decision.

> **Risk Note** If a vendor cannot provide clear, specific answers to data handling questions, that is a disqualifying concern. Responsible vendors know exactly what happens to customer data and can explain it clearly. Vagueness or deflection suggests either inadequate practices or inadequate understanding of their own systems.

Security evaluation also considers the vendor's track record. Have they experienced breaches? How did they respond? Do they have dedicated security personnel? Do they participate in bug bounty programs or third-party security assessments? A vendor's history and security culture indicate how they'll handle future challenges.

For enterprise tools, review the actual contract, not just the marketing materials. Data handling commitments mean nothing unless they're contractually binding. Ensure that the terms you're relying on actually appear in the agreement you're signing, and that they can't be modified unilaterally.

Criterion 3: Cost

Cost analysis for AI tools must extend beyond subscription fees to capture total investment. Implementation, training, integration, and ongoing administration all consume resources. Value projections should be realistic about actual time savings and quality improvements rather than aspirational about theoretical maximum benefit.

Direct costs include subscription or licensing fees, per-user or per-query charges, implementation fees, and training expenses. These costs are usually quantifiable from vendor pricing, though watch for variable pricing that could escalate with usage.

Indirect costs include staff time for implementation and configuration, training time for users, ongoing administration and support, integration with existing systems, and workflow disruption during adoption. These costs are often larger than direct costs but easier to overlook.

Value projections should be grounded in realistic assessment of how the tool will actually be used. If a research tool saves thirty minutes per research task, how many research tasks do you perform monthly? If a drafting tool accelerates document production, does that translate to reduced hours billed, increased capacity, or improved quality? Be specific about where value comes from and skeptical about projections that assume maximum adoption and optimal use.

Cost-benefit analysis should consider alternatives. Is the proposed tool the only way to address the need, or are there cheaper alternatives that would suffice? Would investment in training on existing tools produce similar benefits at lower cost? Is the premium for a specialized tool justified compared to a general-purpose alternative?

Criterion 4: Governance Fit

Governance fit asks whether your organization can actually use the tool responsibly. A tool that exceeds your capacity to implement, train, oversee, or maintain creates risk regardless of its theoretical capability.

Implementation capacity considers whether you can deploy the tool effectively. Do you have technical resources for integration? Do you have personnel to configure the tool for your specific needs? Can you absorb the disruption of a new system? Organizations that adopt tools they can't implement properly end up with shelfware or shadow workarounds.

Training capacity considers whether your people will actually learn to use the tool well. Effective AI use requires skill development, not just tool access. Do you have resources for initial training? For ongoing skill development as tools evolve? For quality oversight that catches misuse before it causes harm? Chapter 3's prompting skills don't develop automatically; they require investment.

Before adopting an AI tool, honestly assess your organization's capacity to implement it properly. The best tool in the market is worthless if you can't configure it, train users, maintain oversight, and keep it current. Governance fit is about your capabilities, not the tool's features.

Oversight capacity considers whether you can maintain appropriate supervision of AI-assisted work. Chapter 4's supervision obligations don't disappear because you're using AI; they intensify. Do you have review processes that catch AI errors? Do you have policies that guide appropriate use? Do you have monitoring that detects misuse or drift from guidelines? Tools that outpace your oversight capacity create accountability gaps.

Compliance requirements may constrain tool selection. If you work in regulated industries, handle certain data types, or operate under specific contractual requirements, some tools may not be permissible regardless of other merits. Governance fit includes confirming that adoption is actually allowed.

Evaluate tools against four criteria: accuracy for your intended uses, security that meets your confidentiality requirements, cost that accounts for total investment, and governance fit with your organizational capacity. Trade-offs are inevitable. A tool might excel on accuracy but raise security concerns. A tool might be affordable but require governance capacity you lack. The framework structures evaluation; judgment determines whether trade-offs are acceptable for your situation.

6.3 Matching Tools to Tasks

Tool selection is only half the challenge. The other half is deployment: which tool for which task? A tool that excels at one task type may perform poorly at another. Matching tools to tasks ensures that deployment aligns with tool strengths while respecting tool limitations.

General AI Assistants: Appropriate Uses

General assistants match tasks where broad language capability matters more than legal-specific accuracy. Strong matches include:

Brainstorming and ideation: generating lists of potential arguments, identifying angles you might not have considered, exploring how different audiences might respond to messaging. The AI's breadth is an asset when you're expanding possibilities rather than narrowing to precision.

Content transformation: converting documents between formats, adjusting tone for different audiences, simplifying complex language for clients, expanding outlines into prose. These tasks leverage language capabilities without requiring legal verification.

Initial drafting of routine content: first drafts of standard letters, email communications, and internal documents where the content isn't legally sensitive and you'll review thoroughly before using.

Explaining concepts: generating explanations of legal concepts for clients or non-lawyer colleagues. The AI can produce clear explanations that you verify for accuracy before sharing.

General assistants should not be used for tasks requiring verified legal content: legal research, citation of authorities, statements of legal standards, or any content where accuracy depends on legal sources the AI cannot access or verify. The hallucination risk is too high, and legal-specific tools are more appropriate.

For general AI assistants, adopt a simple rule: never use the output without verification if it contains legal propositions, citations, or factual claims. The tool is a drafting accelerator, not a source of reliable legal content. Verify everything substantive before relying on it.

Legal Research Platforms: Appropriate Uses

Legal research AI matches tasks where finding and synthesizing legal authorities is central. Strong matches include:

Initial research on unfamiliar topics: getting oriented in a new area, identifying key cases and statutes, understanding the structure of a legal question before diving deeper.

Finding authorities for known propositions: locating cases that support arguments you've already developed, identifying additional authorities to strengthen positions.

Comparing authorities across jurisdictions: understanding how different courts have addressed similar questions, identifying splits or trends.

Summarizing lengthy authorities: getting the key holdings and reasoning from long opinions before deciding whether to read in full.

Research AI requires verification of every authority you rely on. The AI may cite cases correctly but summarize them incorrectly. It may retrieve relevant cases but miss the most important ones. Use research AI to accelerate the finding phase while maintaining the verification rigor that competent research requires.

Drafting Tools: Appropriate Uses

Drafting tools match document creation tasks, but appropriate use depends on the tool type.

Template-based tools match routine documents that fit established patterns: standard contracts with predictable variations, form letters, compliance documents with fixed structures. The template provides quality control; the AI provides customization efficiency.

Generation-based tools match first drafts of documents where you'll invest significant review time. They can produce starting points faster than drafting from scratch, but every generated sentence requires the same scrutiny you'd apply to work from an untested drafter.

Drafting tools generally should not be used as final-output generators. Even well-crafted drafts require human review, revision, and refinement. The tool produces material for your editing, not documents ready for filing or sending.

Contract Analysis Tools: Appropriate Uses

Contract analysis tools match tasks involving processing existing contract text. Strong matches include:

Extracting key terms: pulling out dates, amounts, party names, and specific provisions from lengthy agreements.

Comparing documents: identifying differences between draft versions, deviations from standard forms, or variations across a portfolio of agreements.

Summarizing obligations: creating digests of what parties have committed to across one or many contracts.

Flagging issues: identifying provisions that match patterns associated with risk, such as unusual liability allocations or missing standard protections.

Contract analysis tools should not be used as substitutes for lawyer judgment about significance. The tool can identify what a contract says; you must determine what it means for your client and how to respond.

Discovery Tools: Appropriate Uses

Discovery tools match large-scale document processing tasks where human review of every document is impractical. Strong matches include:

Relevance classification: prioritizing documents likely to be relevant for human review rather than reviewing everything linearly.

Privilege identification: flagging documents that may be privileged for focused attorney review rather than reviewing all documents for privilege equally.

Issue coding: sorting documents by topic or issue to facilitate organized review and production.

Quality control: identifying potentially incorrect human classifications through consistency checking.

Discovery tools require human oversight at multiple stages: training, validation, and final determination on uncertain documents. The AI accelerates review; humans maintain ultimate responsibility for production decisions.

Building a Tool-Task Matrix

For your practice, create a matrix mapping approved tools to appropriate task types. The rows are your tools; the columns are task types you commonly perform. Each cell indicates whether that tool is appropriate for that task, with notes on any conditions or limitations.

This matrix becomes a practical reference guiding daily decisions. When you face a task, consult the matrix to identify which tool to use rather than defaulting to whatever tool is most convenient. Update the matrix as you learn more about tool performance and as your tool portfolio evolves.

Match tools to tasks based on alignment between task requirements and tool capabilities. General assistants match brainstorming and transformation but not legal research. Research AI matches finding authorities but not generating novel analysis. Drafting tools match document creation but require verification scaling with generative freedom. Contract analysis matches extraction but not evaluation. Discovery tools match classification but require human oversight. Create a tool-task matrix for your practice to guide deployment decisions.

6.4 Vendor Diligence: The Questions to Ask

Before adopting any AI tool, conduct diligence that provides answers determining whether adoption is responsible. Vendor demonstrations and marketing materials tell you what the vendor wants you to know. Diligence tells you what you need to know. The questions in this section provide a minimum framework; Appendix E provides a comprehensive assessment protocol for formal evaluations, and a printable vendor diligence checklist is available on the companion site for download.

Data Handling Questions

Data handling questions implement Chapter 5's confidentiality framework at the adoption stage. You need clear answers before any client information enters the system.

What happens to data I enter? Does the provider process it only to generate responses, or is it retained for other purposes? If retained, for how long and in what form? Can I request deletion, and does deletion actually remove all copies?

Is my data used for training? Do my inputs or outputs contribute to improving the model? If so, can I opt out? Is opt-out effective retroactively for data already entered?

Who can access my data? Which provider employees can see my inputs? Under what circumstances? Is access logged and auditable? Are there technical controls limiting access to those with legitimate need?

Are subprocessors involved? Does my data pass through third parties for processing, storage, or analysis? If so, who are they, and what are their data handling commitments?

What happens to my data if the provider is acquired or ceases operations? Are there contractual commitments about data handling in these scenarios?

Get data handling answers in writing and verify they match the actual contract terms. Vendor representatives sometimes describe ideal practices that aren't reflected in binding agreements. What the salesperson says matters less than what the contract requires.

Security Questions

Security questions assess whether the provider maintains protections adequate for the information you plan to enter.

What security certifications does the provider hold? SOC 2, ISO 27001, and similar certifications indicate third-party validation of security practices. What scope do these certifications cover? Are audit reports available for review?

How is data encrypted? Is data encrypted in transit and at rest? What encryption standards are used? Who holds encryption keys?

Has the provider experienced security incidents? If so, what happened, how did they respond, and what changes resulted? Responsible providers are transparent about incidents and improvements.

What is the incident response process? If a breach occurs, how will you be notified, and how quickly? What support will the provider offer?

Does the provider have dedicated security personnel? What is their security organization's structure and leadership?

Accuracy Questions

Accuracy questions probe beneath marketing claims to understand actual performance characteristics.

What training data underlies the tool? For legal tools, does training include legal authorities? From what jurisdictions and time periods? How current is the training data?

What accuracy metrics exist? How has the provider measured performance? On what tasks and datasets? Are metrics independently validated?

What are known limitations? Where does the tool perform less well? What task types should users avoid? What failure modes have been identified?

How does the tool signal uncertainty? When the tool is less confident in outputs, does it indicate uncertainty? How should users interpret confidence signals?

Be skeptical of vendors who claim high accuracy without explaining methodology, who can't articulate limitations, or who dismiss questions about failure modes. Responsible providers understand their tools' weaknesses and communicate them clearly. Vendors who only discuss strengths are either uninformed or misleading you.

Operational Questions

Operational questions address practical aspects of adoption, use, and termination.

What service level commitments apply? What uptime guarantees exist? What happens if the service is unavailable when you need it?

How is pricing structured? What determines cost? Are there usage limits or overage charges? How does pricing change at renewal?

What integrations are available? Does the tool connect with your document management system, research platforms, or other infrastructure? What does integration require?

What training and support are provided? What resources exist for onboarding and skill development? What support channels are available for problems?

What are the termination terms? If you decide to stop using the tool, what happens to your data? What is the notice period? Are there penalties for early termination?

Documenting Diligence

Document your diligence process and findings. Record what questions you asked, what answers you received, and how you evaluated the responses. This documentation serves multiple purposes.

It demonstrates reasonable evaluation if questions arise later about why you adopted a particular tool. It provides a reference when tool practices change and you need to assess the impact. It supports consistent evaluation across tools, enabling comparison. It contributes to organizational learning as you refine your diligence process.

The documentation need not be elaborate, but it should capture the key questions and answers in a form you can reference later.

Vendor diligence asks the questions that determine whether adoption is responsible. Data handling questions implement confidentiality requirements. Security questions assess protection adequacy. Accuracy questions probe beneath marketing claims. Operational questions address practical adoption concerns. Document your diligence to demonstrate reasonable evaluation and support future reference. See Appendix E for a comprehensive diligence protocol.

6.5 Implementation Pitfalls: Shadow AI and Inconsistent Prompting

Even well-selected tools fail to deliver value if implementation undermines them. Two pitfalls are particularly common and particularly damaging: shadow AI, where people use unapproved tools outside organizational visibility, and inconsistent prompting, where variable practices produce unreliable results. Addressing these pitfalls is essential for realizing the benefits of responsible AI adoption.

Shadow AI: The Invisible Risk

Shadow AI refers to AI tool use that occurs outside organizational visibility and governance. People use consumer AI tools for work because they're convenient, because approved tools aren't available or are cumbersome, or because they don't understand why it matters. The organization can't manage risks it doesn't know about, and shadow AI creates exactly those invisible risks.

The confidentiality exposure from shadow AI is obvious. Consumer tools enter confidential information into systems without adequate protection, potentially training AI models on client data, exposing information to breaches, or making data available in ways you can't control. Chapter 5's confidentiality analysis assumes you know what tools are being used; shadow AI defeats that assumption.

Less obvious is the quality risk. Shadow AI use typically lacks the verification processes, supervision structures, and documentation practices that responsible AI use requires. Outputs may enter work product without appropriate review. Errors may propagate without detection. The accountability gaps from Chapter 4 multiply when AI use is invisible.

If you don't know what AI tools people in your organization are using, you can't manage the risks those tools create. Shadow AI is not a minor compliance issue; it's a failure of the governance infrastructure that responsible AI use requires. Discovery typically happens after harm has occurred.

Preventing shadow AI requires understanding why it occurs. People use unapproved tools because approved alternatives don't meet their needs. The solution isn't primarily enforcement; it's meeting legitimate needs through governed channels with minimal friction.

Provide approved tools that actually work. If your approved AI is cumbersome, limited, or unavailable, people will find alternatives. Evaluate whether your approved tools genuinely address the tasks people are trying to accomplish.

Reduce friction in accessing approved tools. If using the approved tool requires multiple authentication steps, switching systems, or navigating complex interfaces, people will choose easier paths. Make approved tools as convenient as consumer alternatives.

Explain why it matters. People are more likely to comply with policies they understand than policies that seem arbitrary. Explain the confidentiality risks, the professional responsibility implications, and the potential consequences. Understanding creates motivation that rules alone don't.

Monitor and address violations constructively. When shadow AI is discovered, treat it as a signal that needs aren't being met rather than purely as a compliance failure. What were they trying to accomplish? Why didn't approved tools suffice? How can you address the underlying need?

> Practice Point Survey your team about what AI tools they are actually using. Create a safe environment for honest disclosure. Use the information not to punish but to understand gaps in your approved tool portfolio and address them. Shadow AI prevention starts with understanding what people need.

Inconsistent Prompting: The Quality Underminer

Even when people use approved tools, inconsistent prompting undermines results. Different people prompting the same tool for similar tasks produce varying outputs depending on their prompting skill, approach, and habits. Variation introduces unreliability that makes supervision harder and quality less predictable.

Chapter 3 established that prompting is a skill that affects output quality. But skills vary across practitioners. Some people have invested in developing prompting capability; others use tools casually without systematic approach. The resulting variation means that AI-assisted work product quality depends on who happened to work on it rather than organizational standards.

The supervision burden increases with inconsistency. When outputs vary unpredictably, supervisors must examine each one freshly rather than applying calibrated review appropriate to known output characteristics. The efficiency gains from AI are consumed by the oversight demands that inconsistency creates.

Addressing inconsistent prompting requires strategies that raise the floor without constraining the ceiling. The goal isn't uniformity that prevents skilled prompters from using their capabilities; it's consistency that ensures minimum quality regardless of individual skill levels.

Shared templates provide starting points for common tasks. Rather than each person constructing prompts from scratch, templates offer structures that embody Chapter 3's prompt engineering principles. Users can customize templates for specific situations while benefiting from tested foundations.

Prompt libraries collect prompts that have proven effective. When someone develops a prompt that works well, capturing it in a shared library lets others benefit. Over time, the library becomes a repository of organizational knowledge about effective prompting for your specific tool portfolio and task types.

Training programs develop baseline capability. Not everyone needs to become a prompting expert, but everyone using AI tools needs minimum competence. Training should cover the principles from Chapter 3, adapted to your specific tools and use cases.

Quality benchmarks establish expectations. What does good AI-assisted work product look like? How should outputs be structured? What verification is required? Benchmarks guide both production and supervision by making expectations explicit.

Supervision that includes prompting feedback addresses skill development. When reviewing AI-assisted work, examine not just the output but the process that produced it. What prompts were used? Could they be improved? Feedback on prompting builds capability over time.

> **Key Point** Implementation pitfalls undermine even excellent tool selection. Shadow AI creates invisible risks that governance cannot address. Prevent shadow AI by meeting legitimate needs through approved channels with minimal friction. Inconsistent prompting produces variable outputs that undermine reliability and complicate supervision. Address inconsistency through templates, libraries, training, benchmarks, and feedback. Implementation quality matters as much as tool quality.

6.6 Tool Cards and Keeping Current

The AI tool landscape a year ago looked different than it does today. The landscape a year from now will look different again. Tools that are cutting-edge today may be superseded. Vendors that seem stable may pivot, merge, or disappear. New categories of tools may emerge that don't fit current frameworks. Capabilities that seem impossible now may become routine. The evaluation you conducted when adopting a tool remains valid only until circumstances change, and circumstances change constantly. Staying current is not optional for responsible AI use; it's an ongoing obligation that your competence duty requires.

The challenge of staying current is managing information flow without drowning in it. AI news is relentless: product announcements, feature updates, research breakthroughs, vendor acquisitions, regulatory developments, court decisions, bar opinions, and commentary on all of the above. Following everything is impossible. Following nothing leaves you uninformed. The goal is a sustainable approach that keeps you adequately informed without consuming time you need for actual legal work.

Tool cards provide a structured way to maintain current knowledge about the specific tools you use. A tool card is a reference document for a single tool that captures the key information you need about that tool in one place. When you adopt a tool, you create its card. When circumstances change, you update the card. When you need to remember what you learned during diligence or how the tool fits your workflows, you consult the card. The card becomes your institutional memory about each tool in your portfolio.

What belongs on a tool card includes the following elements. Basic identification: tool name, vendor, version or tier you use, contract or subscription details, renewal dates. Capabilities: what the tool does, what task types it suits, what the Section 6.3 matching guidance says about appropriate uses. Data handling: the answers to the Section 6.4 diligence questions about retention, logging, training, and access. Security: certifications, encryption, incident history, and your assessment of adequacy for different information tiers. Limitations: known weaknesses, task types to avoid, accuracy concerns. Operational details: how to access the tool, training resources, support contacts. Status: when the card was last updated, what changes have occurred since adoption, any pending concerns.

The format matters less than the discipline. A tool card might be a structured document, a wiki page, a database entry, or a section in a spreadsheet. What matters is that the information exists, is accessible when needed, and gets updated when circumstances change. Choose a format that fits how your organization manages information and that you'll actually maintain.

Create a tool card template and use it consistently for every AI tool you adopt. The template ensures you capture the same information for each tool, making comparison easier and preventing gaps. Store cards where everyone who uses the tools can access them.

Triggers for updating tool cards include vendor communications about changes, contract renewals that prompt re-evaluation, incidents or problems that reveal new information, and periodic reviews you schedule proactively. When a vendor announces a policy change, update the affected card. When you renew a subscription, review the card to confirm it remains accurate. When something goes wrong, document what you learned. When your scheduled review date arrives, verify that nothing has changed since the last update.

Periodic review schedules ensure that cards don't become stale through inattention. For tools central to your practice, quarterly review is appropriate. For tools used less frequently, annual review may suffice. The review need not be elaborate: check the vendor's website for announcements, confirm that your documented understanding matches current terms, and note any changes. If nothing has changed, update the "last reviewed" date to show that you checked. If something has changed, update the substantive content.

Model Rule 1.1 Comment 8's requirement of technological competence is ongoing, not a one-time obligation satisfied at adoption. A lawyer who conducted thorough diligence two years ago but hasn't checked whether the tool's practices have changed may no longer be meeting the competence standard. Tool cards with documented review dates demonstrate ongoing attention to competence.

Staying current beyond your specific tools requires information sources that track the broader landscape. No single source covers everything, but a combination of sources can provide adequate coverage without overwhelming volume.

Legal technology publications cover AI developments relevant to law practice. Publications like Law.com, Above the Law, Legaltech News, and similar outlets report on new tools, vendor changes, and industry trends. Following one or two such publications provides baseline awareness of what's happening in legal AI. The coverage tends toward news and announcements rather than deep evaluation, so it alerts you to developments without replacing your own assessment.

Bar association resources increasingly address AI. State bars publish ethics opinions on AI use, and many have formed task forces or committees tracking AI developments. The ABA has multiple initiatives addressing AI in law practice. These resources provide authoritative guidance on professional responsibility dimensions that general technology coverage doesn't address. Following your state bar's communications and the ABA's AI-related publications keeps you current on the regulatory and ethical landscape.

> **Risk Note** Be cautious about AI information from sources with commercial interests in particular tools. Vendor blogs, sponsored content, and affiliate marketing present information through a lens that favors certain products. This content can be useful but should be evaluated with awareness of its promotional purpose. Independent sources provide more balanced perspective.

Vendor communications about your specific tools deserve attention despite their promotional nature. Vendors announce feature changes, policy updates, and capability expansions through blogs, emails, and release notes. These communications tell you what's changing in tools you actually use, which matters more than general industry news. Subscribe to update notifications from vendors whose tools you rely on, and actually read the communications rather than letting them accumulate unread.

Professional networks provide informal intelligence about what's working and what isn't. Colleagues at other organizations have experiences with tools you're considering or using. Bar association sections and legal technology groups facilitate information sharing. CLE programs on legal technology include practitioner perspectives. These networks surface practical insights that formal sources miss: the tool that looks good on paper but frustrates users in practice, the vendor whose support is excellent or terrible, the workaround that makes a difficult tool usable.

Cultivate relationships with a few colleagues who share your interest in legal technology. Regular conversations about what you're each learning, trying, and discovering compound your individual knowledge. A small network of informed peers provides better ongoing education than any publication.

Continuing legal education on AI is increasingly available and sometimes required. Some jurisdictions have begun requiring technology-related CLE. Even where not required, AI-focused CLE helps maintain competence and often provides practical insights beyond what self-study offers. Prioritize CLE that addresses practical application over theoretical discussion, and that comes from presenters with actual experience using AI in practice rather than those who only study it from outside.

Experimentation builds knowledge that reading cannot provide. Try new tools in low-stakes contexts to develop firsthand understanding of capabilities and limitations. When a new tool generates buzz, spend an hour exploring it rather than relying solely on others' descriptions. Direct experience reveals nuances that summaries miss. The time investment in experimentation pays returns through better-informed tool selection and use.

Information management prevents staying-current efforts from consuming excessive time. Rather than trying to read everything, develop efficient habits. Set aside a regular time, perhaps 30 minutes weekly, for AI-related reading. Use RSS feeds, email newsletters, or news aggregators to bring relevant content to you rather than hunting for it. Skim headlines to identify what warrants deeper reading. Save important items for reference rather than relying on memory. These habits make staying current sustainable rather than burdensome.

Organizational approaches to staying current scale individual efforts. Designating someone to monitor AI developments and share relevant updates with the team distributes the burden and ensures coverage. Creating a shared channel or folder for AI news and resources makes collective knowledge accessible. Scheduling periodic team discussions about AI developments creates space for sharing and learning. These organizational mechanisms build capability that doesn't depend on any single person's attention.

Appendix F provides tool card templates for different tool categories. The templates identify the information most relevant for each category: general AI assistants, legal research platforms, document drafting tools, discovery and review tools, and knowledge and operations tools. Use these templates as starting points, customizing them for your specific needs and organizational context. Editable versions of the templates are available on the companion site for download.

The evolution ahead will bring capabilities we cannot fully anticipate. Tools will become more powerful, more integrated, and more embedded in legal workflows. New risks will emerge alongside new capabilities. Regulatory frameworks will develop and mature. The lawyers who thrive will be those who maintain the ongoing attention that responsible use requires: evaluating new tools thoughtfully, updating their understanding as tools change, and adapting their practices as the landscape evolves. Staying current is not a burden to resent but a professional discipline to cultivate, no different from staying current on legal developments in your practice areas.

The AI tool landscape changes continuously, requiring ongoing attention to maintain competence. Tool cards provide structured documentation of key information about each tool you use, with periodic reviews ensuring cards remain current. Staying current with the broader landscape requires information sources including legal technology publications, bar association resources, vendor communications, professional networks, CLE programs, and direct experimentation. Manage information flow through efficient habits that make staying current sustainable. Organizational approaches distribute the burden and build collective knowledge. Appendix F provides tool card templates for different tool categories. The competence obligation is ongoing; evaluation at adoption is necessary but not sufficient.

Mini-Lab: Building Your AI Tool Portfolio

This exercise guides you through evaluating your current AI tool portfolio, identifying gaps, and conducting diligence on a tool you're considering.

Time Required: 60 to 90 minutes

Step 1: Inventory Your Current Tools

List every AI tool you currently use or have access to for legal work. For each tool, note:

- Tool name and vendor
- Category (general assistant, legal research, drafting, contract analysis, discovery, other)
- Whether it's approved or unapproved (shadow AI)
- What you actually use it for
- Your rough assessment of how well it works

If you're working within an organization, distinguish between tools provided by the organization and tools you've adopted individually.

Step 2: Create Tool Cards for Current Tools

Select the three tools you use most frequently. For each one, create a tool card using the structure from Section 6.6:

- Basic identification (name, vendor, tier, contract details)
- Capabilities and appropriate uses
- Data handling practices
- Security certifications and assessments
- Known limitations
- Operational details
- Current status and last review date

If you don't know answers to some questions, note them as gaps requiring follow-up. Blank tool card templates are available on the companion site to make this step faster.

Step 3: Build a Tool-Task Matrix

Create a matrix with your tools as rows and common task types as columns. Task types might include: brainstorming, legal research, document drafting, contract review, correspondence, internal memos, litigation support, and others specific to your practice.

For each tool-task combination, indicate:

- Green: appropriate match, tool is good for this task
- Yellow: conditional match, tool works with caveats
- Red: poor match, avoid using tool for this task
- Blank: not applicable or haven't evaluated

Review your matrix for gaps: task types where you lack an appropriate tool. A blank tool-task matrix template is available on the companion site for download.

Step 4: Evaluate a Prospective Tool

Select one tool you're considering adopting or one that might address a gap identified in Step 3. Conduct diligence using the Section 6.4 framework:

Prepare questions across the four categories: data handling, security, accuracy, and operations.

Research answers using publicly available information: the vendor's website, terms of service, privacy policy, security documentation, and published reviews.

Note questions you cannot answer from public sources that would require vendor engagement.

Assess whether the tool passes your threshold for adoption based on available information.

Step 5: Create an Improvement Plan

Based on your inventory and analysis, identify:

- Shadow AI that should be eliminated or replaced with approved alternatives
- Gaps in your tool portfolio that limit your AI capabilities
- Tools that should be retired because better alternatives exist
- Training needs to improve how you use existing tools
- Process improvements for maintaining tool cards and staying current

For each item, note the action required and a timeline.

Deliverable

By the end of this Mini-Lab, you should have:

1. A complete inventory of your current AI tools
2. Tool cards for your three most frequently used tools
3. A tool-task matrix showing appropriate uses for each tool
4. Diligence documentation for one prospective tool
5. An improvement plan addressing gaps, shadow AI, and maintenance practices

Bonus Challenge

If you work within an organization, extend this exercise to your team or practice group:

Interview colleagues about what AI tools they use (including shadow AI). Compile a collective inventory.

Identify common gaps and shadow AI patterns across the group.

Propose a practice-group tool portfolio: which tools should be approved, what training is needed, and how tool cards will be maintained.

Present your recommendations to firm leadership or IT, advocating for resources to address identified gaps.

This organizational perspective develops the leadership skills that senior lawyers need as AI adoption scales beyond individual use.

Revisit your tool portfolio and cards quarterly. The landscape changes too quickly for annual review to maintain currency. Block time on your calendar for these reviews rather than waiting until something goes wrong. Proactive maintenance is easier than reactive remediation.

Key Takeaways

1. AI tools for legal work fall into categories with different strengths and appropriate uses. General AI assistants offer versatility but lack legal-specific grounding and verification. Legal research platforms connect AI capabilities to verified authorities, addressing hallucination risks while still requiring verification of analysis. Drafting tools range from template-based systems with vetted foundations to generation-based tools requiring extensive review. Contract analysis tools excel at extraction and comparison but cannot provide contextual judgment about significance. Discovery tools have the longest track record and court acceptance, accelerating human review without replacing it. No single tool excels at everything; responsible practice often requires multiple tools matched to different task types.
2. Selecting tools requires evaluating them against criteria that matter: accuracy, security, cost, and governance fit. Accuracy assessment should focus on performance for your specific intended uses, not general capabilities or vendor demonstrations. Security evaluation implements the Chapter 5 confidentiality framework by examining data handling, retention, logging, training policies, and subprocessor involvement. Cost analysis must include implementation, training, and ongoing administration, not just subscription fees, weighed against realistic value projections. Governance fit considers whether the tool can operate within your policies, training capacity, oversight mechanisms, and compliance requirements. Trade-offs are inevitable; the rubric structures evaluation while judgment determines decisions.
3. Matching tools to tasks ensures that tools are deployed for purposes aligned with their strengths. General assistants match brainstorming, transformation, and initial drafting but not tasks requiring verified legal content. Legal research AI matches finding and summarizing authorities but not strategic judgment. Drafting tools match structured document creation but not novel analysis. Contract analysis tools match extraction and comparison but not contextual assessment. Discovery tools match prioritization and classification but not final determinations. For complex tasks, decompose into components and match each component to an appropriate tool.
4. Vendor diligence answers the questions that determine whether adoption is responsible. Data handling questions address what happens to information you enter: processing, retention, access, training use, and subprocessors. Security questions address encryption, certifications, incident history, and breach response. Accuracy questions address training

data, metrics, known limitations, and uncertainty signaling. Operational questions address service levels, pricing, integrations, training, support, and termination terms. Document your diligence to demonstrate responsible evaluation, and use Appendix E for comprehensive assessments.

5. Implementation pitfalls can undermine even well-selected tools. Shadow AI occurs when people use unapproved tools outside organizational visibility, creating risks that cannot be managed. Prevent shadow AI by meeting legitimate needs through approved channels with minimal friction. Inconsistent prompting produces variable results that undermine reliability and supervision. Address inconsistency through shared templates, prompt libraries, training on prompting principles, quality benchmarks, and supervision that includes prompting feedback. Other pitfalls include over-reliance that erodes verification, under-utilization that wastes investment, and integration failures that create workflow friction.
6. Staying current is an ongoing competence obligation, not a one-time task satisfied at adoption. Tool cards document key information about each tool and provide a framework for tracking changes. Information sources including legal technology publications, bar association resources, vendor communications, professional networks, and CLE programs provide broader landscape awareness. Efficient information management habits make staying current sustainable. Periodic review of tool cards ensures documented understanding remains accurate as tools and policies evolve.

Chapter Checkpoint

Before moving on, make sure you can:

☐ Identify which tool category (general AI, legal research, drafting, contract analysis, discovery) is most appropriate for a given task and explain why that category's strengths match the task requirements

☐ Evaluate a potential AI tool against the four-criterion rubric (accuracy, security, cost, governance fit) and identify the trade-offs involved in adoption

☐ Create a tool-task matching matrix for your practice that shows which approved tools are appropriate for which task types

☐ Conduct vendor diligence using the minimum questions framework, obtaining specific answers about data handling, security, accuracy, and operational terms

☐ Identify shadow AI risks in an organization and design approaches to prevent unapproved tool use by meeting legitimate needs through governed channels

☐ Develop prompting standardization strategies including templates, libraries, and training that improve consistency across a team

☐ Create and maintain tool cards that document key information about AI tools and support ongoing competence through periodic review

☐ Identify information sources and habits that support staying current with AI developments without consuming excessive time

What's Next

Parts I and II have built the foundation: how AI works, how to prompt effectively, what professional responsibilities apply, how to protect confidentiality, and how to select and implement tools responsibly. Part III puts that foundation to work. The next five chapters walk through the core workflows of legal practice, showing how AI assistance fits into the actual tasks lawyers perform daily.

Chapter 7 addresses legal research and analysis, the workflow where AI's promise and peril are both most apparent. You'll learn how to use AI to accelerate research while avoiding the hallucination trap, how to structure research workflows that produce reliable results, and how to build research memos that leverage AI assistance without compromising accuracy. The verification imperative that runs throughout this book becomes concrete in specific techniques for confirming that authorities exist and say what you think they say.

From research, we'll move through drafting and redlining in Chapter 8, litigation and discovery workflows in Chapter 9, client communications in Chapter 10, and practice management in Chapter 11. Each chapter applies the frameworks you've learned to specific task types, providing workflow structures, prompt patterns, and practical checklists you can implement immediately. By the end of Part III, you'll have a comprehensive toolkit for integrating AI into the work you actually do, grounded in the professional responsibility and confidentiality frameworks that make that integration responsible.

Key Terms

General AI assistants — Large language models trained on broad datasets and accessible through chatbot interfaces or embedded in productivity software. Versatile across many task types but lack legal-specific training, verification mechanisms, or grounding in authoritative legal sources.

Contract analysis tools — AI tools that extract, compare, and summarize contract provisions. They accelerate review by surfacing information but cannot replace lawyer judgment about significance or appropriate response.

Total cost of ownership — All costs associated with adopting and operating a tool: licensing/subscription, implementation, training, ongoing administration, and productivity impact during transition. Subscription fees alone underestimate true cost.

Tool-task matching — Deploying specific AI tools for specific task types based on alignment between task requirements and tool capabilities. Effective matching leverages tool strengths while respecting tool limitations.

Prompting standardization — Organizational practices that promote consistent, effective prompting across users: shared templates, prompt libraries, training programs, quality benchmarks, and supervision that includes prompting feedback.

Tool card — A structured reference document capturing key information about a specific AI tool: capabilities, data handling practices, appropriate uses, limitations, and current status. Provides quick reference and a framework for tracking changes over time.

Staying current — The ongoing effort to maintain accurate, up-to-date knowledge about AI tools, practices, and regulatory requirements as the landscape evolves. A component of the technological competence professional responsibility requires.

Part III: Core Workflows

Chapter 7: AI for Legal Research and Analysis

Legal research is where AI assistance feels most natural. The work involves processing large volumes of text, identifying patterns across authorities, and synthesizing findings into usable analysis—exactly the tasks where AI capabilities shine. A research question that once required hours of database searching and case reading can now begin with AI assistance that surfaces relevant authorities, suggests additional search terms, and helps organize what you find. The efficiency potential is genuine, and lawyers who learn to use AI effectively for research gain meaningful advantages.

Yet research is also where the discipline this book teaches matters most. The same AI that accelerates research can corrupt it if used carelessly. *Mata v. Avianca* happened in a research context: lawyers asked AI to find supporting authorities, and it fabricated them. That outcome wasn't random bad luck; it was a predictable consequence of using AI for tasks it cannot perform reliably without the verification workflows that make AI assistance responsible. This chapter teaches you to capture research efficiency while building in the safeguards that prevent research disasters.

What This Chapter Covers

- The specific research tasks where AI adds value and the tasks where AI creates unacceptable risk without intensive verification
- A structured research workflow that sequences AI assistance appropriately: framing questions, identifying authorities, synthesizing findings, and verifying results
- Techniques for extracting accurate holdings from authorities versus the dangers of AI paraphrasing that distorts meaning
- How to use AI to build argument maps and identify counterarguments without accepting AI's framing uncritically
- The process of drafting research memos with AI assistance while maintaining the accuracy standards your professional obligations require
- A Mini-Lab producing a research memo from provided authorities, applying the chapter's techniques to a realistic scenario
- A practice checklist and prompt pack providing ready-to-use resources for research workflows

Chapter Summary

AI can accelerate legal research significantly when used for tasks that play to its strengths: processing large volumes of text, identifying patterns across documents, suggesting search terms and research angles, transforming dense legal language into clearer summaries, and organizing information into useful structures. AI should not be trusted for tasks that require reliability it cannot provide: generating citations, asserting that authorities exist, characterizing holdings without verification, or providing definitive answers to legal questions. The line between these categories is not the line between helpful and unhelpful; it's the line between responsible and reckless.

Effective research workflow sequences AI assistance appropriately. The workflow begins with question framing, where AI can help refine vague research questions into precise legal issues but cannot replace your understanding of what the client actually needs to know. It proceeds to authority identification, where AI's role depends entirely on whether the tool connects to verified legal databases or generates from training patterns. It continues through synthesis, where AI can help organize and summarize authorities you've verified but cannot be trusted to characterize authorities accurately without your review. It concludes with verification, which is not optional and cannot be abbreviated regardless of time pressure. Each stage has appropriate AI uses and inappropriate ones; knowing the difference determines whether AI helps your research or corrupts it.

The authority-handling techniques in this chapter address the specific challenges of working with case law, statutes, and regulations. Extracting holdings requires distinguishing between what an opinion actually decided and what AI thinks it decided, a distinction that requires reading the opinion yourself. Paraphrasing is dangerous because AI confidently restates holdings in ways that subtly or dramatically change their meaning. Building argument maps with AI assistance works well for identifying potential arguments but requires critical evaluation of whether AI's framing reflects how courts actually analyze the issues. Throughout, the principle is consistent: AI accelerates the mechanical aspects of research while human judgment remains essential for the analytical aspects.

The chapter concludes with practical resources: a Mini-Lab that walks through producing a research memo using the chapter's techniques, and a checklist and prompt pack that provide templates for implementing these approaches in your own research. By the end of this chapter, you'll have both the conceptual framework and the practical tools to use AI for legal research responsibly, capturing efficiency gains without creating accuracy risks.

7.1 Where AI Helps Research (and Where It Shouldn't Be Trusted)

The question is not whether AI can help with legal research. It can, substantially. The question is which parts of research it can help with and which parts it cannot. Treating AI as generally useful for research leads to the disasters we've discussed. Treating AI as generally unreliable for research forfeits genuine efficiency gains. The path forward requires granular understanding: this specific task benefits from AI assistance, that specific task requires caution, this other task should be done without AI or with intensive verification. Developing that granular understanding is what this section provides.

AI helps research most reliably when it processes information you provide rather than generating information from its training. This principle, introduced in Chapter 1's discussion of task types, applies directly to research. When you give AI a case opinion and ask it to summarize the holding, the AI is transforming text you supplied. When you ask AI what cases support a particular proposition, the AI is generating content from patterns in its training data. The first task builds on a verified foundation. The second task may produce fabrications indistinguishable from real authorities. The distinction is fundamental.

Research question refinement is a task where AI adds genuine value with minimal risk. Lawyers often begin research with vague questions: "What are our options if the supplier breaches?" AI can help transform that vague question into precise legal issues: What constitutes breach under the contract's terms? What remedies does the contract specify? What gap-filling rules apply under the governing law? What defenses might the supplier raise? This refinement doesn't require AI to know the answers; it requires AI to help you ask better questions. Because you'll verify the answers through actual research, AI's role in shaping the questions creates little risk.

Search term generation similarly leverages AI strengths without depending on AI accuracy. When you're researching an unfamiliar area, you may not know the terminology courts use to discuss the issues. AI can suggest search terms, synonyms, and related concepts that expand your research scope. "What terms do courts use when discussing liability for autonomous vehicle accidents?" produces suggestions you can test in legal databases. If the suggestions lead to relevant authorities, they helped. If they don't, you've lost only the time spent testing them. The downside is limited because you're using AI suggestions as inputs to verified searches, not as final answers.

> Practice Point Use AI to brainstorm search terms before diving into database research. Ask for terminology variations, related concepts, and how different jurisdictions might frame the same issue. This front-end investment often surfaces authorities you would miss using only the terms that first occurred to you.

Issue spotting benefits from AI's ability to identify patterns across large text inputs. Give AI a contract, a fact pattern, or a set of documents and ask what legal issues the materials raise. AI may identify issues you'd have spotted yourself, but it may also surface issues you'd have missed. Because issue spotting produces a list of topics to investigate rather than conclusions to rely on, AI errors in this task are self-correcting: you'll discover during research whether a suggested issue is actually relevant. The cost of a false positive is wasted research time, not a fabricated authority in your brief.

Authority summarization works well when you provide the authority. If you have a case opinion and need a concise summary of its holding, facts, and reasoning, AI can produce that summary quickly. The summary is transformation of text you supplied, not generation from training patterns. You can verify the summary against the source because you have the source. This verification is still necessary since AI may mischaracterize holdings, but the verification is straightforward: read the relevant portions of the opinion and confirm the summary is accurate.

The reliability differs dramatically when AI summarizes authorities you haven't provided. If you ask AI to summarize a case by name without providing the text, the AI may generate a summary from its training data. That summary might be accurate if the case is well-represented in training data. It might be partially accurate with subtle errors. It might describe a completely different case. It might describe a case that doesn't exist. You cannot know which outcome occurred without retrieving and reading the actual opinion, which eliminates any efficiency gain from the AI summary.

> Risk Note Never rely on AI summaries of authorities you have not independently retrieved and reviewed. The summary may be fabricated, partially accurate, or subtly wrong in ways that undermine your analysis. Summaries are useful only when you can verify them against the actual source.

Research organization leverages AI's ability to structure information. After you've gathered authorities through verified means, AI can help organize them: grouping cases by issue, arranging them chronologically, identifying which authorities support which propositions, and flagging apparent conflicts. This organizational work involves classification and transformation of materials you've verified, not generation of new content. The risk is limited because errors in organization are visible when you review the organized output against the sources you know.

Synthesis assistance helps you see patterns across authorities. "Based on these five cases I've provided, what factors do courts consider in determining whether a duty exists?" AI can identify commonalities and differences across the materials you've supplied. This synthesis may reveal analytical frameworks you'd have developed yourself, or it may suggest frameworks that don't actually fit the authorities. Either way, you evaluate the synthesis against sources you have, maintaining the verification capability that makes AI assistance responsible.

The distinction between AI that retrieves from verified databases and AI that generates from training patterns corresponds to different product categories discussed in Chapter 6. Legal research platforms with database connections can provide citations that exist because the platform retrieved them from actual legal databases. General AI assistants generate citations from patterns and may fabricate authorities that don't exist. Know which type of tool you're using.

Citation generation is where AI poses the greatest research danger. When you ask AI to provide cases supporting a proposition, general AI assistants generate responses by predicting what a helpful answer would look like. That prediction may produce real cases, but it may produce fabricated cases with plausible-sounding names, citations, and holdings. The fabrications are not obvious; they look exactly like real citations. Without verification through authoritative databases, you cannot distinguish real from fabricated. The Mata lawyers learned this lesson publicly. You should learn it from their example.

Legal research AI platforms with database connections occupy a middle ground. These platforms can retrieve real cases because they're connected to databases of real cases. When such a platform cites an authority, the authority exists because the platform found it in a verified source. This connection addresses the fabrication problem but doesn't eliminate all accuracy concerns. The platform may retrieve cases that aren't actually relevant to your question. It may characterize holdings inaccurately. It may miss the most relevant authorities. Verification remains necessary, but the verification is qualitative rather than existential: you're checking whether the authorities say what the platform claims, not whether they exist at all.

Definitive legal conclusions should not come from AI regardless of tool type. "Is this contract enforceable?" and "Will we win this motion?" are questions AI cannot reliably answer. Legal conclusions require applying law to facts with judgment about how courts will interpret ambiguous authorities and weigh competing considerations. AI can help you gather and organize the inputs to that judgment. The judgment itself must be yours. Lawyers who ask AI for conclusions and report those conclusions to clients are outsourcing professional judgment to systems incapable of exercising it.

> **Key Point** AI helps research reliably when processing information you provide: refining questions, suggesting search terms, spotting issues, summarizing authorities you have retrieved, and organizing verified materials. AI creates unacceptable risk when generating information from training patterns: producing citations that may be fabricated, characterizing authorities you have not reviewed, and providing definitive legal conclusions. The distinction between processing and generating determines whether AI assistance is responsible. Use grounded legal research platforms when available, but verify even their outputs for accuracy and relevance. Combine AI assistance with traditional research methods to capture efficiency at each stage while maintaining the verification that professional responsibility requires.

The reliability ladder from Chapter 2 maps onto research tasks. At the bottom, low-risk uses include question refinement, search term generation, and organization of verified materials. Moving up, moderate-risk uses include summarizing and synthesizing authorities you've provided, which require verification but build on verified foundations. At the top, high-risk uses include citation generation from general AI and accepting AI characterizations without verification. The ladder provides a framework for calibrating your AI use to appropriate risk levels throughout the research process.

Before using AI for any research task, ask yourself: Is AI processing information I've verified, or is it generating information I'll need to verify? The first category is generally safe with normal review. The second category requires intensive verification that may eliminate any efficiency gain.

Combining AI assistance with traditional research produces better results than either approach alone. Use AI for question refinement and search term generation, then use traditional database research to find authorities. Use AI to summarize and organize the authorities you've found, then verify those summaries against the sources. Use AI to help draft your analysis, then review the draft against your verified understanding of the authorities. This combined approach captures AI efficiency at each stage while maintaining the verification that responsible research requires.

7.2 Research Workflow: Question → Authorities → Synthesis → Verification

Effective legal research follows a structure whether or not AI is involved. You begin with a question, identify authorities that bear on the question, synthesize what those authorities tell you, and verify that your understanding is accurate. AI doesn't change this fundamental structure; it changes how you execute each stage. The workflow presented in this section sequences AI assistance appropriately, showing where AI accelerates the work and where human judgment remains essential. Following this workflow captures AI efficiency while maintaining the accuracy your professional obligations require.

The workflow has four stages: Question, Authorities, Synthesis, and Verification. The stages are sequential in concept but iterative in practice. Research that surfaces new issues sends you back to refine your questions. Authorities that contradict your initial understanding require revised synthesis. Verification that reveals errors requires correction before the research is complete. Think of the stages as a framework for organizing your efforts, not a rigid sequence that prohibits revisiting earlier work.

Stage One: Question

Research quality depends on question quality. A vague question produces scattered research that may or may not address what you actually need to know. A precise question focuses your efforts on authorities that matter and helps you recognize when you've found them. Investing time in question framing before diving into databases pays returns throughout the research process.

AI assists question framing by helping you decompose broad questions into specific legal issues. Start with what you actually need to know, expressed in whatever terms come naturally. Then use AI to refine that expression into researchable questions. A prompt might look like this: "My client is a commercial landlord whose tenant stopped paying rent and abandoned the premises, leaving behind equipment and inventory. The client wants to know their rights regarding the abandoned property and any obligations they have before disposing of it. Help me break this down into specific legal questions I should research."

The AI response will suggest questions you might not have articulated yourself: What constitutes abandonment under applicable law? What notice requirements apply before a landlord can dispose of abandoned property? Does the lease address abandoned property, and do lease terms supersede statutory defaults? What liability exposure exists for improper disposal? Are there differences between how equipment and inventory are treated? These questions give you specific targets for research rather than a general topic to explore.

Question refinement also involves identifying the governing law and any threshold issues that affect how you frame subsequent questions. AI can help here too: "What factors determine which state's law governs abandoned property in a commercial lease context?" The answer helps you focus research on the right jurisdiction's rules rather than conducting broad research you'll later have to narrow.

> **Practice Point** Spend fifteen minutes on question framing before opening any database. Use AI to decompose your initial question, identify the governing law, and surface related issues you might have missed. This investment typically saves more time during research than it costs, and it produces more complete analysis.

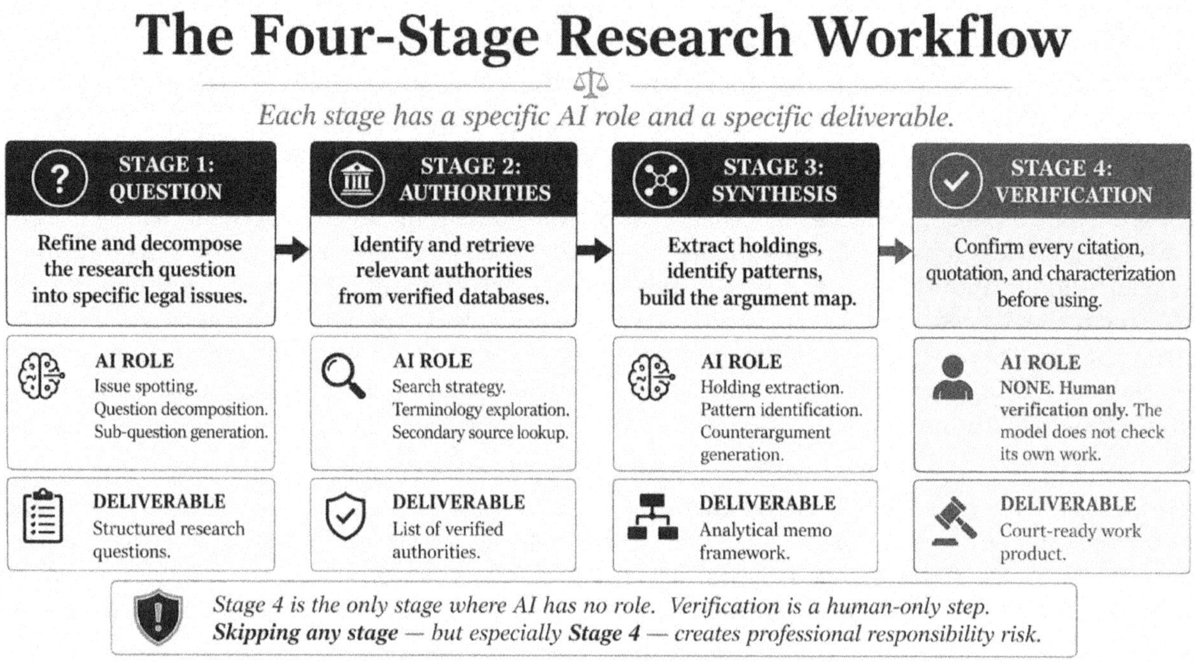

Figure 7.1 — The Four-Stage Research Workflow

The human judgment that AI cannot replace at this stage involves understanding what the client actually needs. The client asked about abandoned property, but do they really need a comprehensive legal analysis, or do they need to know whether they can dispose of specific items by a specific date? Are there business considerations that affect which legal options are practically viable? Does the relationship with the tenant matter for how aggressively to pursue remedies? These questions require understanding the client and the matter context, which AI doesn't have.

Stage Two: Authorities

Authority identification is where tool selection matters most. The distinction from Section 7.1 applies directly: grounded AI tools connected to legal databases can retrieve authorities that exist, while general AI tools may fabricate authorities that don't. Your approach to this stage should differ based on which tools you're using.

With grounded legal research AI, you can prompt the tool with your refined questions and receive citations to potentially relevant authorities. "What cases address a commercial landlord's obligations before disposing of property abandoned by a tenant under California law?" The platform searches its database and returns cases it found. These cases exist because the platform retrieved them from verified sources. Your task is evaluating whether they're actually relevant and accurately characterized, not whether they're real.

Even with grounded tools, AI should complement rather than replace your research judgment. The tool finds what its algorithms determine is relevant, but its relevance assessment may not match yours. It may miss authorities that don't use the terminology it searched. It may return authorities that match keywords but address different issues. It may prioritize recent cases over foundational ones, or vice versa. Use AI results as a starting point, then expand through traditional research techniques: checking citations in relevant cases, searching for related terms, and examining secondary sources that may identify additional authorities.

With general AI tools lacking database connections, authority identification requires different handling. You can still use these tools productively, but not for generating citations. Use them to suggest search terms, identify how courts might frame the issues, and describe the general legal landscape. Then take those suggestions to verified legal databases and conduct traditional searches. The AI helps you know what to look for; the verified database confirms what actually exists.

> **Risk Note** Do not stop researching when the AI tool stops returning results. AI surfaces what matches its search parameters, which may not include everything relevant. Traditional research techniques like following citations, checking secondary sources, and searching alternative terms remain necessary even when AI accelerates the initial search.

A workflow using general AI might proceed like this. First, ask the AI what legal concepts typically govern abandoned property in commercial leases. The response mentions bailment, landlord lien rights, statutory notice requirements, and lease provisions as common frameworks. Second, take those concepts to Westlaw or Lexis and search for California cases addressing each. Third, build your authority set from what the verified database returns, not from anything the general AI specifically cited. The AI suggested productive avenues; the research confirmed what those avenues actually contain.

Courts have accepted technology-assisted research as an appropriate component of legal practice, but the lawyer remains responsible for the accuracy of authorities cited. A citation to a non-existent case reflects on the lawyer regardless of what tool produced the citation. Verification is the lawyer's responsibility, and "the AI told me it was real" is not a defense.

Authority evaluation occurs as you identify sources. Not every case a search returns deserves inclusion in your analysis. Evaluate relevance: does the case actually address the issue you're researching, or does it merely mention related terms? Evaluate precedential weight: is this a binding authority, persuasive authority, or an outlier? Evaluate currency: has subsequent case law modified or overruled the holding? These evaluations require reading the authorities, not just their AI-generated summaries. Skim at first to assess relevance, then read more carefully for authorities you'll rely on.

Organizing authorities as you find them prevents the chaos of an unsorted pile at the synthesis stage. Note which question each authority addresses, whether it helps or hurts your position, and how it relates to other authorities you've found. AI can help with this organization if you provide the authorities: "Here are the cases I've found on landlord notice requirements for abandoned property. Help me organize them by what notice period they required and what factors affected that determination." The AI processes your verified materials into a useful structure.

Stage Three: Synthesis

Synthesis transforms a collection of authorities into understanding. What do these cases collectively tell you about the legal question? Where do they agree, and where do they conflict? What factors drive different outcomes? What principles emerge that you can apply to your client's situation? Synthesis is where research becomes analysis.

AI assists synthesis by helping you identify patterns across the authorities you've gathered. After reading the key cases, you might prompt: "Based on the five cases I've summarized here regarding landlord disposal of abandoned tenant property, what factors appear to determine whether a landlord's notice was sufficient?" The AI identifies commonalities and differences across your summaries, suggesting analytical frameworks you can evaluate against your own reading of the cases.

The danger at this stage is letting AI synthesis replace your own analytical judgment. AI identifies patterns through statistical analysis of text. It may surface patterns that are coincidental rather than meaningful. It may miss patterns that require legal sophistication to recognize. It may frame synthesis in ways that don't reflect how courts actually reason. AI synthesis is a tool for generating hypotheses about what the authorities mean, not a substitute for your own analysis.

Verify synthesis against the sources. When AI suggests that courts consider three factors in determining notice adequacy, check whether the cases actually discuss those factors as determinative. When AI identifies an apparent split in authority, confirm that the cases actually conflict rather than addressing different issues. When AI proposes an analytical framework, test that framework against each authority to see if it holds. This verification catches AI synthesis errors before they corrupt your analysis.

Gaps and conflicts identified during synthesis may require additional research. If your authorities don't address a sub-question you identified at the question stage, you need more research. If authorities conflict without clear resolution, you may need to research how courts handle that conflict or identify additional authorities that clarify the landscape. Synthesis reveals what you know and what you still need to learn.

Write a preliminary synthesis statement before using AI to assist synthesis. "Based on my reading, I think the cases establish X." Then use AI to test that statement: "Here are the cases. Do they support the proposition that X?" AI challenge to your preliminary synthesis forces you to articulate and defend your understanding, improving analysis quality.

Stage Four: Verification

Verification is not a final polish; it's a non-negotiable requirement. Every citation must be confirmed through authoritative sources. Every characterization of a holding must be checked against the actual opinion. Every factual assertion must trace to the record or an authority. Verification catches the errors that, if uncaught, produce sanctions, malpractice exposure, and professional embarrassment.

Citation verification confirms that authorities exist and are cited correctly. For each citation, confirm in a verified legal database that the case exists, that the citation format is accurate, and that the case hasn't been overruled or significantly limited. This verification takes seconds per citation using citator tools, but it must actually occur. Don't assume that because you found a case through a grounded AI platform, the citation format is correct; transcription errors happen.

Holding verification confirms that authorities say what you think they say. For each authority you rely on for a specific proposition, read the relevant portion of the opinion and confirm that your characterization is accurate. If you're claiming that a case held X, the opinion should actually hold X, not something close to X that could be interpreted as X. Subtle characterization errors are more common than fabricated citations and potentially as damaging. AI summaries are particularly prone to these subtle errors, which is why reading the actual authorities remains essential.

The most dangerous errors are subtle mischaracterizations that sound right but aren't quite accurate. A case that "held employers must provide reasonable accommodation" versus one that "held employers may be required to provide reasonable accommodation under certain circumstances" differ meaningfully, but AI might summarize both the same way. Reading the actual holding catches what summaries obscure.

Quotation verification confirms that quoted language actually appears in the source. If you're quoting from a case, the quote should be verbatim from the opinion. AI sometimes generates quotation-like text that doesn't actually appear in the source or that combines language from different parts of an opinion in misleading ways. Verify every quote against the original.

Signal verification confirms that your signals accurately characterize the support each authority provides. "See" and "cf." and "but see" convey different relationships between your proposition and the cited authority. Make sure your signal choices reflect what the authorities actually say, not what you wish they said or what AI suggested they said.

Rule 11 of the Federal Rules of Civil Procedure (Fed. R. Civ. P. 11) requires reasonable inquiry to ensure that legal contentions are warranted by existing law or by a nonfrivolous argument for modification. Verification is not mere formality; it's the inquiry that Rule 11 requires. Documentation of your verification process supports your certification if it's ever questioned.

Verification documentation creates a record that the work was done. Note which database you used to verify each citation, when you verified it, and any issues you discovered. For significant matters, maintain a verification checklist that you complete and retain. This documentation serves the purposes discussed in Chapter 4: protection if questions arise, support for compliance with disclosure requirements, and evidence of the professional process you followed.

Time pressure does not relax your verification obligations. When deadlines loom, the temptation is to abbreviate the review process on the assumption that an AI-generated output is probably accurate. As we have seen over and over again in the reported sanctions cases, that assumption is exactly how lawyers end up in front of judges explaining filings they could not defend. If time constraints prevent thorough verification of the citations and the record, the tool should not be used for that filing. In a crunch, the better move is almost always to file a shorter brief with fewer, fully verified authorities than a longer document carrying unconfirmed text.

Build verification into your timeline from the start. If the brief is due Friday, plan to complete your draft Thursday and spend Friday on verification. Verification compressed into the final hour before filing will be rushed and incomplete. Verification given adequate time catches errors while you can still correct them. Independent verification has always been part of competent practice. What generative AI changes is the nature of the risk you are checking for. Quality control used to mean catching your own mistakes; verification now means catching the machine's mistakes before they become yours.

Workflow Integration

The four stages integrate into a complete research process. You refine your question with AI assistance, identifying the specific legal issues to research. You identify authorities using grounded AI tools where available, complemented by traditional techniques. You synthesize what the authorities tell you, using AI to suggest patterns while maintaining your own analytical judgment. You verify everything: citations, holdings, quotations, and signals.

The time allocation across stages should reflect where human judgment matters most. Question framing deserves meaningful investment because good questions drive good research. Authority identification can proceed quickly with appropriate tools, though breadth matters more than speed. Synthesis requires the careful reading and thinking that AI can supplement but not replace. Verification requires systematic attention that cannot be rushed.

AI's overall contribution is acceleration, not automation. Each stage still requires human judgment, but AI reduces the time required to exercise that judgment. Question decomposition that might have taken an hour of brainstorming happens in minutes. Authority identification that might have required numerous search iterations happens in fewer. Synthesis that required reading every case word-for-word can focus on key passages once AI identifies where those passages are. Verification still takes the time it takes, but you arrive at verification sooner because earlier stages moved faster.

Effective research workflow sequences AI assistance appropriately across four stages. Question framing uses AI to decompose broad questions into specific legal issues. Authority identification uses grounded AI tools to find authorities that exist, complemented by traditional techniques. Synthesis uses AI to help identify patterns across verified materials while maintaining human analytical judgment. Verification confirms that every citation, holding, quotation, and signal is accurate, with documentation that demonstrates the verification occurred. Time pressure does not reduce verification requirements. AI accelerates research without automating the judgment that professional responsibility requires.

7.3 Authority Summaries: Extracting Holdings vs. Paraphrasing

A case holding is a precise thing. It's what the court actually decided on the specific facts before it, expressed in the court's own reasoning and language. When you cite a case for a proposition, you're representing that the case supports that proposition. If your characterization drifts from what the case actually held, your research product becomes unreliable regardless of how many cases you cite. AI creates a particular risk here: it paraphrases fluently, producing summaries that sound authoritative but may subtly or substantially distort what the court decided. This section teaches you to extract holdings accurately, using AI to assist the process while guarding against the paraphrasing errors that corrupt legal analysis.

The difference between extraction and paraphrasing is the difference between capturing what a source says and restating it in your own words. Extraction preserves the source's meaning by staying close to its language and structure. Paraphrasing reprocesses the source through the summarizer's understanding, which may introduce distortions. Both have legitimate uses, but for legal holdings, extraction is safer. A holding extracted from an opinion can be verified against the source. A paraphrased holding may not match any language in the opinion, making verification harder and errors more likely.

AI paraphrases by default. When you ask AI to summarize a case, it generates text that conveys what it interprets the case to mean, not text that mirrors what the case says. This paraphrasing tendency produces several types of errors. AI may generalize holdings beyond their actual scope, turning a narrow ruling on specific facts into a broad legal principle. AI may soften or strengthen the court's language, turning "may" into "must" or "generally" into "always." AI may merge holdings from different cases into composite statements that no single case supports. AI may omit qualifications that limit a holding's applicability. Each type of error can mislead your analysis and, if incorporated into work product, mislead your readers.

> **Practice Area Note: Criminal Defense** The qualification-stripping risk is most consequential in suppression motion practice, where the difference between a granted and denied motion often turns on a single qualifying phrase. A case might hold that "consent to search may be withdrawn if expressed unequivocally and prior to the discovery of contraband." AI might summarize that as "consent to search may be withdrawn." The stripped version loses both qualifications — the unequivocal-expression requirement and the temporal cutoff — which are precisely the elements that determine whether your suppression motion succeeds. When extracting holdings from Fourth Amendment cases, preserve every qualifier the court used. The qualifications are usually where the suppression argument lives or dies.

Over-generalization is perhaps the most common AI paraphrasing error. A case holds that "a landlord who retakes possession of abandoned premises must provide written notice to the tenant's last known address at least fifteen days before disposing of property valued over $500." AI might summarize this as "landlords must provide notice before disposing of abandoned tenant property." The summary is technically accurate but omits crucial details: the retaking-possession trigger, the written form requirement, the address specification, the fifteen-day period, and the $500 threshold. A lawyer relying on the summary might miss issues that the actual holding would have flagged.

Qualification stripping removes the hedges and limitations that courts include for good reason. A court writes that "under the circumstances presented, the landlord's actions were reasonable." AI summarizes: "the court found the landlord's actions reasonable." The stripped version loses "under the circumstances presented," which signals that the holding is fact-specific and may not apply when circumstances differ. Courts write carefully; AI summaries often lose that care.

When reviewing AI case summaries, ask yourself: What qualifications or limitations might the court have included that this summary omits? Then check the actual opinion to find out. The qualifications are often where the real analysis lies.

Confidence mismatch occurs when AI presents uncertain or disputed propositions with unwarranted certainty. A case might note that "courts are divided on whether constructive abandonment triggers the same duties as actual abandonment." AI might summarize: "constructive abandonment triggers the same duties as actual abandonment." The uncertainty has disappeared, replaced by a confident assertion that misrepresents the state of the law. When you cite this case, you may be claiming it supports a proposition it actually identifies as contested.

Temporal confusion involves misrepresenting what law applied when or how the law has evolved. AI might summarize an older case without noting that subsequent legislation changed the rule. Or it might describe a case's discussion of prior law as if it were the case's holding. Or it might combine holdings from cases decided under different statutory regimes as if they established a consistent rule. Legal analysis is inherently temporal; AI summaries often flatten time in ways that mislead.

> **Risk Note** Watch for AI summaries that state rules without citing specific authority for each element of the rule. Composite summaries that merge multiple cases into a single statement may not accurately represent any individual case. Each proposition needs traceable support.

Extraction techniques produce more reliable summaries by staying closer to source language. When you need to characterize a holding, start by identifying the specific passage where the court states what it decided. Quote that passage or paraphrase it minimally, preserving key terms and qualifications. If the court said "may," don't upgrade to "must." If the court said "in these circumstances," preserve the limitation. The goal is capturing what the court said, not translating it into your preferred formulation.

For complex holdings, break them into components and extract each separately. A holding might establish (1) the legal standard, (2) how the standard applies to these facts, and (3) the outcome. Extract each component with appropriate attribution. "The court held that the standard for landlord liability is X (establishing legal standard). Applying this standard, the court found that the landlord's conduct satisfied X because Y and Z (application to facts). The court therefore ruled for the tenant (outcome)." This structured extraction preserves the holding's logic rather than compressing it into a single potentially distorted statement.

Appellate courts frequently address whether lower courts correctly stated the controlling legal standard. Characterization matters because it shapes analysis. A brief that overstates or understates what prior cases held may find its argument rejected not on the merits but on the ground that the precedents don't support the characterization. Accurate extraction is a persuasion strategy, not just a professional obligation.

AI can assist extraction when properly directed. Rather than asking AI to "summarize this case," ask it to perform specific extraction tasks that keep it closer to the source. "Quote the sentence or sentences from this opinion where the court states its holding." "What specific language does the court use to describe the standard it is applying?" "List the factors the court identifies as relevant to its determination, using the court's own terminology." These prompts constrain AI's tendency to paraphrase by directing it to identify rather than restate.

Comparison prompts help verify extraction accuracy. After extracting a holding yourself, ask AI to do the same, then compare. "Based on this opinion, what did the court hold regarding the landlord's duty to provide notice?" Compare the AI response to your own extraction. Discrepancies signal that one of you may have misread the case; investigate to determine which reading is correct. This comparison process uses AI as a check on your extraction rather than as a source of holdings you adopt uncritically.

Use AI as an extraction verifier, not just an extraction generator. Extract the holding yourself first, then ask AI for its extraction, then compare. The comparison reveals ambiguities and potential errors that either approach alone might miss.

Verification of extracted holdings requires reading the relevant passages yourself. There's no shortcut here. If you're going to represent that a case held X, you need to have read enough of the opinion to confirm that it held X. For holdings central to your analysis, read the full opinion or at least the sections addressing the relevant issue. For holdings you're citing for peripheral support, reading the specific passages establishing the holding may suffice. But some reading is always required; AI extraction cannot substitute for your own engagement with the source.

Pin cites support verification and strengthen your work product. Rather than citing a case generally, cite the specific pages where the relevant holding appears. Pin cites force you to identify exactly where the court said what you claim it said. They help readers verify your characterizations. And they demonstrate that your research involved actual engagement with authorities rather than reliance on AI summaries. The discipline of pin citing improves extraction accuracy because it requires finding the specific language that supports your characterization.

Different authority types require different extraction approaches. Case holdings are extracted from the court's statement of its ruling. Statutory provisions are quoted or closely paraphrased with citation to the specific section. Regulatory provisions similarly require close attention to exact language, since regulatory interpretation often turns on precise wording. Secondary sources may be paraphrased more freely since they're analytical rather than authoritative, but their analysis should still be accurately represented.

Multi-case synthesis presents particular extraction challenges. When you're characterizing what "the cases" establish rather than what one case held, you're synthesizing holdings into a broader statement. This synthesis is appropriate and often necessary, but it must be built from accurate extraction of each underlying case. State the synthesized principle, then support it with correctly extracted holdings from each case. "Courts applying California law require written notice at least fifteen days before disposal. Smith v. Jones, 123 Cal. App. 4th 456, 460 (2015) (requiring 'written notice served at least fifteen days prior to any disposal'); Brown v. Green, 234 Cal. App. 4th 567, 572 (2017) (same)." The synthesis is supported by extracted holdings, not by AI's composite paraphrase.

> **Note on Citations** *Smith v. Jones* and *Brown v. Green* are illustrative placeholders used throughout this chapter to demonstrate citation format and synthesis structure. They are not real cases and should not be cited as authority. The same convention is used in Section 7.5's sample drafting prompt. Do not attempt to look these citations up; they do not exist.

When AI synthesis and your reading diverge, trust your reading. AI may suggest that cases establish a principle you don't see in the authorities. If careful reading doesn't support the AI's synthesis, the AI is wrong. AI generates plausible-sounding synthesis by pattern-matching across its training data; it cannot read the specific cases you've gathered and reason about them with genuine understanding. Your reading of the actual authorities trumps AI's pattern-based synthesis.

If you can't find language in the opinion that supports your characterization of its holding, the characterization is probably wrong. Don't force holdings to fit the analysis you want. Extract what the cases actually say, then build your analysis on that foundation.

Practical workflow for holding extraction combines AI assistance with human verification. First, read the relevant portion of the opinion yourself and identify what you believe the holding to be. Second, ask AI to extract the holding from the same text. Third, compare your extraction to the AI extraction. Fourth, resolve any discrepancies by returning to the source. Fifth, formulate your final characterization of the holding, supported by pin cites to the specific language that establishes it. This workflow uses AI to challenge and refine your extraction without delegating the extraction to AI.

Time investment in accurate extraction pays returns in analysis quality and professional protection. Sloppy extraction produces sloppy analysis that may not withstand scrutiny. Accurate extraction produces reliable analysis that supports confident assertions. When opposing counsel reads the same cases, you want your characterizations to hold up. When courts review your citations, you want them to find authorities that say what you claimed they said. The time spent getting holdings right is not overhead; it's the substance of competent legal work.

AI paraphrases by default, producing summaries that may over-generalize, strip qualifications, convey false certainty, or create temporal confusion. Extraction techniques produce more reliable summaries by staying close to source language and preserving scope, qualifications, and limitations. Direct AI to extract rather than paraphrase by asking for specific language and court terminology. Use AI as a verification check against your own extraction rather than adopting AI extractions uncritically. Pin cites force identification of supporting language and demonstrate actual engagement with authorities. For multi-case synthesis, build synthesized principles from accurately extracted individual holdings. When AI synthesis and your reading diverge, trust your reading. Time invested in accurate extraction produces analysis that withstands scrutiny.

7.4 Building an Argument Map and Counterarguments

Research produces authorities. Analysis transforms authorities into arguments. The gap between knowing what cases say and knowing how to use them for your client requires a different kind of thinking: seeing how legal principles connect, understanding which facts matter under which standards, anticipating how opponents will respond, and structuring your position to be persuasive. AI can assist this analytical work in ways that go beyond finding and summarizing cases. It can help you see the argumentative landscape, identify moves you might not have considered, and stress-test your reasoning against potential attacks. But AI assistance at this stage requires particular vigilance because you're no longer checking whether authorities exist; you're evaluating whether AI's analytical framing is sound.

Argument mapping creates a visual or structured representation of how your arguments fit together. At its simplest, a map shows your conclusion, the premises that support it, and the authorities establishing each premise. More sophisticated maps show how arguments relate hierarchically, which arguments are independent versus cumulative, and where vulnerabilities exist. Mapping forces clarity about the structure of your position: what you're actually claiming, why you believe it follows from the law and facts, and what you're assuming along the way.

AI can generate initial argument maps quickly. After you've gathered and extracted your authorities, prompt AI to identify the potential arguments they support. "Based on the cases I've provided regarding landlord duties for abandoned property, what arguments could a landlord make to justify disposing of property without the full statutory notice period?" AI will identify potential arguments: substantial compliance with notice requirements, waiver by tenant conduct, commercial impracticability, estoppel based on tenant representations, and others you might not have considered. This brainstorming accelerates the analytical process by surfacing possibilities you can then evaluate.

The danger is adopting AI's framing without critical evaluation. AI generates arguments based on patterns in its training data. Those patterns reflect how arguments have been structured in legal materials the AI was trained on, which may or may not match how arguments should be structured for your specific matter. AI might suggest arguments that sound sophisticated but don't actually fit your facts. It might frame issues in ways that courts in your jurisdiction have rejected. It might identify arguments that are theoretically available but practically weak. The map AI generates is a hypothesis about argumentative structure, not a blueprint to follow uncritically.

Evaluating AI-suggested arguments requires asking several questions. First, does this argument actually apply to my facts? AI may suggest arguments that require factual predicates your case doesn't satisfy. A waiver argument requires facts suggesting the tenant waived notice rights; if no such facts exist, the argument is unavailable regardless of how elegant it sounds. Second, does this argument have support in controlling authority? AI may suggest arguments that have been made in other jurisdictions or in academic commentary but that haven't been adopted by courts whose precedents bind your matter. Third, is this argument worth making? Strategic judgment about which arguments to pursue involves considerations AI cannot assess: court tendencies, opposing counsel's likely responses, client risk tolerance, and case themes you want to emphasize.

Treat AI argument suggestions as a brainstorming partner's ideas, not as legal advice. Some suggestions will be excellent, some will be mediocre, and some will be unusable. Your judgment determines which is which. The value is in expanding the possibilities you consider, not in automating the selection.

Building your own map from AI suggestions involves filtering, organizing, and supplementing. Start with the AI-generated possibilities, eliminate those that don't fit your facts or lack legal support, organize the remaining arguments by strength and type, and add arguments the AI missed based on your own analysis. The result is a map that reflects your professional judgment, informed by AI suggestions but not dictated by them.

Hierarchical organization improves map utility. Some arguments are foundational: if they fail, everything built on them fails too. Others are independent: they stand or fall on their own without affecting other arguments. Others are alternative: you pursue them in the disjunctive, needing any one to succeed. Understanding these relationships helps you allocate attention appropriately. Foundational arguments deserve the most support because everything depends on them. Independent arguments provide backup if primary theories fail. Alternative arguments require only that one succeed.

AI can help identify these hierarchical relationships. "Looking at the arguments I've identified for the landlord, which ones depend on each other and which ones are independent?" AI analysis of argument relationships can reveal structure you hadn't articulated. But verify the relationships against your own understanding; AI may perceive dependencies that don't actually exist or miss dependencies that do.

Courts respond to arguments differently based on how they're framed. An argument presented as an application of settled law faces a lower persuasive burden than an argument presented as an extension or modification of existing doctrine. AI may frame arguments without attending to these distinctions. When building your map, consider how you'll position each argument relative to existing law.

Counterargument identification is where AI provides particularly valuable assistance. Seeing your own position's weaknesses is difficult because you're invested in your arguments. AI isn't invested; it can identify counterarguments without the psychological resistance that affects human self-assessment. Prompting AI to attack your position surfaces vulnerabilities you might have overlooked or minimized.

Effective counterargument prompts ask AI to argue against you. "I'm arguing that the landlord complied with notice requirements through substantial compliance. What are the strongest arguments against this position?" AI generates opposition arguments: that substantial compliance doesn't apply when statutory language is mandatory, that the deviations weren't minor, that the tenant was actually prejudiced by inadequate notice, that prior cases rejected similar substantial compliance claims. These counterarguments tell you what you'll face and where your position needs strengthening.

> **Practice Area Note: Family Law** Steelmanning is particularly valuable in custody and relocation disputes, where best-interests analysis requires anticipating how opposing counsel will frame the same facts in opposite directions. Use AI to construct the strongest possible argument that a proposed relocation would not serve the child's best interests under your state's specific statutory factors. The steelmanned opposition surfaces evidence gaps, expert-witness needs, and authority you should distinguish before drafting your motion. Family law practitioners often face the same opposing counsel repeatedly; knowing the strongest version of their argument lets you address it preemptively rather than reactively.

Steelmanning takes counterargument identification further by asking AI to present the opposition's best possible case. "Make the strongest argument you can that the landlord violated the notice statute and should be liable for wrongful disposal." Steelmanning reveals not just what counterarguments exist but how effectively they could be deployed. The steelmanned opposition argument may be more formidable than what your actual opponent will produce, preparing you for the strongest version of the challenge you face.

Counterargument evaluation applies the same critical scrutiny you applied to your own arguments. Not every counterargument AI identifies is actually threatening. Some counterarguments rest on facts the opponent can't establish. Some rely on authority that doesn't control your matter. Some are theoretically possible but practically unlikely to be raised. Assess each counterargument's seriousness based on factual applicability, legal support, and strategic likelihood. Focus your preparation on counterarguments that are both valid and probable.

Response development addresses the counterarguments you've identified as serious. For each significant counterargument, determine your best response. Can you distinguish the facts? Can you argue that the legal rule the counterargument invokes doesn't apply? Can you show that even accepting the counterargument's premise, your client still prevails? AI can assist here too: "The other side will argue that the notice statute is mandatory and substantial compliance doesn't apply. What are the strongest responses to this argument?" AI suggestions give you response options to evaluate, not finished rebuttals to adopt.

Don't let AI counterargument identification paralyze your analysis. AI will generate counterarguments even for strong positions because you asked it to. The existence of counterarguments doesn't mean your position is weak; it means legal arguments can be made on multiple sides. Evaluate counterargument strength rather than simply counting counterarguments.

Argument map refinement incorporates counterargument analysis. A complete map shows not just your arguments but the challenges they face and your responses to those challenges. This enriched map clarifies where your position is strongest, where it's most vulnerable, and where you need to develop additional support. The map becomes a strategic tool for allocating research and drafting effort.

The structured representation helps you see gaps that unstructured thinking might miss. If your map shows an argument supported by only one authority that's easily distinguishable, you've identified a weakness to address. If your map shows three independent arguments, any of which would be sufficient, you've identified strength you might not have appreciated. If your map shows a devastating counterargument with no good response, you've identified a risk to evaluate before committing to your position.

> **Practice Point** Build your argument map before you start drafting, then use the map to structure your draft. A map built in advance prevents the disorganized drafts that result from discovering your argument structure as you write. The map is a planning tool, not just an analytical exercise.

Iterative refinement improves maps over time. Your first map represents your initial understanding, which will evolve as you work with the material. Arguments that seemed strong may prove weaker upon closer examination. Counterarguments you dismissed may prove more serious. New arguments may emerge from deeper engagement with authorities. Update your map as your understanding develops so that it remains a reliable representation of your analytical thinking.

Client communication benefits from argument mapping. Explaining to a client why their position is strong or weak often requires showing how arguments fit together and where vulnerabilities exist. A map provides a structure for that explanation, allowing you to walk through the analysis systematically rather than offering conclusory assessments. Clients who understand the argument structure can participate more meaningfully in strategic decisions about which arguments to emphasize and how aggressively to pursue the matter.

Different map formats suit different purposes. A simple outline may suffice for straightforward arguments. A flowchart showing logical dependencies works well for complex multi-step reasoning. A matrix comparing your arguments to opposing arguments and responses provides comprehensive strategic overview. A visual diagram with color-coded strength assessments offers quick reference. Choose or create a format that matches your analytical needs and working style.

Judicial clerks often create argument maps when analyzing briefs, tracking what each party argues and how the arguments relate. A well-organized brief that makes its argument structure clear is easier for clerks and judges to follow. The mapping you do for your own analysis can inform how you structure your written presentation.

AI limitations in argument mapping include inability to assess argument weight without context, tendency to generate arguments without regard for strategic considerations, and potential to miss creative arguments that aren't well-represented in training data. AI helps you see possibilities but cannot tell you which possibilities matter most for your particular situation. That judgment requires understanding your client, your court, your opponent, and the broader litigation or transactional context that AI cannot access.

The human judgment that remains essential includes factual assessment of which arguments your facts support, strategic judgment about which arguments to emphasize, predictive judgment about how courts and opponents will respond, and practical judgment about resource allocation. AI expands the possibilities you consider; your judgment narrows them to those worth pursuing. This division of labor captures AI's analytical assistance while preserving the professional judgment that responsible practice requires.

AI can help build argument maps by identifying potential arguments from gathered authorities and surfacing possibilities you might not have considered independently. Evaluate AI-suggested arguments critically: do they fit your facts, have support in controlling authority, and merit inclusion in your strategy? Hierarchical organization reveals which arguments are foundational, independent, or alternative. Counterargument identification and steelmanning prepare you for opposition by surfacing challenges and vulnerabilities. Response development addresses serious counterarguments systematically. Refine maps iteratively as your understanding deepens. AI expands analytical possibilities while human judgment determines which possibilities to pursue. The map structures your analysis and subsequently guides your drafting.

7.5 Drafting a Research Memo

The research memo is where analysis becomes work product. Everything you've done so far in this chapter, the question framing, authority gathering, holding extraction, and argument mapping, converges into a document that communicates your conclusions to whoever needs them: a supervising partner, a colleague on the matter, or your future self returning to the question later. AI can assist memo drafting significantly, accelerating the transformation of your analysis into polished prose. But drafting with AI requires the same vigilance that research with AI requires. The memo represents your work; errors in the memo are your errors regardless of what assistance produced them.

The research memo serves specific purposes that should shape how you draft it. A memo answers a question. It explains the reasoning behind that answer. It provides enough information for the reader to evaluate whether the analysis is sound. It anticipates follow-up questions and addresses them where possible. It acknowledges uncertainty where certainty isn't warranted. It serves as a reference that allows the reader to locate underlying authorities for deeper investigation. A memo that doesn't serve these purposes fails regardless of how polished its prose.

Memo structure typically follows a predictable pattern: issue statement, brief answer, facts, analysis, and conclusion. Variations exist depending on organizational preferences and matter requirements, but the core elements recur. Issue statement frames what question the memo addresses. Brief answer provides the bottom-line conclusion upfront. Facts summarizes the relevant circumstances the analysis applies to. Analysis presents the legal rules, applies them to the facts, and supports the conclusion. Conclusion restates the answer and may address implications or next steps. This structure helps readers navigate the memo and find what they need.

AI can assist each structural component, but the assistance differs by component. Issue statements benefit from AI's ability to distill questions into precise legal formulations, building on the question refinement work from the research workflow's first stage. Brief answers require your judgment about what the analysis actually shows, though AI can help articulate that judgment clearly. Facts sections involve selecting which circumstances matter and may benefit from AI's summarization capabilities, but selection judgment remains yours. Analysis sections are where AI assistance is most substantial and most dangerous, combining the synthesis and argument mapping work into connected prose. Conclusions typically flow from everything preceding them and require minimal AI assistance beyond drafting refinement.

Drafting workflow should sequence AI assistance to maximize value while minimizing risk. Begin with your argument map and extracted holdings as inputs rather than asking AI to generate the memo from scratch. The map provides structure; the extracted holdings provide content. Your task is connecting them into flowing analysis. This approach keeps you grounded in the research you've verified rather than generating fresh content that might introduce errors.

Never prompt AI to write a research memo without providing the authorities you've gathered. "Write a memo on landlord duties for abandoned property under California law" invites fabrication. "Based on the following cases I've researched, write a memo analyzing landlord duties for abandoned property" builds on verified foundation.

The source-bounded prompting technique from Chapter 3 applies directly to memo drafting. Your prompt should provide the authorities you're relying on, the facts you're analyzing, and the structure you want the memo to follow. AI works with what you provide rather than generating from training patterns. A comprehensive drafting prompt might include: your issue statement, your brief answer, the facts of your matter, each authority you're relying on with your extracted holding, and your argument map showing how the authorities support your conclusion. AI transforms these inputs into connected prose while you retain control over the substance.

A sample drafting prompt structure demonstrates the approach:

> "I need to draft a research memo with the following parameters:
>
> **Issue:** *[your precise issue statement]*
>
> **Brief Answer:** *[your bottom-line conclusion]*
>
> **Relevant Facts:** *[the facts of your matter relevant to the analysis]*
>
> **Authorities (with my extracted holdings):**
>
> - *Smith v. Jones, 123 Cal. App. 4th 456 (2015): [your extraction of the holding]*
> - *Brown v. Green, 234 Cal. App. 4th 567 (2017): [your extraction of the holding]*
> - *Cal. Civ. Code § 1946.1: [the relevant statutory text]*
>
> **Argument Structure:** *[your argument map showing how authorities support the conclusion]*
>
> Please draft an analysis section that applies these authorities to these facts, following this argument structure. Use proper legal citation format and include pin cites where I've provided them."

This prompt constrains AI to work with your verified materials rather than generating independently. The resulting draft builds on your research rather than supplementing or replacing it.

Watch for AI adding authorities you didn't provide. If the draft cites a case you don't recognize, AI has generated beyond your inputs. These additions must be verified through the full authority verification process or removed entirely. A memo built on verified authorities shouldn't incorporate unverified ones.

Iterative drafting improves results more than attempting to generate a complete memo in one pass. Start with the analysis section, which is typically the most substantial. Review the AI draft against your argument map: does the draft follow the structure you specified? Does it accurately represent the authorities you provided? Does it apply the law to the facts correctly? Identify problems and prompt AI to revise specific passages rather than regenerating the entire section. This iterative approach maintains tighter control than wholesale generation.

Section-by-section drafting extends the iterative approach. Rather than generating the full analysis at once, generate each major analytical component separately. If your analysis has three parts, address the standard, apply the standard to facts, and address counterarguments, generate and refine each part before combining them. This granular approach allows closer verification of each component and prevents errors in one section from propagating through AI-generated text in subsequent sections.

Research memos may become exhibits in malpractice litigation or disciplinary proceedings if the advice they provided is later questioned. A memo that accurately characterizes authorities provides evidence of competent research. A memo with errors provides evidence of the opposite. The verification discipline throughout this chapter protects not just the current matter but your future ability to defend your work.

Voice and tone in research memos should be professional and objective. AI tends toward the tone of its training data, which may not match your firm's conventions or the appropriate register for internal work product. You may need to adjust AI-generated prose to match expected style: more or less formal, more or less hedged, longer or shorter sentences. Style adjustments are routine drafting refinements, not substantive changes, but they affect how the memo reads and whether it matches organizational expectations.

Qualification calibration is a specific voice issue that affects accuracy. Legal analysis involves uncertainty that the memo should reflect. AI sometimes generates overconfident assertions that your research supports more weakly than the prose suggests. If your research shows that courts "have generally" adopted a particular rule, the memo shouldn't say courts "always" follow that rule. If the analysis depends on fact questions not yet resolved, the memo should acknowledge that contingency. Review AI drafts specifically for qualification accuracy, adjusting hedges to match what the authorities actually support.

Read AI-generated analysis with a skeptic's eye toward qualifications. Where the draft says "courts hold," ask yourself: is that universally true, or is there variation? Where the draft says "clearly," ask yourself: is it actually clear, or is that rhetorical overstatement? Adjust qualifications to match the actual strength of support.

Citation integration requires particular attention. AI drafts may place citations in ways that don't accurately reflect what each authority supports. A draft might cite three cases together for a proposition that only one of them supports. It might place a citation at the end of a paragraph when the cited authority supports only one sentence within that paragraph. It might use signals incorrectly, citing a case with "see" when "cf." would be more accurate. Review citation placement and signals against your extracted holdings to ensure each citation accurately represents its authority's relevance.

Pin cite insertion may not survive AI drafting even when you provided pin cites in your prompt. AI might drop pin cites to general citations or might generate pin cites that don't match what you provided. Verify that pin cites in the draft match your extraction work. If pin cites are missing or wrong, add or correct them. The discipline of pin citing discussed in Section 7.3 should carry through to the final memo.

Fact-law integration is where legal analysis lives. The analysis section shouldn't be a law summary followed by a fact summary; it should integrate law and facts to show how the law applies to these specific circumstances. AI can struggle with this integration, sometimes producing rule discussions disconnected from the facts or fact discussions disconnected from the rules. Review the draft for integration quality. Does the analysis show why the facts satisfy or fail to satisfy each legal element? Does it use the facts to address the specific questions the legal rules pose? Prompt revisions if the integration is weak.

The brief answer and conclusion deserve specific attention even though they're shorter sections. These sections state your bottom line, which is the most important information in the memo. AI drafts of these sections should accurately reflect what the analysis demonstrates. If the analysis shows that the landlord "likely" complied with notice requirements, the brief answer shouldn't say the landlord "definitely" complied. Check brief answer and conclusion language against the analysis to ensure consistency in how the conclusion is characterized.

Final verification before memo completion revisits everything the chapter has covered. Citation verification confirms that every authority cited exists, is cited correctly, and hasn't been overruled or significantly limited. Holding verification confirms that authorities are characterized accurately where they're cited. Quotation verification confirms that quoted language is verbatim from the source. Signal verification confirms that signals accurately describe the relationship between authorities and propositions. Fact verification confirms that factual assertions in the memo match the record. Consistency verification confirms that the brief answer, analysis, and conclusion characterize the result consistently.

> **Risk Note** Do not let AI-generated confidence inflate your brief answer beyond what the analysis supports. Readers often skip to the brief answer and may rely on it without reading the full analysis. An overconfident brief answer misleads even when the analysis is appropriately hedged.

Verification documentation for the memo might be as simple as a checklist you complete and retain with your research files. For each citation, check that you verified existence, accuracy, and good law status. For key holdings, note that you confirmed the characterization against the source. This documentation serves the purposes discussed in Chapter 4: protection if questions arise and evidence of the professional process you followed.

Build a memo verification checklist you use for every research memo. The checklist should include citation verification, holding verification, quotation verification, and consistency checks. Complete the checklist before treating any memo as final. Keep completed checklists with your matter files.

Template development for recurring research needs improves efficiency and consistency over time. If you frequently write memos on similar topics, create memo templates that incorporate your firm's preferred structure, citation conventions, and standard caveats. AI drafting within templates produces more consistent results than generation without structural guidance. Templates also reduce the risk of omitting important elements because the template reminds you what to include.

Learning from each memo improves future performance. After completing a memo, note what worked well and what required significant correction in the AI drafting process. Did certain prompt structures produce better results? Did particular types of analysis require more manual revision? Were there recurring AI errors you should watch for in future memos? This reflection builds personal knowledge about effective AI-assisted drafting that improves your efficiency over time.

Supervising attorneys reviewing AI-assisted memos should verify the work as rigorously as they would verify any junior associate's work. Supervision responsibility discussed in Chapter 4 applies fully. The fact that AI assisted the drafting doesn't change the human review obligation; if anything, it reinforces the need for careful review given AI's known limitations.

The complete drafting workflow integrates with the research workflow from Section 7.2. You frame questions with AI assistance, gather authorities through appropriate tools, extract holdings carefully, build argument maps to structure your analysis, then draft the memo using AI assistance constrained by your verified inputs. Each stage builds on the previous stage's work. Each stage maintains the verification discipline that prevents AI errors from corrupting your work product. The result is a memo that captures AI efficiency at every stage while maintaining the accuracy your professional obligations require.

AI can significantly accelerate research memo drafting when used appropriately. Provide AI with your extracted holdings, argument map, and matter facts rather than generating from scratch. Use source-bounded prompting to keep AI working with verified materials. Draft iteratively, reviewing and refining each section before proceeding. Verify citation placement, signals, and pin cites against your extraction work. Check qualification calibration to ensure hedges match actual support strength. Ensure fact-law integration shows how rules apply to specific circumstances. Verify that brief answer and conclusion language matches the analysis characterization. Complete final verification covering citations, holdings, quotations, and consistency. Document your verification process. Build templates and learn from each memo to improve future efficiency.

7.6 Mini-Lab: Produce a Memo Using Provided Authorities

Objective: Apply the research workflow and memo drafting techniques from this chapter to produce a research memo using pre-selected authorities. Practice holding extraction, argument mapping, source-bounded drafting, and verification in a controlled exercise where the authorities are provided rather than discovered through research.

The complete Mini-Lab packet, including the scenario, authority summaries, and analysis guide, is available on the companion site for download.

Time Required: 90 to 120 minutes

What You'll Need:

- The research workflow from Section 7.2 (Question → Authorities → Synthesis → Verification)
- The holding extraction techniques from Section 7.3
- The argument mapping approach from Section 7.4
- The memo drafting workflow from Section 7.5
- Access to an AI tool for drafting assistance
- A method for documenting your process

Instructions

Step 1: Review the Scenario and Authorities

Read the following scenario and the authority summaries provided. In actual practice, you would have full case opinions and statutory text. For this exercise, the summaries provide the essential information you need to work with. Treat the summaries as if they were your notes from reading the actual authorities.

Scenario

Your client is Greenfield Property Management ("Greenfield"), a commercial property management company. Greenfield manages a small strip mall in Sacramento, California. One of the tenants, Digital Solutions Inc. ("Digital Solutions"), a computer repair business, stopped paying rent four months ago. After multiple failed attempts to contact Digital Solutions, Greenfield's property manager visited the premises and found the business closed, the signage removed, and the space empty except for several items left behind: two commercial computer workstations (estimated value $3,000 each), a small safe (contents unknown), cardboard boxes containing what appear to be customer computers left for repair (approximately 15 units), and miscellaneous office furniture.

Greenfield served a three-day notice to pay rent or quit, which was posted on the door and mailed to Digital Solutions' registered agent address. No response was received. Greenfield then filed an unlawful detainer action and obtained a default judgment restoring possession. The judgment was entered 30 days ago. Greenfield now wants to dispose of the remaining property to re-lease the space. The property manager sent a letter to Digital Solutions' registered agent address two weeks ago stating that the property would be disposed of in 15 days if not claimed.

The supervising partner wants to know: Has Greenfield satisfied its legal obligations such that it can now dispose of the abandoned property without liability?

Authority Summaries

For this exercise, treat the following as your research results from reading the actual authorities:

> **Note on Authorities** All five authorities below are teaching constructions designed to interact with the Greenfield Property Management facts. Authority 1 is loosely based on California Civil Code § 1993, the real California statute governing disposition of personal property remaining on commercial premises, but has been simplified and adapted for this exercise. The actual statute differs in several material respects, including the notice period, the mailing requirements, and the value threshold for public sale. Authorities 2 through 5 are constructed hypothetical cases, named *Hypothetical v. [Legal Issue]* to make clear they are teaching examples. None of these authorities should be cited or relied upon in actual practice. Always verify current law using primary research tools before relying on any authority in actual matters.

Authority 1: Adapted Abandoned-Property Statute (loosely based on California Civil Code § 1993)

This statute governs disposition of personal property remaining on commercial premises after a tenancy terminates. Key provisions:

- The landlord must give written notice to the tenant describing the property in reasonable detail, stating that the property will be disposed of if not claimed, and providing at least 15 days for the tenant to claim the property.
- Notice must be personally delivered or sent by first-class mail and certified mail to the tenant's last known address.
- If the property is reasonably believed to be worth less than $700 total, the landlord may dispose of it in any manner. If reasonably believed to be worth $700 or more, the landlord must sell it at public sale and hold the proceeds (minus costs) for the tenant for one year.
- The notice must include specific information: a description of the property, the address where the property may be claimed, the date before which the property must be claimed (at least 15 days from notice), and a statement that the property will be disposed of if not claimed.
- The landlord is liable to the tenant and any other owner of the property for damages caused by the landlord's failure to comply with this section.

Authority 2: Hypothetical v. Statutory Notice

Holding: The statutory notice requirements under the predecessor to Section 1993 are mandatory, not permissive. A landlord who fails to provide the required notice before disposing of tenant property is liable for conversion regardless of whether the tenant was otherwise in default or had abandoned the premises.

The court emphasized that the statute exists to protect property rights separate from the tenancy itself. The tenant's breach of the lease does not forfeit their rights in personal property left on the premises.

Authority 3: Hypothetical v. Dual Mailing

Holding: The dual notice requirement (first-class mail and certified mail) is a substantive requirement, not merely a procedural formality. Where the landlord sent only first-class mail without certified mail, the notice was defective and the landlord was liable for wrongful disposition even though the tenant actually received the notice.

The court rejected the argument that actual notice should excuse technical noncompliance, reasoning that the statutory requirements create a safe harbor: landlords who comply are protected, while those who don't bear the risk that their disposition was wrongful.

Authority 4: Hypothetical v. Express Abandonment

Holding: A landlord is not liable for disposing of property without following statutory procedures where the tenant expressly abandoned the property in writing. The statutory notice requirements apply to property the tenant has left behind, not to property the tenant has affirmatively relinquished.

The court distinguished cases involving implied abandonment through conduct from cases involving express abandonment through written communication, holding that only the latter excuses compliance with the statutory notice requirements.

Authority 5: Hypothetical v. Mailing Date

Holding: The 15-day notice period runs from the date of mailing, not the date of receipt. A landlord who mailed compliant notice on March 1 could dispose of property on March 16, regardless of when (or whether) the tenant actually received the notice.

The court also held that where the tenant's whereabouts are unknown, the landlord satisfies the notice requirement by sending notice to the last address actually known to the landlord, even if the landlord has reason to believe the tenant may no longer be at that address.

Step 2: Extract Holdings and Identify Issues

Before using AI assistance, work through the authorities yourself.

For each authority, write a one-sentence extraction of the key holding relevant to your client's question. Stay close to the language in the summaries; resist the urge to paraphrase broadly.

Identify the specific legal requirements your client must satisfy based on Authority 1 (the statute).

Note which requirements Greenfield appears to have satisfied and which are potentially problematic based on the facts provided.

List any factual gaps that affect your ability to provide a definitive answer.

Step 3: Build an Argument Map

Create a simple argument map addressing the question: Has Greenfield satisfied its obligations to dispose of the property?

Your map should identify:

- Arguments supporting the conclusion that Greenfield has satisfied its obligations
- Arguments suggesting Greenfield may not have satisfied its obligations
- The authorities that support each position
- The factual predicates each argument requires

Use AI assistance to brainstorm additional arguments you might have missed, but evaluate each suggestion critically against the authorities you've been provided. Don't adopt arguments that lack support in these specific authorities.

Step 4: Draft the Memo

Using the source-bounded drafting approach from Section 7.5, draft a research memo addressing the partner's question.

Your memo should include:

- Issue: A precise statement of the legal question
- Brief Answer: Your bottom-line conclusion with appropriate qualification
- Facts: A summary of the relevant facts from the scenario
- Analysis: Application of the legal requirements to Greenfield's situation
- Conclusion: Restatement of your answer with any recommendations

When prompting AI to assist with drafting, provide your extracted holdings, your argument map, and the facts. Constrain the AI to work with the authorities you've provided. Watch for any AI additions of authorities you didn't provide.

Step 5: Verify and Refine

Review your draft against the verification requirements from Section 7.5:

- Does every citation accurately represent what the authority says?
- Are your holdings characterized accurately, without over-generalization or qualification stripping?
- Do your qualifications match the strength of support in the authorities?
- Is your brief answer consistent with your analysis?
- Are there any AI-generated additions that you didn't provide and therefore cannot verify?

Revise the draft to correct any problems you identify.

Step 6: Document Your Process

Create a brief documentation record of your process:

- What AI tool did you use?
- At what stages did you use AI assistance?
- What verification steps did you complete?
- What changes did you make between the AI draft and your final version?

This documentation practices the habits discussed in Chapter 4 and provides a record of your professional process.

Scenario Analysis Guide

After completing your own analysis, use the following guide to check your work. The guide identifies the key issues the scenario raises; reasonable lawyers might reach somewhat different conclusions on some points.

Statutory Requirements Analysis

Under Section 1993, Greenfield must satisfy the following requirements:

Written notice: Greenfield sent a letter, satisfying the written requirement.

Description of property: The scenario doesn't specify whether the letter described the property in reasonable detail. This is a factual gap. If the letter merely said "your property" without description, it may be defective.

Statement that property will be disposed of: The scenario states the letter said property would be "disposed of," which appears to satisfy this requirement.

15-day notice period: The letter stated property would be disposed of in 15 days, and 15 days have not yet passed (letter was sent two weeks ago). This is sufficient under Hypothetical v. Mailing Date, which held the period runs from mailing.

Dual mailing requirement: This is the critical issue. The scenario says Greenfield sent "a letter" but doesn't specify the mailing method. Under Hypothetical v. Dual Mailing, both first-class and certified mail are required. If Greenfield sent only first-class mail, the notice is defective regardless of actual receipt.

Address requirement: Greenfield sent notice to the registered agent address, which appears to be the last known address. Under Hypothetical v. Mailing Date, this satisfies the address requirement even if Greenfield suspects Digital Solutions may no longer receive mail there.

Disposition Method Analysis

The property value affects the required disposition method. The workstations alone ($6,000 estimated) exceed the $700 threshold. Under Section 1993, property reasonably believed to be worth $700 or more must be sold at public sale, not simply disposed of. Additionally, the customer computers present a particular concern: they may belong to third parties, not Digital Solutions, complicating Greenfield's right to dispose of them.

Key Uncertainties

Two factual gaps prevent a definitive answer:

1. Did the notice letter adequately describe the property?
2. Did Greenfield use both first-class and certified mail?

If either answer is no, Greenfield has not satisfied its obligations and faces conversion liability under *Hypothetical v. Statutory Notice.*

Express Abandonment Inapplicability

The Hypothetical v. Express Abandonment exception doesn't help Greenfield. Digital Solutions never provided written abandonment of the property. The factual circumstances suggest implied abandonment through conduct (closing the business, removing signage), but Hypothetical v. Express Abandonment specifically held that implied abandonment through conduct doesn't excuse statutory compliance.

The customer computers require separate analysis. These may not be Digital Solutions' property at all. Disposing of property belonging to third parties creates direct liability to those third parties regardless of compliance with the landlord-tenant statutory scheme. Greenfield should make reasonable efforts to identify and notify the computer owners.

Sample Brief Answer

Based on the authorities provided, Greenfield likely has not yet satisfied all statutory requirements for disposing of the abandoned property, though this conclusion depends on factual details not provided. Specifically, California Civil Code Section 1993 requires notice by both first-class and certified mail, and the dual mailing requirement is substantive, not merely procedural. If Greenfield sent notice only by first-class mail, the notice is defective and Greenfield faces conversion liability for disposing of the property. Additionally, the property value appears to exceed $700, requiring public sale rather than simple disposal. Finally, the customer computers may belong to third parties whose interests are not addressed by the landlord-tenant statutory scheme. I recommend confirming the mailing method used, providing compliant notice if necessary, and addressing the customer computer issue before disposing of any property.

Deliverable

By the end of this Mini-Lab, you should have:

1. Extracted holdings for each of the five authorities
2. An argument map showing supporting and opposing positions with authority support
3. A complete research memo with Issue, Brief Answer, Facts, Analysis, and Conclusion
4. Documentation of your AI-assisted drafting process
5. A revised final memo reflecting verification and corrections

This deliverable demonstrates the complete research-to-memo workflow using AI assistance while maintaining the verification discipline that professional responsibility requires.

Bonus Challenge

Extend your analysis in two directions.

First, draft a follow-up email to the partner identifying the additional information you need to provide a more definitive answer. The email should explain why each piece of information matters and what difference it would make to your conclusion. This exercise practices the client communication skills that Chapter 10 will address more fully.

Second, assume the partner responds that Greenfield did use only first-class mail, not certified mail. Draft a supplemental memo analyzing whether Greenfield has any viable defenses or arguments that the technical noncompliance should be excused. Consider what arguments Hypothetical v. Dual Mailing rejected and whether any factual distinctions might allow a different result. This exercise practices the counterargument and response development skills from Section 7.4.

Research memos often reveal that the question cannot be answered definitively without additional facts. This is not a failure of research; it's a success of analysis. Identifying what you need to know is as valuable as answering what you can answer. The ability to distinguish what's certain from what's contingent is a hallmark of sophisticated legal analysis.

7.7 Practice Checklist and Prompt Pack for Research

This section provides ready-to-use resources for implementing AI-assisted legal research in your practice. The checklist consolidates the workflow requirements from this chapter into a single reference. The prompt pack provides templates for common research tasks that you can adapt to your specific needs. These resources translate the chapter's concepts into practical tools you can use immediately.

Research Workflow Checklist

Use this checklist for any significant AI-assisted research project. For routine research, the checklist serves as a mental framework even if you don't complete it formally. For research that will inform court filings or significant client advice, complete the checklist and retain it with your matter files. A printable version of this checklist is available on the companion site for matter files.

Stage 1: Question Framing

☐ Identified the core question the research must answer

☐ Used AI to decompose the question into specific legal issues

☐ Determined the governing jurisdiction and any choice-of-law considerations

☐ Identified threshold issues that affect how subsequent questions are framed

☐ Confirmed understanding of what the client actually needs (not just what they asked)

☐ Documented the refined research questions before proceeding to authority identification

Stage 2: Authority Identification

☐ Selected appropriate research tools based on task requirements (grounded AI for citation retrieval, general AI only for search term brainstorming)

☐ Conducted initial searches using AI-suggested terms and concepts

☐ Expanded research beyond initial AI results using traditional techniques (following citations, checking secondary sources, searching alternative terms)

☐ Evaluated each authority for relevance, precedential weight, and currency

☐ Organized authorities by issue and by whether they support or undermine the client's position

☐ Identified gaps where additional research is needed

☐ Confirmed that all authorities were retrieved from verified legal databases, not generated by AI

Stage 3: Synthesis

☐ Read the relevant portions of each authority (not just AI summaries)

☐ Extracted holdings using language close to the court's own words

☐ Preserved qualifications and limitations in extracted holdings

☐ Created pin cites identifying where key holdings appear

☐ Used AI to help identify patterns across authorities, then verified suggested patterns against the sources

☐ Built argument map showing how authorities support conclusions

☐ Identified counterarguments and evaluated their strength

☐ Developed responses to serious counterarguments

☐ Noted uncertainties and factual contingencies that affect conclusions

Stage 4: Verification

☐ Verified existence of every cited authority through authoritative database

☐ Verified citation format accuracy for every citation

☐ Checked good law status (not overruled or significantly limited) for every authority

☐ Verified that holding characterizations accurately reflect the cited opinions

☐ Verified that quoted language appears verbatim in the source

☐ Verified that signals accurately describe the relationship between authorities and propositions

☐ Documented verification process (which database, when verified, any issues identified)

Memo Drafting (if applicable)

☐ Provided AI with extracted holdings, argument map, and facts rather than generating from scratch

☐ Reviewed AI draft for accuracy against provided authorities

☐ Removed or verified any authorities AI added that you didn't provide

☐ Checked qualification calibration (hedges match actual support strength)

☐ Confirmed fact-law integration (analysis shows how rules apply to specific facts)

☐ Verified consistency between brief answer, analysis, and conclusion

☐ Completed final verification of all citations, holdings, and quotations

Prompt Pack for Legal Research

The following prompts are templates you can adapt for your specific research needs. Replace bracketed placeholders with your actual information. These prompts implement the source-bounded approach that keeps AI working with verified materials rather than generating from training patterns. All prompts in this section are available as copy-paste templates on the companion site.

Question Refinement Prompts

Basic question decomposition: "My client needs to know [general question in natural language]. Help me break this down into specific legal questions I should research. For each question, identify what area of law it implicates and what type of authority (statute, case law, regulation) is most likely to address it."

Jurisdiction-focused refinement: "I'm researching [general topic] under [jurisdiction] law. What are the key legal issues I should investigate? What statutory provisions or common law doctrines typically govern this area? What terminology do courts in this jurisdiction use when addressing these issues?"

Issue spotting from facts: "Here are the facts of my matter: [fact summary]. What legal issues do these facts raise? For each issue, briefly explain why these particular facts trigger that issue and what legal questions I should research."

Customize these prompts by adding context about your practice area, jurisdiction, or the type of matter. More specific prompts produce more useful results. "Break down this contract dispute question under Delaware law for a motion to dismiss" produces better output than "break down this question."

Search Term Generation Prompts

Terminology exploration: "I'm researching [topic]. What terms and phrases do courts typically use when discussing this issue? Include synonyms, related concepts, and any technical terminology I should search for. Focus on terms likely to appear in [jurisdiction] case law."

Expanding search scope: "My initial searches for [terms you've used] returned limited results. What alternative search terms might surface additional relevant authorities? What related legal concepts might courts discuss in cases that don't use my initial terminology?"

Secondary source identification: "What treatises, law review articles, or practice guides typically address [topic]? Which secondary sources are most respected in this area and most likely to identify leading cases?"

Authority Summarization Prompts

Holding extraction (provide the opinion text): "Here is the text of [case name]. Quote the sentence or sentences where the court states its holding on [specific issue]. Then explain in one sentence what the court decided, staying close to the court's own language."

Structured case summary (provide the opinion text):

> Based on the following opinion, provide a structured summary with these elements:
>
> 1. **Procedural posture** *(one sentence)*
> 2. **Key facts** *(2–3 sentences focusing on facts relevant to the holding)*
> 3. **Issue** *(the specific legal question the court addressed)*
> 4. **Holding** *(what the court decided, using language close to the opinion)*
> 5. **Reasoning** *(the court's key rationale in 2–3 sentences)*
> 6. **Pin cite** *(the page number where the holding appears)*

Comparison extraction (provide multiple opinion texts): "Here are excerpts from three cases addressing [issue]. For each case, extract the specific standard or test the court applied. Then identify: Where do these courts agree? Where do they differ? Are the differences genuine conflicts or are they addressing different fact patterns?"

Never use summarization prompts without providing the actual opinion text. Asking AI to summarize a case by name without providing the text invites fabrication. Always retrieve the opinion from a verified database first, then provide the text to AI for summarization assistance.

Synthesis Prompts

Pattern identification (provide your case summaries): "Based on the following case summaries I've prepared, what factors appear to determine [outcome at issue]? List each factor and identify which cases discuss it. Note any apparent conflicts between how different courts weigh these factors."

Rule synthesis (provide your authority extractions): "Using only the authorities I've provided below, synthesize the legal rule for [issue] in [jurisdiction]. State the rule, then show which authorities support each element of the rule. Identify any gaps where the authorities don't clearly establish the rule."

Application framework (provide authorities and facts): "Based on the legal standards from the authorities I've provided and the facts of my matter below, create a framework for analyzing [issue]. For each element of the standard, identify what facts are relevant and what additional facts I would need to reach a definitive conclusion."

Argument Mapping Prompts

Argument brainstorming (provide your authorities): "Based on the authorities I've provided regarding [issue], what arguments could be made that [proposition client wants to establish]? For each argument, identify which authorities support it and what facts it requires."

Counterargument identification: "I'm arguing that [your position]. Based on the authorities I've provided, what are the strongest arguments against this position? For each counterargument, identify which authorities might support it and how an opponent might frame the argument."

Steelmanning: "Present the strongest possible argument that [opposing position]. Use only the authorities I've provided. Make the argument as compelling as you can, identifying the best authorities for each point and acknowledging any weaknesses only where necessary to maintain credibility."

Response development: "My opponent will likely argue [counterargument]. Based on the authorities I've provided, what are the strongest responses to this argument? For each response, identify supporting authorities and explain why the response should be persuasive."

Save your most effective prompts in a personal prompt library. When a prompt produces excellent results, note what made it effective and keep it for future use. Over time, your library becomes a collection of proven approaches for your specific practice needs.

Memo Drafting Prompts

Analysis section draft (provide all inputs):

> Draft an analysis section for a research memo with the following parameters:
>
> **Issue:** *[your precise issue statement]*
>
> **Legal standard:** *[the rule synthesized from your authorities]*
>
> **Authorities with extracted holdings:**
>
> - *[Authority 1]: [your extraction]*
> - *[Authority 2]: [your extraction]*
> - *[continue for each authority]*
>
> **Facts to apply:** *[relevant facts from your matter]*
>
> **Argument structure:** *[your argument map or outline]*
>
> Draft the analysis applying these authorities to these facts. Use proper legal citation format with pin cites where I've provided them. Include appropriate qualifications where the conclusion depends on factual determinations or where authorities provide less than definitive support.

Brief answer draft:

> Based on the following analysis [paste your analysis section], draft a brief answer that:
>
> **1.** States the bottom-line conclusion in the first sentence
>
> **2.** Identifies the key reasons supporting that conclusion
>
> **3.** Notes the most significant qualifications or contingencies
>
> **4.** Uses language consistent with the qualifications in the analysis
>
> Keep the brief answer to one paragraph.

Conclusion draft:

> Based on the following issue, brief answer, and analysis [paste each], draft a conclusion section that:
>
> **1.** Restates the answer to the question presented

> **2.** Summarizes the key supporting points
>
> **3.** Identifies any recommended next steps or additional research
>
> **4.** Notes any factual investigations that would strengthen the analysis
>
> Keep the conclusion concise, approximately one-half to one page.

Prompts that specify citation format and request pin cites produce more professionally useful output than prompts that leave formatting to AI defaults. Specify what you need in the prompt rather than correcting format issues afterward.

Verification Prompts

Holding verification check: *"I've characterized [case name] as holding [your characterization]. Here is the relevant excerpt from the opinion. Does my characterization accurately capture what the court held? If not, what is inaccurate and how should I revise the characterization?"*

Consistency check: *"Here is my brief answer: [brief answer]. Here is my analysis section: [analysis]. Here is my conclusion: [conclusion]. Are these three sections consistent in how they characterize the conclusion and its strength? Identify any inconsistencies."*

Qualification audit:

> Review the following analysis section for qualification accuracy. Flag any places where the language (such as "clearly," "always," "must") may overstate the certainty of the legal conclusion based on the authorities cited. Suggest more accurate qualifications where appropriate.
>
> *[paste analysis section]*

Customization Guidance

These prompts are starting points. Customize them based on your practice area, jurisdiction, preferred drafting style, and organizational conventions.

For specialized practice areas, add context that improves AI performance. Employment lawyers might add standard references to Title VII, the ADA, or state equivalents. Corporate lawyers might specify the state of incorporation. Litigators might specify the court level and procedural posture. This context helps AI provide more relevant assistance.

For firm-specific conventions, add style guidance. If your firm uses particular citation formats, heading styles, or structural conventions, incorporate those requirements into your drafting prompts. AI can follow style guidance when you provide it.

For recurring research types, create specialized versions of these prompts. If you frequently research the same issues, pre-populate prompts with the governing statutory framework and leading cases so you do not have to include them each time.

Quick Reference: Research Risk Levels

Use this quick reference to calibrate AI assistance to task risk. A printable desk card version of this reference is available on the companion site.

Lower risk (AI assistance appropriate with normal review):

- Question decomposition and refinement
- Search term brainstorming
- Summarizing authorities you've provided
- Organizing verified materials
- Pattern identification across provided authorities
- Drafting from verified inputs

Higher risk (intensive verification required):

- Any AI characterization of holdings
- Synthesis of rules across multiple authorities
- Application of law to facts
- Counterargument identification and evaluation

Unacceptable risk (do not rely on AI):

- Citation generation from general AI (not connected to verified databases)
- Accepting authority summaries without reading the source
- Legal conclusions without independent analysis
- Any output going to clients or courts without verification

Post this quick reference where you can see it when conducting research. The risk calibration becomes intuitive with practice, but the reference helps until the calibration is habitual.

Documentation Template

Use this template to document your AI-assisted research process for significant matters. A fillable version of this template is available on the companion site.

RESEARCH DOCUMENTATION

Matter: ____________________ Date: _____________ Researcher: _____________

TOOLS USED:

- ☐ Grounded legal research AI: _________________
- ☐ General AI assistant: _________________
- ☐ Traditional legal databases: _________________

AI ASSISTANCE STAGES:

- ☐ Question refinement
- ☐ Search term generation
- ☐ Authority summarization

☐ Synthesis assistance

☐ Argument mapping

☐ Memo drafting

VERIFICATION COMPLETED:

☐ All citations verified in: ________________ Date: ______

☐ All holdings verified against source opinions

☐ All quotations verified verbatim

☐ Good law status confirmed for all authorities

NOTES: [Document any issues discovered during verification, corrections made, or limitations of the analysis]

Verification completed by: ________________ Date: ________________

This template creates a record demonstrating that AI assistance was used responsibly with appropriate verification. Retain completed templates with your research files.

The checklist consolidates research workflow requirements into a single reference for ensuring quality and completeness. The prompt pack provides adaptable templates for common research tasks, implementing source-bounded prompting that keeps AI working with verified materials. Customize prompts for your practice area, jurisdiction, and organizational conventions. Save effective prompts in a personal library for reuse. Use the risk reference to calibrate AI assistance to task requirements. Document your process using the template to demonstrate responsible AI use. These resources translate the chapter's concepts into practical tools for immediate implementation.

Key Takeaways

1. AI assists legal research most reliably when processing information you provide rather than generating information from training patterns. Question refinement, search term generation, issue spotting, authority summarization, research organization, and synthesis assistance all work well because AI transforms verified inputs rather than creating unverified outputs. Citation generation from general AI, characterizing authorities you haven't read, and providing definitive legal conclusions create unacceptable risk because they depend on AI accuracy that cannot be assumed. The distinction between processing and generating determines whether AI assistance is responsible.
2. Effective research workflow sequences AI assistance across four stages. Question framing uses AI to decompose broad questions into specific legal issues suitable for systematic investigation. Authority identification uses grounded AI tools connected to verified databases, complemented by traditional research techniques that expand beyond initial AI results. Synthesis uses AI to help identify patterns across authorities you've gathered while maintaining your own analytical judgment about what the authorities actually establish. Verification confirms that every citation exists, is formatted correctly, hasn't been overruled, and is characterized accurately. Time pressure does not reduce verification requirements.

3. Holding extraction requires staying close to source language rather than accepting AI paraphrasing. AI paraphrases by default, producing summaries that may over-generalize holdings, strip qualifications, convey false certainty, or create temporal confusion. Extraction techniques preserve the court's meaning by using language close to the opinion's own words and carefully preserving scope and limitations. Pin cites force identification of supporting language and demonstrate actual engagement with authorities. When AI synthesis and your reading diverge, trust your reading.
4. Argument mapping structures your analysis by showing how legal principles connect, which facts matter under which standards, and how authorities support conclusions. AI can brainstorm potential arguments from gathered authorities, but each suggestion requires critical evaluation: does it fit your facts, have support in controlling authority, and merit inclusion in your strategy? Counterargument identification and steelmanning prepare you for opposition by surfacing challenges and vulnerabilities. Response development addresses serious counterarguments systematically. The map structures your analysis and subsequently guides your drafting.
5. Memo drafting with AI assistance requires source-bounded prompting that constrains AI to work with your verified materials. Provide extracted holdings, argument map, and matter facts rather than generating from scratch. Draft iteratively, reviewing and refining each section before proceeding. Verify citation placement, signals, and pin cites against your extraction work. Check qualification calibration to ensure hedges match actual support strength. Ensure fact-law integration shows how rules apply to specific circumstances. Complete final verification before treating any memo as final.
6. Verification is non-negotiable regardless of time pressure or confidence in AI outputs. Every citation must be confirmed through authoritative databases. Every holding characterization must be checked against the actual opinion. Every quotation must be verified verbatim against the source. Every signal must accurately describe the relationship between authority and proposition. Documentation of verification creates a record demonstrating responsible process. If you don't have time to verify, you don't have time to use AI-generated content in work product.

Chapter Checkpoint

Before moving on, make sure you can:

- ☐ Distinguish between research tasks where AI assistance is appropriate with normal review and tasks where AI creates unacceptable risk without intensive verification
- ☐ Execute the four-stage research workflow (Question → Authorities → Synthesis → Verification), using AI assistance appropriately at each stage
- ☐ Extract holdings from authorities using language close to the court's own words while preserving qualifications, limitations, and scope
- ☐ Identify the common AI paraphrasing errors (over-generalization, qualification stripping, confidence mismatch, temporal confusion) and catch them during review
- ☐ Build an argument map showing how authorities support your conclusion, including hierarchical relationships among arguments
- ☐ Use AI to identify counterarguments and develop responses without adopting AI's framing uncritically

☐ Draft a research memo using source-bounded prompting that constrains AI to work with your verified materials

☐ Complete comprehensive verification of citations, holdings, quotations, and signals before finalizing any research product

☐ Document your AI-assisted research process in a manner that demonstrates responsible use

What's Next

Research produces understanding. Drafting transforms understanding into documents that accomplish legal work: contracts that allocate risk, motions that persuade courts, letters that communicate positions, and agreements that memorialize deals. Chapter 8 applies the workflow thinking from this chapter to document drafting and redlining, showing how AI can accelerate the creation and revision of legal documents while maintaining the quality and accuracy your professional obligations require.

The drafting context presents different challenges than research. Research asks whether authorities exist and what they say. Drafting asks whether documents accomplish their intended purposes, whether language creates the legal effects clients need, and whether provisions adequately address risks that may materialize in the future. AI assistance in drafting must account for these differences. The verification focus shifts from confirming that citations are real to confirming that language means what you intend it to mean.

Chapter 8 covers the full drafting lifecycle: from initial instructions through first draft, revision, and final verification. You'll learn how to use AI for contract clause generation, comparison, and risk identification. You'll see how AI can assist with redlining by analyzing proposed changes, suggesting responses, and identifying implications you might miss. You'll address the particular challenges of litigation drafting, where facts must trace to the record and arguments must build on verified authorities. And you'll develop approaches for maintaining style consistency when AI assists drafting across an office or practice group. The drafting workflow builds on everything you've learned while adapting to the specific demands of document creation.

Key Terms

Research question refinement — Transforming vague or broad research questions into precise legal issues suitable for systematic investigation. AI can suggest issue breakdowns and sub-questions and help structure the research plan.

Grounded AI — AI systems connected to authoritative data sources that constrain outputs to information actually present in those sources. For legal research, grounded AI connects to verified legal databases, reducing fabrication risk.

Question decomposition — Breaking a broad research question into specific, discrete legal issues that can be researched independently and then synthesized into a complete answer.

Synthesis (in legal research) — Analyzing authorities collectively to identify principles, patterns, and frameworks that address the research question. AI can suggest synthetic frameworks; evaluation requires lawyer judgment.

Holding extraction — Capturing what a court actually decided by using language close to the opinion's own words and preserving the scope and limitations of the ruling. Contrasts with paraphrasing, which may introduce distortions.

Pin cite — A citation identifying the specific page or pages where supporting material appears within a longer source. Pin cites facilitate verification and demonstrate engagement with the source.

Argument mapping — Creating a structured representation of how legal arguments connect premises, authorities, facts, and conclusions. Maps clarify structure, reveal dependencies, and identify where support is strong or weak.

Steelmanning — Presenting an opposing position in its strongest possible form rather than attacking a weakened version. AI-assisted steelmanning helps identify the best counterarguments to your position.

Research memo — A written document presenting legal analysis of a specific question, typically including issue, brief answer, facts, analysis, and conclusion. Internal work product that informs strategy and decision-making.

Signal (in legal citation) — Indicates the relationship between a cited authority and the proposition it supports. Signals like "see," "see also," "cf.," and "but see" convey different degrees and types of support.

Practice checklist — A standardized list of steps to complete for a recurring task type, ensuring consistent quality and preventing omission of important elements.

Prompt library — A personal or organizational collection of effective prompts saved for reuse, customized for specific practice needs and refined based on experience.

Chapter 8: AI for Drafting and Redlining

Drafting is where lawyers create value. The research memo informed your understanding; the contract, motion, or agreement creates legal reality. A well-drafted document allocates risk appropriately, accomplishes the client's objectives, and anticipates problems before they arise. A poorly drafted document creates ambiguity, fails to address foreseeable contingencies, and may not accomplish what the client needed at all. The stakes of drafting quality are high, which makes AI assistance in drafting both appealing and consequential.

AI can accelerate drafting substantially. Tasks that once required starting from scratch or hunting through precedent files can now begin with AI-generated first drafts. Clause variations that required manual comparison can be generated and analyzed in minutes. Style consistency that demanded painstaking manual review can be supported by AI that identifies deviations and suggests corrections. The efficiency potential is real, and lawyers who learn to use AI effectively for drafting gain meaningful advantages in speed without sacrificing quality.

The research workflow from Chapter 7 asked whether authorities exist and say what you think they say. The drafting workflow asks different questions. Does this language accomplish what the client needs? Does it create the legal effects intended? Does it adequately address risks that might materialize? Does it fit with the rest of the document and the broader transaction or matter? These questions require judgment that AI cannot provide, but AI can accelerate your path to exercising that judgment by handling mechanical aspects of document creation while you focus on the substantive decisions that require professional expertise.

What This Chapter Covers

- A structured drafting workflow that sequences AI assistance appropriately: from instructions through first draft, revision, and verification
- How to use AI for contract clause work, including generating clause options, comparing alternative language, and identifying risks in proposed provisions
- AI assistance for redlining, including analyzing counterparty proposals, developing negotiation positions, and identifying fallback options
- The particular challenges of litigation drafting, where facts must trace to the record and arguments must build on verified legal foundations
- Fact discipline techniques that prevent AI from inventing or embellishing facts that don't exist in the record
- Style systems that maintain consistency when AI assists drafting across an office or practice group
- A Mini-Lab improving a contract clause and producing a risk memo, applying the chapter's techniques to a realistic scenario
- A practice checklist and prompt pack providing ready-to-use resources for drafting workflows

Chapter Summary

Drafting with AI assistance follows a workflow parallel to the research workflow: structured stages that sequence AI assistance appropriately. The workflow begins with instructions, where you define what the document must accomplish, what constraints apply, and what context AI needs to produce useful output. It proceeds to first draft generation, where AI creates initial content you'll revise rather than final content you'll file. It continues through revision, where you transform AI output into work product that reflects your professional judgment. It concludes with verification, which for drafting focuses on whether the document accomplishes its purposes rather than whether citations exist.

Contract work offers particular opportunities for AI assistance. Clause generation produces options you evaluate rather than language you adopt uncritically. Clause comparison helps you understand how different formulations allocate risk and identify which approach best serves your client. Risk identification surfaces issues you might have missed, though AI risk assessment requires the same critical evaluation as AI argument suggestions in research. Redlining assistance helps you analyze counterparty proposals efficiently, identify what changed and why it matters, and develop responses that advance your negotiating position.

Litigation drafting presents distinct challenges because facts and authorities must trace to verified sources. A contract clause can be evaluated on its own terms: does this language accomplish the objective? A motion's factual recitation must match the record, and its legal arguments must build on authorities that exist and say what you claim. The fact discipline techniques in this chapter prevent AI from inventing facts that would undermine your credibility and potentially violate your professional obligations.

Style consistency matters when AI assists drafting across multiple documents, drafters, or matters. Without attention to style, AI produces generic output that doesn't match your firm's conventions, your client's preferences, or the voice appropriate to the document type. Style systems establish parameters that AI follows, producing drafts that require less revision and that maintain consistency across work product. These systems also support supervision by creating expectations that reviewers can apply.

Throughout, the principle from earlier chapters applies: AI accelerates mechanical work while human judgment remains essential for substantive decisions. Drafting involves countless judgment calls about word choice, structure, emphasis, and risk allocation. AI can generate options and surface considerations, but selecting among options and weighing considerations remains your professional responsibility. The verification that matters for drafting is different from research verification, focusing on whether the document works rather than whether citations are real, but verification remains essential. A document that doesn't accomplish its purpose fails regardless of how efficiently it was produced.

8.1 Drafting Workflow: Instructions → First Draft → Revise → Verify

Drafting without a workflow produces inconsistent results. Sometimes the document comes together smoothly; sometimes it requires painful revision after revision. Sometimes AI assistance helps; sometimes it creates more work than it saves. The inconsistency stems from approaching each drafting task ad hoc rather than following a structured process that sequences work appropriately. A drafting workflow imposes that structure, ensuring that each stage builds on the previous stage and that AI assistance occurs where it adds value without introducing problems.

The drafting workflow has four stages: Instructions, First Draft, Revise, and Verify. Like the research workflow, these stages are sequential in concept but iterative in practice. A first draft may reveal that your instructions were incomplete, sending you back to refine them. Revision may surface issues that require additional drafting. Verification may identify problems that require substantive changes rather than minor corrections. The stages provide a framework for organizing your efforts while accommodating the back-and-forth that complex drafting requires.

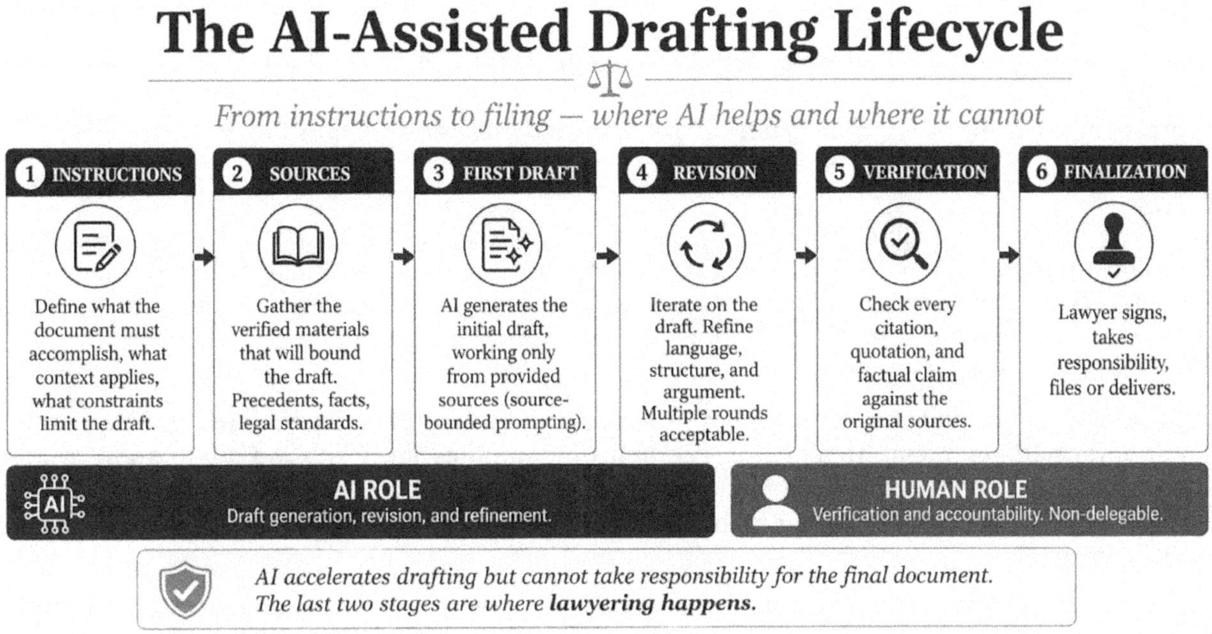

Figure 8.1 — The AI-Assisted Drafting Lifecycle

Stage One: Instructions

The quality of AI-assisted drafting depends heavily on instruction quality. Vague instructions produce generic output that requires extensive revision. Precise instructions produce drafts that are closer to what you need, reducing the gap between AI output and final work product. Time invested in crafting good instructions pays returns throughout the drafting process.

Instructions for AI drafting should address five elements: purpose, context, constraints, structure, and output specifications. Purpose tells the AI what the document must accomplish. Context provides background the AI needs to draft appropriately. Constraints identify limitations the draft must respect. Structure specifies how the document should be organized. Output specifications describe the format, length, and style of the desired output. These elements parallel the Prompt Blueprint from Chapter 3, adapted for drafting tasks.

Purpose clarity prevents drafts that miss the point. "Draft an indemnification clause" is less useful than "Draft an indemnification clause that protects the buyer from third-party claims arising from the seller's pre-closing conduct, with carve-outs for claims the buyer knew about at signing." The second instruction tells AI what the clause must accomplish, not just what type of clause to generate. AI can draft toward a purpose; it cannot infer purpose from a label.

Context provision gives AI the background needed to draft appropriately. Relevant context might include the type of transaction, the parties' relative bargaining positions, industry-specific considerations, related provisions in the same document, and any unusual circumstances that affect how the document should be drafted. A confidentiality clause for a joint venture between competitors requires different drafting than a confidentiality clause for a routine vendor agreement. AI cannot know the difference unless you provide the context.

Constraints identify what the draft cannot do or must avoid. Legal constraints might include mandatory statutory language or regulatory requirements. Negotiation constraints might include positions the counterparty has already rejected or terms outside your client's authority. Practical constraints might include length limitations, formatting requirements, or integration with existing document sections. Making constraints explicit prevents AI from producing drafts that violate them.

> **Practice Point** Before prompting AI to draft anything, spend five minutes writing out what the document must accomplish, what context matters, and what constraints apply. This preparatory thinking improves your instructions and often clarifies your own understanding of what you need.

Structure specification tells AI how to organize the output. Should the contract clause be a single paragraph or multiple subsections? Should the motion follow a particular argument sequence? Should the letter have specific sections? AI will impose some structure if you don't specify, but that default structure may not match what you need. Specifying structure in your instructions produces output that fits your requirements.

Output specifications address format, length, tone, and style. A demand letter requires different tone than an informational memo to a client. A complex indemnification provision might appropriately span a page; a simple acknowledgment clause should be a few sentences. Formal contract language differs from conversational client communications. Specify what you expect so AI can calibrate its output accordingly.

The relationship between instruction investment and revision time is roughly inverse. Detailed instructions require more upfront effort but produce drafts closer to final form. Minimal instructions are faster to create but produce drafts requiring extensive revision. For simple, short documents, minimal instructions may be efficient enough. For complex, consequential documents, detailed instructions typically save time overall despite higher upfront investment.

Documents filed with courts or regulatory agencies often must follow specific format requirements. Instructions to AI should incorporate those requirements so that the first draft is compliant rather than requiring reformatting. Review applicable rules before drafting instructions.

Stage Two: First Draft

The first draft is a starting point, not an ending point. This orientation is essential for responsible AI-assisted drafting. AI produces plausible-seeming text that may or may not accomplish your purposes. Treating the first draft as raw material for your professional judgment rather than as nearly-finished work product maintains the quality standards your obligations require.

Generating the first draft with AI follows from your instructions. The prompt incorporates the five instruction elements, providing AI with what it needs to produce useful output. For source-bounded drafting, the prompt also includes materials the AI should draw from: precedent clauses, fact summaries, legal standards from your research, or other inputs that ground the draft in verified content rather than AI generation from training patterns.

Source-bounded drafting for documents parallels source-bounded prompting for research. When drafting a contract, you might provide sample clauses from your precedent files that AI should use as starting points. When drafting a motion, you might provide the legal standards from your research memo and the facts from the record. When drafting client correspondence, you might provide the key points you want to communicate. AI works with what you provide, reducing reliance on its training patterns for substantive content.

First draft review should be critical, not accepting. Read the draft with the question: does this accomplish what I need? Not "is this plausible-seeming text" but "does this language actually serve my client's interests." The distinction matters because AI produces fluent prose that may not accomplish the underlying purpose. A confidentiality clause that reads well but contains loopholes doesn't serve the client regardless of how professional it sounds.

Common first draft problems to watch for include: - Gaps where the draft doesn't address issues your instructions specified. - Additions where the draft includes content you didn't request and don't want. - Misalignment where the draft addresses the right topics but not in the way your client's interests require. - Generic language where the draft uses boilerplate that doesn't fit your specific circumstances. - Style mismatches where the draft's tone or formality doesn't fit the document type or recipient. Read AI first drafts with a pen in hand, marking every passage that requires attention. Don't passively read and conclude "that looks okay." Actively evaluate each sentence against what you need. The discipline of marking forces engagement that passive reading doesn't.

Multiple drafts may be appropriate before proceeding to revision. If the first draft misses the mark substantially, generating a revised draft with refined instructions may be more efficient than manually revising the existing draft. AI can iterate quickly, and sometimes the fastest path to a good starting point is generating several AI drafts rather than extensively revising one poor draft.

Comparison across drafts helps when you've generated alternatives. "Here are three versions of this indemnification clause I've generated. Compare them and identify the key differences in risk allocation." AI analysis of its own outputs can surface distinctions you might not have noticed and help you select the best starting point for revision. Stage Three: Revise Revision transforms AI output into work product that reflects your professional judgment. The draft provided raw material; revision shapes that material into something you can stand behind. How much revision is required depends on draft quality, document complexity, and the stakes involved. Some drafts need light editing; others need substantial rewriting. All need review through the lens of professional responsibility.

Revision priorities should address substance before style. First, ensure the document accomplishes its purpose. Does each provision do what it needs to do? Does the overall document achieve the client's objectives? Second, ensure accuracy. Are all statements correct? Are any facts asserted that you haven't verified? Are legal characterizations accurate? Third, ensure completeness. Does the document address everything it should? Are there gaps that create risk? Only after substance, accuracy, and completeness are addressed should you focus on style, word choice, and polish.

Substantive revision asks whether the draft works. For contracts, does each clause allocate risk appropriately? Does the language create the intended legal effects? Are the obligations clear and enforceable? For litigation documents, does the argument logically support the conclusion? Are the facts presented accurately and persuasively? Are the authorities used correctly? For correspondence, does the message communicate what the recipient needs to know? Will the recipient understand and respond as intended?

> **Risk Note** Do not let fluent AI prose lull you into accepting substantive problems. AI produces professional-sounding language regardless of whether the underlying content is sound. A beautifully drafted clause that does not protect your client is worse than an awkwardly drafted clause that does, because the beautiful language may obscure the substantive failure.

Accuracy revision ensures that everything the document asserts is true and supportable. In litigation documents, facts must trace to the record. In research memos, holdings must be accurately characterized. In client communications, representations about law and procedure must be correct. AI may generate plausible-sounding assertions that aren't accurate. Your revision must catch these before the document leaves your hands.

Completeness revision ensures that nothing important is missing. AI drafts may omit issues you expected to be addressed. They may fail to include standard provisions appropriate to the document type. They may address the main issue while ignoring subsidiary matters that also require attention. Review against a mental or written checklist of what the document should contain.

Iterative AI assistance during revision can be valuable. If you've identified problems in the draft, you can prompt AI to address them specifically. "This indemnification clause doesn't address environmental liabilities. Add a subsection specifically covering environmental claims with appropriate definitions and limitations." Targeted revision prompts produce focused improvements without regenerating content that was already acceptable.

For complex documents, revise in multiple passes with different focus areas. First pass for substance and purpose. Second pass for accuracy and completeness. Third pass for style and consistency. This layered approach catches more issues than trying to address everything in a single read-through.

Track changes can document your revision process. Working in track changes shows what the AI draft contained and what you modified. This documentation can be valuable if questions arise about how the document was produced and can demonstrate the professional judgment you exercised in transforming AI output into final work product. Stage Four: Verify Verification for drafting differs from verification for research. Research verification asks whether authorities exist and say what you think. Drafting verification asks whether the document accomplishes its purpose and whether all assertions it contains are accurate. The verification mindset is the same, but the specific checks differ.

Purpose verification confirms that the final document does what it needs to do. This is the substantive check you applied during revision, now conducted on the final product. Read the document as if you were the client asking whether your interests are protected, or as opposing counsel looking for weaknesses. Does the document hold up under that scrutiny?

Accuracy verification confirms that all factual and legal assertions are correct. For litigation documents, trace each factual assertion to its record support. For legal characterizations, confirm accuracy against the authorities. For representations about procedures or requirements, verify correctness. This verification catches any AI-introduced errors that survived revision.

Consistency verification confirms that the document is internally coherent. Defined terms should be used consistently. Cross-references should point to the correct sections. Representations in one part shouldn't contradict representations in another. AI drafts sometimes introduce inconsistencies, particularly in longer documents assembled from multiple prompts. Verify that the whole document hangs together.

Integration verification matters when the drafted document is part of a larger set. Does this schedule align with the agreement it supplements? Does this motion brief match the statement of facts filed separately? Does this amendment integrate properly with the original document it modifies? Documents don't exist in isolation, and verification should confirm that the drafted document fits with related materials.

Courts regularly encounter documents with internal inconsistencies: defined terms used incorrectly, cross-references to non-existent sections, and provisions that contradict each other. These errors reflect poorly on the lawyers involved and can create ambiguity that undermines the document's purpose. Verification prevents these preventable errors.

Style verification confirms that the document meets applicable style requirements. Does citation format comply with court rules or firm conventions? Does formatting match requirements? Is tone appropriate for the document type and recipient? Style errors are less serious than substantive errors, but professional work product should be professionally presented.

Verification documentation creates a record that the verification occurred. For significant documents, note what verification you performed and when. This documentation demonstrates professional process and provides protection if questions arise later about how the document was prepared. Workflow Integration The four stages integrate into a complete drafting process. You craft detailed instructions that define what the document must accomplish. You generate a first draft using AI assistance, treating the output as a starting point for your professional work. You revise the draft to address substance, accuracy, completeness, and style, transforming AI output into work product you can stand behind. You verify that the final document accomplishes its purpose and contains no errors.

Time allocation across stages should weight revision heavily. Instructions require upfront investment but can be developed relatively quickly once you understand the document's purpose. First draft generation is fast with AI assistance. Revision is where you add the professional judgment that makes the document yours. Verification ensures that revision caught all problems. A rough allocation might be 15% instructions, 10% first draft, 55% revision, 20% verification, though this varies with document complexity and draft quality.

The workflow applies across document types with adaptation. Contract clauses, full agreements, motions, briefs, letters, memos, and other documents all benefit from the structured approach, though the specific content of each stage varies. A demand letter needs different instructions than an asset purchase agreement. A motion to dismiss requires different verification than a client engagement letter. The workflow provides the framework; you adapt the content to the document at hand.

> **Key Point** Effective drafting workflow sequences AI assistance across four stages. Instructions define purpose, context, constraints, structure, and output specifications, with instruction quality heavily influencing draft quality. First draft generation treats AI output as a starting point for revision, not a final product. Revision addresses substance, accuracy, completeness, and style, transforming AI output into work product reflecting professional judgment. Verification confirms the document accomplishes its purpose and contains no errors. Time allocation should weight revision heavily, with verification receiving adequate attention to catch problems before the document is final.

8.2 Contract Clauses: Generate, Compare, and Risk-Spot

Contracts are built from clauses, and clause work is where transactional lawyers spend much of their drafting time. Generating appropriate language for a particular deal, comparing alternative formulations to understand their implications, and identifying risks in proposed provisions are core transactional skills. AI can assist each of these activities substantially, but the assistance requires the same critical evaluation applied throughout this book. A clause that reads professionally may not protect your client. A comparison that identifies differences may miss the differences that matter. A risk assessment that surfaces concerns may overlook the most significant ones. AI accelerates clause work; your judgment determines whether the results serve your client.

Clause generation produces language options you evaluate rather than language you adopt uncritically. The generation process follows the drafting workflow from Section 8.1: clear instructions that specify what the clause must accomplish, AI generation of draft language, revision to ensure the language actually serves your client, and verification that the final clause works as intended. Generation is the easy part; evaluation is where professional skill matters.

Effective clause generation instructions specify the clause's function, the interests it must protect, the context within the broader agreement, and any constraints on acceptable approaches. "Draft an indemnification clause" produces generic output. "Draft an indemnification clause for an asset purchase agreement where the buyer is acquiring a manufacturing business from a family-owned seller. The clause should protect the buyer against third-party claims arising from the seller's pre-closing operations, including environmental and employment matters. The indemnification period should extend three years for general representations and six years for fundamental representations and tax matters. Include a basket with a tipping mechanism, a cap at the purchase price for general matters and no cap for fraud. Address the exclusive remedy question and survival of claims." This instruction tells AI what the clause must accomplish, producing output that addresses your specific needs rather than generic possibilities.

Source-bounded generation improves clause quality by grounding AI output in precedent you've verified. Rather than generating from training patterns alone, you provide sample clauses from your precedent files that represent good drafting for similar situations. "Using the following indemnification clause from a recent deal as a starting point, draft an indemnification clause for this transaction that addresses the following differences..." AI adapts known-good language rather than creating from scratch, reducing the risk of problematic formulations.

Build a personal clause library of well-drafted provisions from your practice. When you encounter or draft a particularly effective clause, save it for future reference. Your clause library becomes source material for AI-assisted generation, improving output quality by grounding it in vetted precedent.

Multiple variations often serve better than a single draft. Ask AI to generate two or three versions of the same clause with different approaches, then compare them. "Generate three versions of this limitation of liability clause: one that is strongly pro-seller, one that is balanced, and one that favors the buyer. For each version, explain what makes it favorable to that party." The variations show you the range of possibilities and help you select or synthesize an approach appropriate to your negotiating position.

Clause comparison helps you understand how different formulations allocate risk. In negotiations, you frequently receive counterparty language that differs from your preferred approach. Understanding what changed and why it matters is essential for developing an informed response. AI can accelerate this analysis, identifying differences and explaining their implications.

Effective comparison prompts provide both versions and ask for specific analysis. "Compare these two indemnification clauses: [your version] and [counterparty version]. Identify every substantive difference between them. For each difference, explain the practical effect and which party benefits from each formulation." The AI response catalogs the differences and characterizes their significance, giving you a structured basis for your own evaluation.

Comparison limitations require recognition. AI identifies differences in text but may not fully understand their significance in context. A change that AI characterizes as minor might be significant given your transaction's specific circumstances. A change that AI characterizes as significant might be acceptable given other protections in the agreement. AI comparison produces a starting point for your analysis, not a final assessment of which version is better.

> **Risk Note** Do not let AI comparison substitute for reading both versions yourself. AI may miss subtle differences in phrasing that have significant legal effects. It may mischaracterize the importance of identified differences. Use AI comparison to accelerate your review, not to replace it.

Comparison chaining addresses complex provisions with multiple moving parts. Rather than comparing entire clauses at once, break the comparison into components. Compare the indemnification coverage separately from the indemnification procedure. Compare the basket and cap provisions separately from the survival provisions. This component-by-component approach produces more detailed analysis than whole-clause comparison.

Tracking changes through negotiation rounds benefits from AI assistance. "Here is the original indemnification clause, the counterparty's first markup, our response, and the counterparty's second markup. Track how each substantive issue has evolved through these versions. Identify issues that have been resolved, issues that remain open, and any new issues introduced." This analysis helps you see negotiation progress and identify where attention is still needed.

Courts interpreting contracts look to the specific language used, not the parties' unexpressed intentions. A clause that reads one way to you may read differently to a court applying principles of contract interpretation. AI comparison can help identify ambiguities that might support readings different from what you intended.

Risk identification surfaces potential problems in contract language. Every clause creates risks for both parties; the question is whether the risks are acceptable given the transaction economics and the available alternatives. AI can help identify risks you might overlook, though AI risk assessment has the same limitations as AI argument suggestions in litigation: the assessment may surface genuine concerns, miss important issues, or flag concerns that aren't actually significant in context.

Risk identification prompts ask AI to critique proposed language. "Review this limitation of liability clause from the buyer's perspective. What risks does this language create for the buyer? What scenarios could leave the buyer inadequately protected? What changes would better protect the buyer's interests?" The AI response identifies potential concerns, each of which you evaluate for actual significance.

Balanced risk assessment considers both sides even when you represent only one. Understanding what the counterparty gains from proposed language helps you assess negotiability and develop strategic responses. "Analyze this warranty provision from both the buyer's and seller's perspectives. What does each party gain and risk under this language?" Understanding both perspectives supports more effective negotiation.

After AI identifies risks in a clause, prioritize them based on your knowledge of the transaction. Some AI-identified risks will be genuinely important. Others will be theoretical concerns that don't apply to your specific circumstances. Your prioritization, not AI's risk list, should guide your revision and negotiation strategy.

Risk categorization helps organize AI analysis into actionable insights. Ask AI to sort identified risks by severity, by likelihood, or by type. "Categorize the risks you identified as: risks that are likely and significant, risks that are significant but unlikely, risks that are likely but minor, and theoretical risks that are neither likely nor significant in most transactions." This categorization helps you focus attention where it matters most.

Interaction effects between clauses often create risks that neither clause creates alone. A limitation of liability provision interacts with an indemnification provision. A warranty interacts with the disclosure schedules that qualify it. A termination provision interacts with the survival provisions that extend certain obligations. AI can help identify these interactions. "Considering this indemnification provision together with this limitation of liability provision, are there any ways these provisions could interact to create unexpected results or gaps in protection?"

Practical clause workflow combines generation, comparison, and risk identification into an iterative process. You generate initial clause language based on your instructions. You compare your draft to precedent or counterparty language to understand differences. You identify risks in your current language and in proposed alternatives. You revise based on what comparison and risk analysis reveal.

You verify that your final language accomplishes your purposes. The workflow is cyclical rather than linear, with each activity informing the others.

Negotiation support extends clause work into the back-and-forth of deal-making. AI can help you understand counterparty positions, develop responses, and prepare for negotiation conversations. "The counterparty has rejected our unlimited liability for fundamental representations. What arguments might support our position? What compromises might be acceptable? What alternatives exist that partially address our concerns while giving the counterparty something they value?" This strategic assistance supports your negotiating efforts without replacing your judgment about what positions to take.

The substantive judgment that remains yours includes whether AI-identified risks actually matter for your deal, what risk allocation is appropriate given the transaction economics, what positions are worth fighting for versus conceding, and how to balance competing client interests. AI accelerates your analysis and surfaces considerations you might have missed. The decisions about what those considerations mean for this transaction are yours to make.

> **Risk Note** AI cannot assess whether a risk is worth accepting for this particular deal. The acceptable risk level depends on transaction economics, client risk tolerance, relationship dynamics, and factors AI does not know. AI identifies possibilities; you determine significance.

Client counseling on clause issues benefits from the analysis AI helps you develop. When you explain to a client why you recommend certain language or why counterparty language is problematic, the structured analysis from AI-assisted comparison and risk identification provides clear explanations. "This clause creates three main concerns: first, it limits your ability to recover for X; second, it requires you to provide notice within Y days; third, it caps recovery at Z. I recommend we push back on the first issue, accept the second with a minor modification, and negotiate an increase on the third." The AI-assisted analysis supports this kind of clear client communication.

Documentation of clause work creates records useful for the current deal and future transactions. Save your clause generation prompts and the analysis that led to your final language. Note what worked well and what required significant revision. This documentation builds institutional knowledge about effective AI-assisted clause drafting and provides precedent for future similar transactions.

Clause generation produces language options you evaluate rather than adopt uncritically. Effective generation requires detailed instructions specifying what the clause must accomplish. Source-bounded generation grounded in verified precedent improves output quality. Clause comparison identifies differences between alternative formulations and explains their implications, though AI comparison doesn't replace reading both versions yourself. Risk identification surfaces potential problems in proposed language, with AI-identified risks requiring prioritization based on your transaction knowledge. Interaction effects between clauses may create risks that neither clause creates alone. The practical workflow combines generation, comparison, and risk identification iteratively, supporting negotiation while preserving your judgment about what positions to take and what risks to accept.

8.3 Redlining Assistance: Negotiation Positions and Fallbacks

Contract negotiation proceeds through rounds of markups. You send your draft; the counterparty returns a redline. You analyze their changes, develop responses, and send your own markup. They do the same. The process continues until the parties reach agreement or walk away. Each round requires understanding what the other side proposed, assessing whether their changes are acceptable, and crafting responses that advance your client's interests. AI can assist throughout this process, accelerating analysis while you retain control over the strategic decisions that shape the negotiation.

Redline analysis begins with understanding what changed. A complex agreement markup might contain dozens or hundreds of changes, ranging from minor wording adjustments to fundamental restructuring of deal terms. Before you can respond, you need to know what you're responding to. AI can help categorize and prioritize changes so you focus attention where it matters most.

Effective redline analysis prompts provide context and ask for structured output. "Here is our original indemnification article and the counterparty's markup. Identify every change the counterparty made. Categorize each change as: substantive (affecting rights or obligations), clarifying (making meaning clearer without changing substance), or stylistic (word choice or formatting without legal effect). For each substantive change, explain what the change does and which party it favors." This structured analysis gives you a map of the markup before you dive into the details.

Change prioritization helps you allocate attention efficiently. Not every change deserves equal consideration. Some changes affect deal economics or fundamental risk allocation. Others address unlikely contingencies or make distinctions without practical difference. Knowing which is which helps you focus negotiation energy where it matters. "Of the substantive changes you identified, rank them by potential impact on the buyer. Which changes most significantly affect the buyer's risk position or economic outcome?"

Pattern recognition across a markup can reveal negotiation strategy. A counterparty who systematically expands exceptions and narrows obligations is taking an aggressive posture. A counterparty making targeted changes to specific risk areas may have particular concerns you could address directly. A counterparty proposing balanced changes throughout may be seeking efficient resolution rather than advantage. "Looking at the pattern of changes in this markup, what does the counterparty appear to be most concerned about? Are there themes across their changes that suggest a particular negotiating focus?"

Before responding to a markup, step back and assess the overall pattern. Is the counterparty's markup reasonable given their position in the deal? Are they making targeted changes or fighting on every front? Understanding the pattern helps you calibrate your response appropriately.

Developing responses to counterparty changes requires balancing multiple considerations. Some changes are acceptable and should be agreed. Some are unacceptable and must be rejected or modified. Some represent reasonable positions you might accept with adjustments. Some reveal counterparty concerns you could address differently. AI can help generate response options for each category.

Response generation for unacceptable changes asks AI to help you articulate why and propose alternatives. "The counterparty has proposed capping indemnification at $1 million on a $50 million deal. This is unacceptable to my client. Draft language explaining why this cap is inadequate and proposing a cap at the full purchase price with a tiered structure based on claim type." AI generates explanatory language and alternative proposals you can refine.

Response options for changes requiring judgment benefit from seeing multiple approaches. "The counterparty has proposed limiting consequential damages for both parties. Generate three possible responses: one that rejects this limitation entirely, one that accepts it with carve-outs for key claim types, and one that accepts it but adds a minimum recovery floor. For each response, explain the tradeoffs." Multiple options help you select or synthesize an approach appropriate to your negotiating position.

Courts interpreting negotiation history may consider the positions parties took during contract formation. Document your negotiating positions clearly and avoid making representations during negotiation that could be used to interpret ambiguous language against your client.

Fallback positions prepare you for negotiation impasses. When your initial position is rejected, you need alternatives that partially address your concerns while giving the counterparty something they value. Developing fallback positions in advance helps you negotiate dynamically rather than scrambling when the counterparty pushes back. AI can help identify potential fallbacks and assess their adequacy.

Fallback identification prompts ask AI to generate alternatives when primary positions fail. "My primary position is unlimited liability for fraud. If the counterparty rejects this, what fallback positions could partially protect my client while giving the counterparty some limitation they value? For each fallback, explain what my client gains and gives up compared to the primary position." AI generates options you evaluate for acceptability and negotiating leverage.

Fallback adequacy assessment considers whether a fallback position still serves your client's core interests. A fallback that gives away too much isn't really a fallback; it's a concession that undermines your negotiating position. AI can help stress-test fallback positions. "If we accept a liability cap of $10 million instead of our preferred full purchase price cap, what scenarios could leave our client underprotected? How likely are those scenarios given what we know about the business?" This analysis helps you understand what you'd be giving up.

Don't develop fallback positions that give away more than you need to. Fallbacks are for genuine impasses, not for early concessions. If you develop a generous fallback, you may anchor yourself to it psychologically or signal weakness if the counterparty senses you have extensive backup positions.

Negotiation preparation integrates redline analysis with strategic planning. Before the negotiation call or session, you should understand every open issue, your position on each, your fallback positions, and your assessment of counterparty priorities. AI can help synthesize this information into a negotiation guide.

Preparation prompts ask AI to organize your negotiation position. "Based on the analysis we've done of this markup, create a negotiation preparation guide that includes: a prioritized list of open issues from most to least important; our position on each issue with key talking points; our fallback position on each issue if our primary position is rejected; our assessment of what the counterparty likely cares most about; and potential trades where we could give ground on lower priorities to gain on higher priorities." This guide structures your preparation without replacing your judgment about strategy.

Issue trading identifies opportunities for mutual gain through package deals. If the counterparty cares intensely about one issue and you care intensely about a different issue, trading concessions can leave both parties better off. AI can help identify these trading opportunities. "Looking at our analysis of counterparty priorities and our own priorities, are there issues where we have differing levels of concern that might support a package deal? What trades might be mutually beneficial?"

Prepare a one-page negotiation sheet before any substantive call. The sheet should list: your must-haves (non-negotiable), your want-to-haves (important but flexible), your nice-to-haves (would take but won't fight for), and your can-give items (concessions available for trading). This framework organizes your thinking regardless of how the conversation flows.

Real-time negotiation support has limits. During a negotiation call, you may not have time to consult AI for every counterparty proposal. The preparation you've done should equip you to respond to expected issues. For unexpected proposals, you can note them for analysis rather than responding immediately. "That's an interesting approach. Let me take that back and think about whether it works for my client." There's no shame in taking time to consider a proposal properly.

Post-negotiation documentation captures what was agreed and what remains open. After each negotiation round, document the current state of play. AI can help organize these notes. "Based on the call notes I've provided, summarize: issues that are now fully agreed, issues where we reached conceptual agreement but need to finalize language, issues that remain open with current positions noted, and new issues that arose during the call." This documentation ensures nothing falls through the cracks and provides a record of negotiation progress.

Turn tracking across multiple negotiation rounds maintains clarity about how issues have evolved. In complex negotiations spanning weeks or months, it's easy to lose track of where specific issues stand. AI can help maintain this history. "Here is the negotiation history for the indemnification cap: our original position, their first counter, our second position, their second counter, and the current language. Summarize how this issue has evolved and where we stand now." Turn tracking prevents reopening issues you thought were resolved and helps you assess whether progress is being made.

Escalation assessment helps you evaluate when to elevate an impasse. Some issues warrant escalation to principals when negotiators can't reach agreement. Others should be conceded rather than escalated. AI can help think through escalation decisions. "We've reached impasse on the liability cap after three rounds of negotiation. What factors should I consider in deciding whether to escalate this to the principals versus finding a compromise at my level? What are the risks of escalation versus the risks of accepting a less favorable position?"

Negotiation tone calibration matters for relationship management. A markup that comes across as aggressive may provoke an aggressive response even if the substance is reasonable. AI can help calibrate tone in your markup language. "Review the explanatory comments in my proposed markup. Are any of them likely to come across as aggressive or dismissive? Suggest revisions that maintain our position firmly but professionally." Tone review helps maintain productive negotiating relationships.

Judges and arbitrators sometimes review negotiation correspondence when resolving contract disputes. Maintain professional tone throughout negotiations, and avoid representations you wouldn't want quoted back to you in a dispute proceeding.

Client reporting on negotiation status keeps clients informed without overwhelming them. AI can help translate complex negotiation developments into client-friendly summaries. "Draft a client update summarizing our negotiation progress on the purchase agreement. Explain the three most significant open issues in plain English, our position on each, and your assessment of likely resolution. Keep it to one page." These summaries support informed client decision-making about negotiation strategy.

The judgment that remains yours throughout redline assistance includes what positions to take, what fallbacks are acceptable, when to push hard versus accommodate, and how to balance relationship preservation with client protection. AI helps you analyze the markup, generate options, and prepare for negotiation. The strategic choices that shape outcomes are yours to make with client input and approval.

Confidentiality considerations apply to negotiation materials as they do to other client work. Counterparty markups and negotiation correspondence may contain confidential information about both sides' positions. Apply the Chapter 5 framework when deciding what materials to provide to AI and which tools are appropriate for that information. Enterprise tools with appropriate protections are generally suitable; consumer tools may not be, depending on the sensitivity of the negotiation.

Redline analysis identifies, categorizes, and prioritizes counterparty changes so you can focus attention where it matters. AI can catalog changes and explain effects, though evaluating acceptability requires your judgment. Response generation produces options for addressing unacceptable changes, with multiple alternatives helping you select appropriate approaches. Fallback positions prepare you for impasses, with AI helping identify and assess alternatives. Negotiation preparation synthesizes analysis into actionable guides. Issue trading identifies opportunities for mutual gain through package deals. Post-negotiation documentation captures agreements and open items. Throughout, AI accelerates analysis while you retain control over strategic decisions that shape negotiation outcomes.

8.4 Litigation Drafting: Letters, Pleadings, Motions

Litigation drafting differs from transactional drafting in a fundamental way. A contract clause can be evaluated on its own terms: does this language accomplish the intended purpose? Litigation documents make assertions about the world: what happened, what the law says, what conclusions follow. These assertions must be true. Facts must trace to the record. Legal claims must build on authorities that exist and say what you claim. The drafting workflow from Section 8.1 applies, but litigation's accuracy requirements demand additional discipline, particularly when AI assists the drafting.

The fact constraint is paramount. Every factual assertion in a litigation document must have a source. In a complaint, facts must be alleged in good faith with evidentiary support or a reasonable expectation that investigation will reveal support. In a motion, facts must cite to the record: deposition transcripts, document exhibits, declarations, or other evidence before the court. In a brief, factual statements must be supported by citations that allow the court to verify what you've claimed. AI doesn't know your record. It generates plausible-sounding factual assertions that may not correspond to anything that actually happened. Your drafting process must prevent these fabrications from reaching your work product.

The authority constraint parallels the fact constraint. Every legal assertion must build on authorities you've verified through the research process in Chapter 7. AI-assisted litigation drafting should draw on your research memo, not on AI's independent generation of legal arguments. When AI drafts a legal argument, it may cite authorities that don't exist, mischaracterize holdings, or assert propositions that the law doesn't support. The source-bounded approach confines AI to working with authorities you've verified rather than generating its own.

Demand letters and litigation correspondence initiate many disputes and set the tone for what follows. These letters assert facts about what happened, state legal claims that arise from those facts, and demand action or threaten consequences. AI can help draft these letters efficiently, but the factual assertions must come from your investigation, and the legal claims must reflect your research.

Effective demand letter prompts provide the facts and law rather than asking AI to generate them. "Draft a demand letter based on the following facts from my client interview and document review: [fact summary]. The letter should assert claims for [legal theories you've researched] under [applicable law you've verified]. The letter should demand [specific relief] and set a response deadline of [timeframe]. Tone should be firm but professional, leaving room for negotiation while making clear we are prepared to litigate." This prompt constrains AI to work with your verified inputs while leveraging its drafting capabilities for organization and prose.

Before drafting any demand letter, create a fact summary and legal theory outline from your own work. These become the inputs for AI-assisted drafting. The discipline of creating these summaries first ensures that the letter builds on verified foundations rather than AI generation.

Complaints and petitions initiate lawsuits and define the scope of the dispute. Pleading requirements vary by jurisdiction and claim type, but all complaints must state facts sufficient to support the claims asserted. Notice pleading jurisdictions require only enough facts to put the defendant on notice.

Plausibility pleading jurisdictions require facts sufficient to make the claims plausible on their face. Either standard requires actual facts, not AI-generated possibilities.

Complaint drafting prompts should provide your factual investigation results and the elements of each claim. "Draft a complaint alleging breach of contract and fraud based on the following facts: [your fact summary]. The breach of contract claim must allege: formation (facts X), performance by plaintiff (facts Y), breach by defendant (facts Z), and damages (facts W). The fraud claim must allege: misrepresentation (facts A), scienter (facts B), reliance (facts C), and damages (facts D). Include a prayer for relief seeking compensatory damages, punitive damages, and attorney fees." This structured prompt ensures each element has factual support from your investigation.

Verification for complaints requires confirming that every factual allegation traces to your investigation. Read through the AI draft and for each factual assertion, identify where that fact came from. Your client interview? A document you reviewed? Information from a third party? If you can't identify the source, the fact may be AI-generated rather than real. Either find support for it or remove it.

AI sometimes embellishes facts to make the narrative more compelling. A complaint that adds details you never provided, even details that seem to strengthen the case, contains fabrications that could undermine your credibility and violate Rule 11 of the Federal Rules of Civil Procedure (Fed. R. Civ. P. 11). Remove any facts you cannot independently verify.

Answers and responsive pleadings respond to the opposing party's allegations. These documents require paragraph-by-paragraph engagement with the complaint: admitting, denying, or stating insufficient information to admit or deny each allegation. AI can help structure this response, but the determination of what to admit or deny must come from your factual investigation.

Response drafting benefits from providing AI with both the complaint and your factual assessment. "Here is the complaint and my paragraph-by-paragraph factual analysis indicating which allegations I believe are accurate, which are inaccurate, and which I cannot assess without discovery. Draft an answer that admits allegations I've marked as accurate, denies allegations I've marked as inaccurate, and states insufficient information for allegations I've marked as uncertain. Include affirmative defenses for [defenses you've identified through research]." This approach ensures the answer reflects your investigation rather than AI assumptions about what should be admitted or denied.

Judges notice when pleadings contain inconsistencies between factual allegations and the record as it develops. A complaint that alleges facts contradicted by later-produced documents creates credibility problems. Ensure that your pleadings reflect careful factual investigation, not creative drafting.

Motions present arguments to the court requesting specific relief. The motion itself is typically brief, stating what relief is sought and the grounds. The supporting memorandum provides the detailed argument. Both require accuracy in facts and law, with the memorandum demanding particular attention because it presents the substantive legal arguments the court will evaluate.

Motion memorandum structure typically includes an introduction, statement of facts, argument, and conclusion. AI can help draft each section, but the content constraints differ. The introduction summarizes your position and can benefit from AI assistance with framing and persuasive organization. The statement of facts must cite to the record for every assertion. The argument must build on verified legal authorities. The conclusion restates the requested relief.

Statement of facts drafting requires record citation for every factual assertion. Your prompt should provide the facts with their record citations, not raw facts for AI to present. "Draft a statement of facts for a summary judgment motion based on the following facts and record citations: [Fact 1, citing Deposition of Smith at 45:12-46:3]; [Fact 2, citing Exhibit A at Bates 00123]; [Fact 3, citing

Declaration of Jones at paragraph 7]." AI drafts the narrative; you ensure every fact has record support. Verify the final draft by confirming that each citation actually supports the assertion made.

Argument section drafting should draw on your research memo rather than asking AI to generate legal arguments independently. The research workflow in Chapter 7 produced verified authorities with accurately extracted holdings. Those authorities and holdings are the inputs for argument drafting. "Draft the argument section for this motion to dismiss based on the following legal authorities and holdings from my research: [Authority 1 with extracted holding]; [Authority 2 with extracted holding]. The argument should establish [legal standard], apply the standard to [these facts], and conclude that dismissal is required because [conclusion]."

Persuasive framing is where AI drafting assistance can add particular value. Given verified facts and authorities, AI can help you present the argument in its most compelling form: choosing effective organization, crafting strong topic sentences, building momentum toward your conclusion, and anticipating counterarguments. This framing assistance leverages AI capabilities without depending on AI for substantive accuracy.

Separate the accuracy function from the persuasion function in litigation drafting. You ensure accuracy by providing verified facts and authorities as inputs. AI assists with persuasion by organizing and presenting those inputs effectively. Keeping these functions distinct maintains quality control over substance.

Opposition briefs respond to the moving party's arguments. Effective opposition requires understanding what the other side argued, identifying weaknesses in their position, and presenting your response persuasively. AI can help with each component while you maintain responsibility for substantive accuracy.

Opposition analysis prompts ask AI to identify weaknesses you can exploit. "Here is the defendant's motion to dismiss and supporting memorandum. Identify the weakest points in their argument. Where are their legal authorities distinguishable? Where do their facts not support their legal conclusions? Where do they fail to address issues that favor the plaintiff?" This analysis surfaces possibilities you evaluate for actual merit.

Response drafting follows the same source-bounded approach as motion drafting. Provide your factual record support and legal authorities, then prompt AI to draft the responsive argument. "Draft an opposition to this motion to dismiss. My response will argue that [your legal position based on research]. The key authorities supporting my position are [verified authorities with holdings]. The facts supporting my position are [facts with record citations]." AI drafts the response; you ensure the inputs are accurate.

Reply briefs have the final word and should address the opposition's strongest points directly. AI can help identify which opposition arguments require response and draft focused rebuttals. "The plaintiff's opposition raises three main arguments against dismissal: [Argument 1], [Argument 2], [Argument 3]. Draft a reply that addresses each argument directly, explaining why it fails. Keep the reply focused and within the page limit."

Courts appreciate briefs that engage directly with opposing arguments rather than repeating initial positions as if the opposition didn't exist. AI-assisted drafting can ensure your reply addresses what the other side actually argued, not a strawman version of their position.

Discovery motions and responses involve both procedural and substantive arguments. Motions to compel must explain what discovery was sought, why it wasn't provided, and why the court should order production. Motions for protective orders must explain what discovery was sought, why it's objectionable, and what protection is appropriate. AI can help structure these arguments while you ensure factual accuracy about what actually happened in the discovery process.

Proportionality arguments in discovery disputes require particular attention. Courts balance the burden of discovery against its likely benefit, considering factors like relevance, need, and availability from other sources. AI can help articulate proportionality arguments, but the factual assessments of burden and benefit require your input. "Draft a proportionality argument opposing this discovery request. The burden includes [specific burden facts you've gathered]. The likely benefit is limited because [relevance limitations you've identified]. Less burdensome alternatives exist, specifically [alternatives you've identified]."

Discovery motions often involve factual disputes about what happened during the discovery process. Ensure that your description of the discovery history is accurate based on the actual correspondence, meet-and-confer communications, and discovery responses. AI cannot know this history unless you provide it accurately.

Verification for litigation documents requires attention to both facts and law. Fact verification confirms that every factual assertion cites to appropriate record support and that the cited support actually says what you claim. Law verification applies the Chapter 7 framework: every authority must exist, be cited correctly, remain good law, and be characterized accurately. Both verifications must occur before filing.

Citation and record formatting varies by court. Local rules specify citation formats, page limits, font requirements, and other technical requirements. AI drafts may not comply with these requirements automatically. Verification should include format review against applicable rules, correcting any AI-generated formatting that doesn't comply.

Rule 11 certification applies to every litigation document you sign. You certify that factual contentions have evidentiary support or will likely have support after reasonable investigation, that legal contentions are warranted by existing law or nonfrivolous argument for changing the law, and that the filing is not for improper purpose. AI assistance doesn't diminish these obligations. Your verification process demonstrates the reasonable inquiry that Rule 11 requires.

Litigation document templates can standardize your AI-assisted drafting process. Create templates for common document types that specify what inputs are needed and what structure the output should follow. A motion for summary judgment template might specify: introduction section with framing, statement of facts section with record citation requirements, legal standard section, argument section built on researched authorities, and conclusion with specific relief requested. These templates ensure consistent quality across matters.

Supervision of AI-assisted litigation drafting requires particular attention. If junior lawyers or staff use AI to draft litigation documents, supervisors must verify that the source-bounded approach was followed. Did the drafter provide verified facts and authorities as inputs? Did the drafter verify the output against those inputs? Did the final product remove any AI-generated content that wasn't supported by verified sources? Supervision questions should address the process, not just the output.

Litigation drafting differs from transactional drafting because facts must trace to the record and authorities must be verified. AI doesn't know your record and cannot satisfy the fact constraint. Source-bounded drafting provides AI with verified facts and authorities as inputs, confining AI to organization and presentation rather than substance generation. Every factual assertion must cite to record support; every legal assertion must build on verified authorities. Verification confirms both factual and legal accuracy before filing. Rule 11 certification applies regardless of AI assistance. Templates and supervision ensure consistent application of source-bounded principles across matters.

8.5 Fact Discipline: Drafting Without Inventing the Record

AI generates text by predicting what should come next based on patterns in its training data. When you ask AI to draft a complaint or motion, it predicts what a good complaint or motion would contain. That prediction includes facts, because good complaints and motions have detailed, compelling facts. The problem is that AI's predicted facts come from patterns, not from your record. AI doesn't know what your client told you, what the documents show, or what witnesses said in depositions. It knows what facts typically appear in similar documents, and it generates accordingly. The result can be plausible-sounding fabrications that undermine your credibility if they reach your filed documents.

Fact invention is not AI malfunction; it's AI functioning as designed. The system generates coherent, contextually appropriate text. When context suggests facts would be helpful, AI supplies facts. When a narrative would benefit from specific details, AI provides details. When a timeline would clarify events, AI constructs a timeline. None of this comes from your record. All of it comes from AI's prediction of what a well-drafted document would contain. Your fact discipline must prevent these predictions from becoming assertions you make to courts.

The danger is subtle because AI-generated facts often seem right. A complaint about a car accident might include details about weather conditions, the speed of vehicles, or the sequence of impacts that you never provided. These details make the complaint more vivid and compelling. They also may be completely fabricated. Because they fit the narrative and seem plausible, you might not notice they weren't in your inputs. The fabricated facts become part of your filing, attributed to you as the attorney who signed it.

Embellishment is fact invention's more subtle cousin. Rather than generating entirely new facts, AI may add details, specificity, or characterization to facts you provided. You input that your client was injured in a fall. AI drafts that your client "suffered severe injuries requiring emergency medical treatment and extensive rehabilitation." Maybe that's true; maybe it's not. The embellishment sounds better than the bare fact you provided, which creates temptation to keep it even if you're not sure it's accurate. Embellishment is still fabrication if the added details aren't supported by your investigation.

> **Practice Area Note: Personal Injury** Fact embellishment risk is highest in personal injury complaints, where medical records, injury descriptions, and damages calculations are exactly the content AI embellishes most aggressively. AI might transform your client's report of "back pain after the accident" into "severe, debilitating spinal injuries requiring emergency surgical intervention and months of intensive physical rehabilitation." That embellishment may not match the medical records — and opposing counsel will have those records. In PI practice, every injury description and treatment characterization in your complaint must trace to specific medical documentation. The gap between what your client told you and what the records show is where embellishment lives.

The source-bounded approach from Section 8.4 is the primary defense against fact invention. By providing AI with the specific facts you want included, drawn from your investigation and record, you constrain what AI has to work with. AI should organize, present, and articulate the facts you provide rather than generating facts it predicts would be helpful. But source-bounded prompting is necessary, not sufficient. Even with good inputs, AI may add facts beyond what you provided. Verification remains essential.

Structured fact inputs reduce invention risk by making clear exactly what facts AI should use. Rather than providing a narrative summary that AI might embellish, provide facts in a structured format that signals completeness. "The complaint should allege the following facts and only these facts: [Fact 1]; [Fact 2]; [Fact 3]." The explicit instruction to use only these facts signals to AI that invention is not wanted. Some AI systems respect this constraint better than others, but the explicit instruction improves results.

> **Practice Point** Read every AI-drafted sentence with the question: where did this come from? If you cannot identify a specific source in your inputs for every factual assertion, the AI may have invented or embellished. Remove or verify any fact you cannot trace to your own work.

Fact sheets prepared before drafting create the authoritative source for what facts exist. Before engaging AI for drafting assistance, compile a fact sheet listing every fact you can support from your investigation. Note the source for each fact: client interview, document, witness statement, public record. This fact sheet becomes your input for AI drafting and your reference for verification. Any fact in the draft that isn't on your fact sheet needs scrutiny.

A draft that reads better than your inputs should trigger suspicion. If AI produces a vivid, detailed narrative and your inputs were sparse, the improvement came from AI invention. Compelling prose that adds specificity you didn't provide is a warning sign, not a benefit.

Negative instructions explicitly tell AI what not to do. "Draft this complaint using only the facts I've provided. Do not add any facts, details, dates, times, descriptions, or characterizations that I haven't specifically included. If information is missing that would normally appear in this type of document, note the gap rather than filling it." Negative instructions address AI's tendency to be helpful by completing what seems incomplete. You're telling AI that incompleteness is acceptable; invention is not.

Gap flagging instructs AI to identify where facts are missing rather than supplying them. "Where this narrative would typically include information I haven't provided, insert [FACT NEEDED: description of what's missing] rather than generating the information." This approach surfaces gaps in your inputs that you can address through further investigation. It also creates visible markers in the draft that prevent you from accidentally filing AI's placeholder for missing facts.

Courts have sanctioned lawyers for including fabricated facts in pleadings, even when the lawyer claimed reliance on client representations. As the signing attorney, you're responsible for ensuring factual allegations have support. AI-generated facts are not client representations; they're fabrications unless your investigation independently supports them.

Verification techniques for fact accuracy differ from verification for legal accuracy. Legal verification confirms that authorities exist and say what you claim. Fact verification confirms that factual assertions trace to your record. Both are essential; neither substitutes for the other.

Line-by-line fact verification reviews each sentence containing factual assertions. For each assertion, identify the source in your fact sheet or record. If you can't identify a source, investigate whether the fact is accurate or remove it. This line-by-line review catches invention that broader review might miss. It's time-consuming but essential for documents with significant factual content.

Comparison verification compares the AI draft against your inputs. "Here is my fact sheet with all the facts I provided. Here is the draft the AI produced. Identify any factual assertions in the draft that don't appear in my fact sheet." AI can help with this comparison, flagging potential additions for your review. But AI checking its own work has obvious limitations; the comparison is a supplement to, not a replacement for, your own verification.

Create a verification checklist specifically for factual accuracy. The checklist should prompt you to: compare the draft against your fact sheet, identify the source for each factual assertion, flag any facts you can't source, and remove or verify flagged facts before finalizing. Use this checklist for every document with factual content.

Record citation as you draft builds verification into the process. Rather than drafting the narrative and adding citations later, draft with citations from the start. Each factual sentence gets its record cite as you write it. If you can't cite a fact, you don't include it. This approach prevents accumulation of unsourced facts that require later review. It also produces drafts with citations already in place, reducing finishing work.

The assertion-source matrix provides a systematic way to track fact verification. Create a two-column table: assertions in the left column, sources in the right column. As you verify the draft, fill in the matrix. Any assertion without a source gets flagged for removal or additional investigation. The completed matrix becomes documentation of your verification process, useful if fact accuracy is later questioned. A fillable assertion-source matrix template is available on the companion site.

Temporal fact discipline addresses dates, times, and sequences. AI often generates temporal details to make narratives more concrete: "On March 15," "at approximately 3:00 p.m.," "three weeks after the initial meeting." These details may be accurate or fabricated. Temporal facts are particularly important because they often matter for legal analysis: statutes of limitations, notice periods, sequence of events. Verify every temporal assertion against your record.

Quantitative fact discipline addresses numbers and amounts. AI generates dollar figures, percentages, quantities, and measurements to add specificity. "Damages exceeding $500,000," "a 30% decrease in revenue," "traveling at approximately 45 miles per hour." Unless your investigation established these specific numbers, they may be AI fabrications. Verify every quantitative assertion.

Characterization discipline addresses how facts are described. The difference between "discussed" and "agreed," between "considered" and "decided," between "concerned" and "alarmed" can be significant. AI may characterize facts more strongly or weakly than your evidence supports. A witness who said she "had some concerns" becomes one who was "alarmed and distressed" in AI's embellished version. Verify that characterizations match your evidence.

Superlatives and intensifiers are embellishment warning signs. "Devastating," "severe," "blatant," "egregious," "massive," "prolonged," "repeated" are characterizations that may not match your evidence. When you see strong language in an AI draft, verify that your evidence actually supports that characterization.

Client communication about fact discipline helps ensure your inputs are accurate. Your fact sheet is only as good as the information your client provided. If your client's account is incomplete or inaccurate, your fact sheet will be too. Explain to clients why accurate, complete information matters: facts alleged in pleadings can be contradicted by later evidence, creating credibility problems and potential sanctions. Encourage clients to tell you what they don't know as well as what they do know.

Witness statements and declarations present particular fact discipline challenges. AI can draft declarations based on interview notes, but the drafter wasn't the witness. The declaration must accurately reflect what the witness said, not an AI-enhanced version of their testimony. When drafting declarations, provide AI with the witness's actual statements and instruct it to use the witness's language rather than generating more polished versions.

When drafting witness declarations, use the witness's own words where possible. A declaration that sounds too polished or uses sophisticated legal terminology may lose credibility. AI can help organize and structure the declaration while preserving the witness's authentic voice.

Document-based facts should cite to the actual documents. When your complaint alleges what a contract says, quote the contract. When your motion describes what an email stated, cite the email. AI may paraphrase documents in ways that don't match the actual text. For important documents, provide AI with the actual text and instruct it to quote rather than paraphrase where feasible.

Investigation gaps should be acknowledged, not filled. If your investigation hasn't established certain facts, your pleading should either omit those facts or allege them on information and belief where appropriate. AI's instinct is to create complete narratives; your discipline requires accepting incomplete narratives when you can't support more. "The complaint should allege only facts I've confirmed through investigation. For facts I'm alleging on information and belief, I'll mark them as such: [list]. For facts I don't have, leave them out rather than inventing them."

Allegations made "on information and belief" signal to courts that the attorney hasn't confirmed the facts but has a reasonable basis to believe them. This is different from fabricating facts and hoping they turn out to be true. Use information and belief allegations appropriately, and never use them to dress up AI-generated facts as somehow acceptable.

Ongoing fact discipline maintains accuracy as drafts evolve. Through multiple revision rounds, facts can creep in through AI assistance at each round. Each time AI touches the document, it has an opportunity to add facts. Verification should occur after each significant revision, not just at the end. A fact that wasn't in the original draft might appear in revision three.

Training others on fact discipline spreads these techniques throughout your organization. If junior lawyers and staff use AI for drafting, they need to understand the fact invention risk and the verification processes that address it. Training should include examples of AI fact invention so people recognize the problem, and it should teach the specific verification techniques that catch it.

AI invents facts by generating plausible-seeming assertions from training patterns rather than your record. Embellishment adds details and characterization beyond what your inputs support. Source-bounded prompting with structured fact inputs reduces invention risk but doesn't eliminate it. Negative instructions and gap flagging tell AI not to invent and to flag where facts are missing. Verification techniques include line-by-line fact review, comparison against your fact sheet, and assertion-source matrices that document verification. Temporal, quantitative, and characterization assertions require particular scrutiny. Accept incomplete narratives rather than filling gaps with fabrications. Maintain fact discipline through multiple revision rounds, verifying after each significant AI interaction.

8.6 Style Systems: Consistency Across the Office

Style matters more than many lawyers acknowledge. A brief written in crisp, direct sentences conveys confidence. A contract with consistent terminology avoids interpretation disputes. A client letter with appropriate tone maintains relationships. When style varies unpredictably across documents, within documents, or between lawyers working on the same matter, the inconsistency signals disorganization and undermines credibility. AI-assisted drafting creates particular style challenges because AI defaults to generic patterns that may not match your firm's voice, your client's expectations, or the conventions appropriate to specific document types.

Style encompasses multiple dimensions. Tone ranges from formal to conversational, aggressive to conciliatory, technical to accessible. Sentence structure can favor short, punchy sentences or longer, more complex constructions. Word choice preferences affect whether you use contractions, how you handle legal jargon, and whether you prefer Anglo-Saxon or Latinate vocabulary. Formatting conventions govern headings, citations, paragraph structure, and visual presentation. Each dimension can be specified, and AI can follow specifications when you provide them. Without specifications, AI imposes its own defaults, which may not serve your purposes.

The generic AI voice presents a particular problem. AI trained on vast text corpora develops a default style that represents a kind of average across that training data. This average style is competent but undistinctive. Documents drafted with AI assistance can all start to sound alike, losing the individual voice that distinguishes excellent legal writing from merely adequate legal writing. A firm whose documents all sound like AI wrote them has lost something valuable even if the substance is correct.

Voice preservation requires explicit attention when using AI for drafting. Your firm has a voice, whether you've articulated it or not. Clients recognize it. Opposing counsel recognizes it. Judges who see your filings repeatedly recognize it. That voice reflects accumulated decisions about how your firm communicates: the level of formality, the degree of advocacy, the balance between precision and accessibility. AI doesn't know your voice unless you teach it through explicit instructions and examples.

Style guides provide the foundation for consistent AI-assisted drafting. If your firm has a written style guide, it should inform your AI prompts. The guide's preferences for citation format, heading structure, defined term conventions, and other specifics become instructions you include when prompting AI to draft. If your firm doesn't have a written style guide, developing one provides benefits beyond AI consistency: it forces articulation of preferences that may have been implicit and provides a reference for training new lawyers.

If your firm lacks a written style guide, create one. Start with a single page covering your most important preferences: citation format, heading conventions, defined term style, tone for different document types. Expand over time as you identify additional preferences worth standardizing. The guide becomes a resource for both human and AI-assisted drafting. A style guide starter template is available on the companion site.

Document-type conventions recognize that different documents call for different styles. A client email differs from a formal opinion letter. A demand letter differs from an informational memo. A motion brief differs from a transactional document. Your style system should specify conventions for each document type you produce regularly. AI can follow document-type specifications when you provide them in your prompts.

Tone calibration for document types might specify approaches like these. Client communications: accessible language, minimal jargon, warm but professional tone. Demand letters: firm and confident, not aggressive or threatening, leaving room for resolution. Opinion letters: formal, precise, appropriately hedged, comprehensive. Motion briefs: direct, persuasive, confident without overstatement. Contracts: precise, consistent terminology, plain language where possible without sacrificing legal effect. These calibrations guide AI output toward appropriate register for each document type.

Style prompts translate your style guide into AI instructions. Rather than including your entire style guide in every prompt, create style prompts for different contexts that you can add to drafting prompts. A litigation style prompt might read: "Write in direct, active voice. Keep sentences under 25 words where possible. Use plain English; avoid unnecessary Latin phrases. Be confident but not overreaching; use 'clearly' sparingly. Use contractions only in less formal documents. Follow Bluebook citation format. Structure arguments with clear topic sentences and logical progression."

Style examples improve AI output by showing rather than just telling. Including an example of well-styled writing from your firm helps AI understand what you're looking for. "Here is a sample brief introduction from our firm that exemplifies our preferred style. Match this style in the document you draft." Examples communicate nuances that explicit instructions may not capture. They show AI how your preferences combine in actual practice.

Build a collection of exemplary documents from your practice. When you produce or encounter particularly well-written work product, save it as a style example. These examples become resources for training both AI and junior lawyers about what good work looks like in your practice.

Practice Area Note: Family Law Tone calibration is critical in custody and parenting plan drafting, where the document must serve the child's interests even when the parents' relationship is adversarial. AI defaults to either aggressive advocacy language or neutral boilerplate — neither fits. Custody agreements and parenting plans require collaborative, child-centered tone that courts expect and that supports co-parenting relationships. Instruct AI explicitly: "Draft in a tone that assumes both parents are acting in good faith toward the child's welfare. Avoid adversarial framing. Use 'the parents shall' rather than 'Father/Mother must.'" Family law practitioners who calibrate AI tone for these documents produce drafts that require far less revision and that judges receive more favorably. Judges develop impressions of lawyers based partly on their writing. Briefs that are clear, direct, and professional enhance credibility. Briefs that are pompous, aggressive, or sloppy undermine it. Your style system should produce documents that present you and your client favorably to the tribunals where you practice.

Terminology consistency matters particularly for contracts and complex matters. When a defined term is introduced, it should be used consistently throughout the document. When concepts are described, the description should use consistent language. AI sometimes varies terminology in ways that create ambiguity or suggest distinctions that don't exist. Style systems should address terminology consistency, and verification should check that drafts maintain it.

Defined term protocols specify how your organization handles defined terms in contracts. Do you use initial caps for all defined terms? Do you bold or quote terms when first defined? Do you use parenthetical definitions or a separate definitions section? Where do you define terms that appear in multiple transaction documents? These choices should be specified so AI follows your conventions rather than its own defaults.

Multi-drafter coordination becomes essential when AI assists work on the same matter. If two lawyers each use AI to draft different sections of a brief, the sections may have different styles unless both lawyers apply the same style specifications. Style systems provide the common reference that ensures coordination. Both lawyers use the same style prompt, the same examples, and the same verification criteria.

Style review in supervision catches inconsistencies before documents are finalized. Supervisors reviewing AI-assisted work should attend to style as well as substance. Does the draft match firm conventions? Is the tone appropriate for the document type and recipient? Are there style inconsistencies within the document? Adding style review to supervision ensures that consistency standards are maintained across the team.

Style inconsistencies within a single document signal AI-assisted drafting more clearly than any other marker. When the introduction is written in one voice and the argument section in another, readers notice. When defined terms are used inconsistently, it suggests careless assembly. Style review should specifically check for internal consistency.

Client-specific style requirements add another layer of complexity. Some clients have their own style guides and expect outside counsel to follow them. Government clients may have specific format requirements. Corporate clients may have preferences about how their company name is used, how products are referenced, or how certain topics are addressed. Your style system should accommodate client-specific requirements, with separate style prompts for clients with distinct preferences.

Practice area conventions recognize that different areas of practice have different style norms. Litigation briefs follow certain conventions about argumentation structure. Corporate documents follow conventions about defined terms and conditions. Regulatory submissions follow conventions about how to address agency staff. Real estate documents follow conventions specific to that practice. Your style system should account for practice area differences, with style prompts adapted to each area where your firm works.

Create practice-area-specific style prompts that address the conventions of each area. A litigator's style prompt differs from a corporate lawyer's style prompt. Maintaining separate prompts for each practice area ensures that AI-assisted drafting follows appropriate conventions rather than generic patterns.

Training on style systems ensures that everyone in the organization can apply them. Style systems only work if people use them. Training should explain why style consistency matters, what the firm's style preferences are, how to apply style prompts in AI-assisted drafting, and how to verify that drafts meet style standards. Without training, style systems remain aspirational rather than operational.

Style feedback in document review reinforces learning and improves consistency over time. When supervisors review drafts, they should provide style feedback along with substantive feedback. "The analysis is correct, but the tone is too aggressive for this client. Soften the language in paragraphs three and four." This feedback teaches what style expectations actually mean in practice and helps lawyers internalize preferences they can then apply in their own drafting and AI instructions.

Legal writing programs increasingly emphasize plain English and clear communication. Courts appreciate briefs that are easy to read and understand. Style systems should generally favor clarity over complexity, accessibility over obscurity. AI can help achieve plain English when explicitly instructed to do so.

Iteration improves style systems over time. Your initial style guide and style prompts may not perfectly capture your preferences. As you use them and review results, you'll identify gaps and adjustments needed. Treat your style system as a living document that evolves based on experience. Note what works well, what produces inconsistent results, and what's missing. Update your style prompts accordingly.

Style verification confirms that final documents meet your standards. Verification for style is less rigorous than verification for facts and law, but it should still occur. Before finalizing any significant document, review it specifically for style: Does it match firm conventions? Is the tone appropriate? Is it internally consistent? Are defined terms used correctly? Style errors are less serious than substantive errors, but professional work product should meet professional standards in all dimensions.

Quality control sampling helps maintain style consistency at scale. You can't review every document in detail for style, but you can review a sample. Periodic sampling of AI-assisted work product reveals whether style systems are being followed and whether style prompts are producing desired results. Quality control findings inform training needs and style system refinements.

Automation opportunities exist for some style elements. Citation format checking can be partially automated. Defined term consistency can be checked with search functions. Formatting compliance can be verified against templates. While AI currently cannot fully evaluate style quality, certain mechanical style elements can be checked systematically. Incorporating these checks into your workflow reduces style errors in final work product.

The balance between consistency and judgment acknowledges that style rules have exceptions. Your style guide provides defaults, but exceptional circumstances may warrant departures. A matter with unusual sensitivity may require different tone than your default. A client with strong preferences may override firm conventions. A creative argument may benefit from unconventional presentation. Style systems should guide without constraining judgment. When lawyers depart from style defaults, they should do so consciously and for good reasons, not accidentally because they forgot to apply the style system.

Style systems translate organizational preferences into specifications that guide AI-assisted drafting toward consistent voice, tone, and formatting. Style guides provide the foundation; style prompts translate guides into AI instructions. Document-type conventions recognize that different documents call for different styles. Multi-drafter coordination requires shared style references. Supervision should include style review to catch inconsistencies. Client-specific and practice-area conventions add additional layers to the system. Training ensures people can apply the system; iteration improves it over time. Style verification confirms that documents meet standards before finalization. Balancing consistency with judgment allows appropriate departures from defaults while maintaining coherence across the organization's work product.

8.7 Mini-Lab: Clause Improvement + Risk Memo

Objective: Apply the drafting workflow and clause techniques from this chapter to improve a problematic contract clause and produce a risk memo advising a client on the improved language. Practice clause analysis, generation, comparison, and risk identification in an integrated exercise.

The complete Mini-Lab packet, including the scenario, original clause, analysis guide, and revised clause elements, is available on the companion site for download.

Time Required: 75 to 90 minutes

What You'll Need

- The drafting workflow from Section 8.1 (Instructions → First Draft → Revise → Verify)
- The clause techniques from Section 8.2 (Generate, Compare, Risk-Spot)
- The style considerations from Section 8.6
- Access to an AI tool for drafting assistance
- A method for documenting your process

Step 1: Review the Scenario and Problem Clause

Read the following scenario and the contract clause that requires improvement.

Scenario

Your client is Riverside Manufacturing, Inc. ("Riverside"), a mid-sized company that manufactures industrial components. Riverside is negotiating an agreement to supply components to Apex Systems, Inc. ("Apex"), a larger company that assembles the components into finished products sold to end users. The supply agreement is worth approximately million annually over a three-year term.

Apex has sent its standard form supply agreement, which contains the following indemnification clause. Riverside's general counsel has reviewed the clause and is concerned that it's too one-sided. She has asked you to analyze the clause, propose improved language, and prepare a short memo explaining the risks in the original clause and the benefits of your proposed revisions.

Original Clause (Apex's Form)

> *"INDEMNIFICATION. Supplier shall indemnify, defend, and hold harmless Buyer, its affiliates, and their respective officers, directors, employees, agents, successors, and assigns from and against any and all claims, damages, losses, costs, and expenses (including reasonable attorneys' fees) arising out of or relating to (a) any products supplied by Supplier under this Agreement, (b) any breach of this Agreement by Supplier, (c) any negligent or wrongful act or omission of Supplier or its employees, agents, or subcontractors, or (d) any violation of applicable law by Supplier. Supplier's indemnification obligations shall survive termination or expiration of this Agreement indefinitely. Buyer shall have no obligation to indemnify Supplier for any reason."*

Step 2: Analyze the Original Clause

Before using AI assistance, conduct your own analysis of the original clause. Identify the specific problems with the clause from Riverside's perspective. Consider the following questions:

- What is the scope of indemnification coverage? Is it appropriately limited or overbroad?
- Does the clause include any limitations on Riverside's exposure (baskets, caps, time limits)?
- Does the clause require fault on Riverside's part, or does it impose strict liability for product-related claims?
- Is the indemnification mutual, or does it flow only one direction?
- What procedural protections exist for the indemnifying party?
- How does the survival provision affect Riverside's long-term exposure?
- Are there any carve-outs for claims caused by Apex's conduct?

Write a brief list of the problems you've identified before proceeding. This list becomes your reference for evaluating AI analysis and your input for the risk memo.

Step 3: Use AI to Analyze the Clause

Prompt AI to analyze the clause from Riverside's perspective. Compare AI's analysis to your own.

Sample prompt:

> *"Analyze the following indemnification clause from the perspective of the Supplier (Riverside Manufacturing). Identify every provision that is unfavorable to the Supplier. For each unfavorable provision, explain the risk it creates and rate its severity as high, medium, or low. Suggest how the provision might be modified to better balance the parties' interests."*

Review AI's analysis critically:

- Did AI identify the same problems you identified?
- Did AI surface additional problems you missed?
- Did AI miss problems you identified?
- Are AI's severity ratings appropriate given the transaction context?
- Are AI's suggested modifications reasonable and commercially realistic?

Note where AI analysis was helpful and where it required correction or supplementation.

Step 4: Generate Improved Clause Language

Using the drafting workflow from Section 8.1, generate improved indemnification language that addresses the problems you've identified while remaining commercially reasonable.

First, develop your instructions. What should the improved clause accomplish? Consider: mutual indemnification so both parties have obligations; appropriate scope limitations (arising from breach, negligence, or willful misconduct rather than strict product liability); fault-based coverage with carve-outs for claims caused by the indemnified party's conduct; reasonable cap on indemnification exposure; defined survival period rather than indefinite exposure; and procedural protections for the indemnifying party (notice, control of defense, cooperation).

Second, generate a first draft using AI with source-bounded prompting. Provide AI with your analysis, your desired outcomes, and any precedent language you want to use as a starting point.

Third, revise the AI draft. Does the generated language actually address your concerns? Is it internally consistent? Does it use defined terms correctly? Does the tone and style match appropriate contract drafting conventions? Make revisions to address any problems.

Fourth, verify the final clause. Does it accomplish what you intended? Are there any provisions that create unintended effects? Is the language clear and enforceable?

Generate multiple versions of the improved clause with different approaches (aggressive, moderate, conservative). Having options helps you advise the client on tradeoffs and prepares you for negotiation by identifying fallback positions.

Step 5: Compare Original and Revised Clauses

Use AI to generate a detailed comparison between the original Apex clause and your revised clause.

Sample prompt:

> *"Compare the following two indemnification clauses: [original Apex clause] and [your revised clause]. For each substantive difference, explain: what changed, why the change matters, and which party benefits from the change. Organize the comparison by topic (scope, limitations, survival, procedures)."*

Review the comparison to ensure it accurately captures the differences and their significance. This comparison becomes source material for your risk memo.

Step 6: Draft the Risk Memo

Prepare a short memo to Riverside's general counsel that: explains the key risks in the original Apex clause; describes your proposed revisions and why they matter; identifies any risks that remain even in the revised clause; and recommends a negotiation approach for presenting the revisions to Apex.

Use AI to assist with drafting, but ensure the memo reflects your analysis rather than AI's independent generation. Provide AI with your clause analysis, your revised language, and your comparison as inputs.

Verify the memo for accuracy. Does it correctly describe the original clause's problems? Does it accurately explain your revisions? Does it provide useful guidance to the client?

Step 7: Document Your Process

Create a brief record of how you used AI in this exercise:

- What AI tool did you use?

- At what stages did you use AI assistance?
- What inputs did you provide to constrain AI to your analysis?
- What verification did you perform on AI outputs?
- What changes did you make between AI drafts and final work product?

This documentation practices the habits discussed in Chapter 4 and reinforces the source-bounded drafting approach.

Scenario Analysis Guide

After completing your own analysis, use the following guide to check your work. The guide identifies the major issues; reasonable lawyers might prioritize them somewhat differently.

Problems in the Original Clause

Overbroad scope in subsection (a): "Any products supplied by Supplier" imposes what amounts to strict liability for product-related claims regardless of fault. Riverside would be liable even if Apex's improper use, modification, or integration of the components caused the claim. This should be narrowed to defects in the products or claims arising from Riverside's manufacturing.

No carve-out for Apex's conduct: The clause provides no exception for claims caused by Apex's negligence, misuse, improper integration, or failure to follow specifications. Riverside should not indemnify Apex for claims that Apex caused.

One-way indemnification: Apex has no obligation to indemnify Riverside for anything. Reasonable supply agreements include mutual indemnification, with each party covering claims arising from its own conduct.

No limitation on exposure: There's no cap, basket, or other limitation on Riverside's indemnification obligation. On a million annual contract, unlimited indemnification exposure creates disproportionate risk.

Indefinite survival: The obligation survives "indefinitely," meaning Riverside could face claims decades after the relationship ends. A reasonable survival period might be three to five years, or tied to applicable statutes of limitation.

No procedural protections: The clause doesn't address notice of claims, control of defense, or cooperation obligations. Standard indemnification provisions give the indemnifying party procedural rights that affect its ability to manage and defend claims.

The combination of strict product liability scope plus no cap plus indefinite survival creates potentially catastrophic exposure. A single serious product liability claim arising from Apex's integration of Riverside's components could exceed the entire value of the contract relationship.

Revised Clause Elements

A well-drafted revision might include:

Mutual structure: Each party indemnifies the other for specified categories of claims.

Fault-based Supplier coverage: Riverside indemnifies for claims arising from (a) defects in products caused by Riverside's manufacturing, (b) Riverside's breach of the agreement, (c) Riverside's negligence or willful misconduct, and (d) Riverside's violation of law.

Carve-out: Riverside's indemnification doesn't apply to the extent claims arise from Apex's negligence, misuse, modification, improper integration, or failure to follow specifications.

Buyer coverage: Apex indemnifies Riverside for claims arising from Apex's negligence or willful misconduct, Apex's breach, and Apex's use, integration, or modification of products in ways that cause harm.

Cap: Riverside's aggregate indemnification liability capped at [the greater of or one year's purchase volume], excluding claims for indemnification arising from gross negligence, willful misconduct, or third-party intellectual property infringement.

Survival: Indemnification obligations survive for [three/four/five] years following termination or expiration, except for claims arising from gross negligence or willful misconduct, which survive until the expiration of the applicable statute of limitations.

Procedures: The indemnified party must provide prompt written notice of claims, allow the indemnifying party to control defense (with consent to settlement not unreasonably withheld), and cooperate reasonably in the defense.

Courts interpret indemnification clauses according to their terms. Ambiguous language may be construed against the drafter. Clear, specific language that addresses foreseeable scenarios reduces litigation risk and produces predictable results.

Risk Memo Elements

A complete risk memo should address:

Executive Summary: The original clause imposes disproportionate risk on Riverside through broad scope, no limitations, one-way obligations, and indefinite duration. Recommend revising to mutual, fault-based indemnification with appropriate limitations before signing.

Key risks identified: Strict product liability exposure, no carve-out for Apex's conduct, unlimited exposure on a finite-value contract, indefinite duration, and no procedural protections.

Proposed changes and benefits: Each change should be explained in business terms the general counsel can relay to company leadership. "The cap limits our worst-case exposure to an amount proportionate to the contract value."

Remaining considerations: Even the revised clause creates some exposure. Product liability claims can be expensive to defend even when meritorious defenses exist. The carve-out for Apex's conduct may be contested in any actual claim situation. Recommend adequate product liability insurance regardless of contract language.

Negotiation approach: Present revisions as industry-standard mutual protections rather than as rejection of Apex's form. Emphasize that Apex benefits from the mutual structure too. Identify which revisions are must-haves versus negotiable preferences. Suggest offering something Apex values (perhaps favorable pricing terms or extended warranty coverage) in exchange for indemnification improvements.

Risk memos should help clients make decisions, not just identify problems. Always include a recommendation. "Here are the risks" is less useful than "Here are the risks, here's what I recommend, and here's why."

Deliverable

By the end of this Mini-Lab, you should have:

1. A written analysis of problems in the original clause
2. Documentation of AI's analysis and your evaluation of its accuracy
3. A revised indemnification clause addressing identified problems
4. A comparison between original and revised clauses

5. A risk memo advising the client on the issues and recommending an approach
6. Documentation of your AI-assisted drafting process

This deliverable demonstrates the clause improvement workflow from analysis through drafting and client communication.

Bonus Challenge

Extend your work in two directions.

First, prepare for negotiation pushback. Assume Apex's counsel rejects your proposed cap, arguing that Apex needs full protection against product liability claims because it faces the end users. Draft talking points responding to this objection and identifying potential compromises: a higher cap, a tiered cap with different limits for different claim types, or insurance requirements that provide Apex protection without unlimited Riverside exposure.

Second, draft a fallback clause. If Apex won't accept your preferred revision, what's the minimum acceptable position? Draft a fallback indemnification clause that addresses your client's most critical concerns while making concessions on less essential points. Identify what this fallback gives up compared to your preferred revision and explain why the fallback is still acceptable.

Know your walk-away point before negotiation begins. If Apex insists on the original clause without meaningful modification, advise your client on whether the business relationship is worth the legal risk. A million annual contract may not be worth potentially unlimited liability exposure.

8.8 Practice Checklist and Prompt Pack for Drafting

This section provides ready-to-use resources for implementing AI-assisted drafting in your practice. The checklist consolidates the workflow requirements from this chapter into a single reference. The prompt pack provides templates for common drafting tasks that you can adapt to your specific needs. These resources translate the chapter's concepts into practical tools you can use immediately.

Drafting Workflow Checklist

Use this checklist for any significant AI-assisted drafting project. For routine documents, the checklist serves as a mental framework. For high-stakes documents, complete the checklist formally and retain it with your matter files. A printable version of this checklist is available on the companion site for matter files.

Stage 1: Instructions

- ☐ Identified the document's purpose and what it must accomplish
- ☐ Gathered context the AI needs: transaction background, party relationships, related documents, relevant constraints
- ☐ Identified constraints: legal requirements, negotiation limitations, client instructions, format rules
- ☐ Specified structure: sections, organization, length expectations
- ☐ Defined output specifications: tone, formality level, style conventions
- ☐ For source-bounded drafting, assembled inputs: precedent clauses, verified facts, legal standards from research
- ☐ Documented instructions before generating first draft Stage 2: First Draft

☐ Provided AI with comprehensive instructions incorporating all five elements (purpose, context, constraints, structure, output specifications)

☐ For source-bounded drafting, included verified source materials in the prompt

☐ Reviewed first draft critically, not acceptingly

☐ Identified gaps where draft doesn't address required topics

☐ Identified additions where draft includes unrequested content

☐ Identified misalignments where draft addresses topics but not as needed

☐ Considered generating multiple draft versions for comparison

☐ Selected best starting point for revision or refined instructions for regeneration Stage 3: Revision

☐ Addressed substance first: Does each provision accomplish its purpose? Does the overall document achieve client objectives?

☐ Addressed accuracy: Are all factual assertions verified? Are legal characterizations correct?

☐ Addressed completeness: Does the document address everything it should? Are there gaps creating risk?

☐ Addressed style: Does tone match document type and recipient? Is terminology consistent? Does formatting follow conventions?

☐ For litigation documents, verified every fact traces to record support

☐ For litigation documents, verified every legal assertion builds on researched authorities

☐ Removed or verified any content AI added beyond provided inputs

☐ Used iterative AI assistance for targeted revisions where helpful

☐ Tracked changes to document revision process Stage 4: Verification

☐ Purpose verification: Does the final document accomplish what it must?

☐ Accuracy verification: Are all factual and legal assertions correct and supported?

☐ Consistency verification: Are defined terms used consistently? Do cross-references point correctly? Are there internal contradictions?

☐ Integration verification: Does this document fit with related documents in the matter?

☐ Format verification: Does the document comply with applicable rules, client requirements, and firm conventions?

☐ Style verification: Does the document meet organizational style standards?

☐ Documented verification process for significant documents

Prompt Pack for Drafting

The following prompts are templates you can adapt for your specific drafting needs. Replace bracketed placeholders with your actual information. These prompts implement the source-bounded approach that keeps AI working with your verified inputs rather than generating from training patterns. All prompts in this section are available as copy-paste templates on the companion site.

Clause Generation Prompts

Detailed clause generation

> Draft a [clause type] for a [transaction type] where [party description and relationship]. The clause should [specific objectives]. Include [specific provisions required]. Address [specific risks or scenarios]. Constraints: [any limitations on acceptable approaches, such as positions already rejected or regulatory requirements].

Multiple variations

> Generate three versions of this [clause type]: one that is strongly pro-[your client's role], one that is balanced, and one that favors [counterparty's role]. For each version, explain what makes it favorable to that party and identify the key tradeoffs.

Source-bounded generation

> Using the following [clause type] from a recent deal as a starting point, draft a [clause type] for this transaction that addresses the following differences: [list specific differences between the precedent deal and the current transaction]. Preserve the structural approach of the precedent while adapting the substance to these new requirements.

Clause Comparison Prompts

Substantive comparison

> Compare these two [clause type] clauses: [your version] and [counterparty version]. Identify every substantive difference between them. For each difference, explain the practical effect and which party benefits from each formulation. Organize the comparison by topic.

Interaction analysis

> Considering this [clause type A] together with this [clause type B], are there any ways these provisions could interact to create unexpected results or gaps in protection? Identify any interaction effects and explain their practical significance.

Risk Identification Prompts

Client-perspective risk review

> Review this [clause type] from the [your client's role] perspective. What risks does this language create for [your client]? What scenarios could leave [your client] inadequately protected? What changes would better protect [your client's] interests?

Balanced risk assessment

> Analyze this [clause type] from both [party A] and [party B] perspectives. What does each party gain and risk under this language? Categorize the risks as: likely and significant, significant but unlikely, likely but minor, and theoretical.

Redlining Prompts

Markup analysis

> Here is our original [section/article] and the counterparty's markup. Identify every change the counterparty made. Categorize each change as: substantive (affecting rights or obligations), clarifying (making meaning clearer without changing substance), or stylistic (word choice or formatting without legal effect). For each substantive change, explain what the change does and which party it favors.

Response generation

> The counterparty has proposed [specific change]. This is unacceptable to my client because [reason]. Draft language explaining why this proposal is inadequate and proposing [alternative approach] as a substitute.

Fallback identification

> My primary position is [your position]. If the counterparty rejects this, what fallback positions could partially protect my client while giving the counterparty some [concession type] they value? For each fallback, explain what my client gains and gives up compared to the primary position.

Negotiation preparation

> Based on our analysis of this markup, create a negotiation preparation guide that includes: a prioritized list of open issues from most to least important; our position on each issue with key talking points; our fallback position if our primary position is rejected; and potential trades where we could give ground on lower priorities to gain on higher priorities.

Litigation Drafting Prompts

Demand letter

> Draft a demand letter based on the following facts from my investigation: [fact summary with sources]. The letter should assert claims for [legal theories] under [applicable law]. Demand [specific relief] with a response deadline of [timeframe]. Tone should be firm but professional, leaving room for negotiation.

Motion argument section

> Draft the argument section for this [motion type] based on the following legal authorities and holdings from my research: [Authority 1 with extracted holding]; [Authority 2 with extracted holding]. The argument should establish [legal standard], apply the standard to [these facts with record citations], and conclude that [requested relief] is required because [conclusion].

Statement of facts with record citations

> Draft a statement of facts for a [motion type] based on the following facts and record citations: [Fact 1, citing source]; [Fact 2, citing source]; [Fact 3, citing source]. Present the facts in [chronological/thematic] order. Every factual assertion must cite to the record support I have provided.

Fact Discipline Prompts

Structured fact input

```
Draft this [document type] using only the facts I have provided below. Do not
add any facts, details, dates, times, descriptions, or characterizations that
I haven't specifically included. If information is missing that would normally
appear in this type of document, insert [FACT NEEDED: description of what's
missing] rather than generating the information. Facts: [your fact list].
```

Fact verification comparison

```
Here is my fact sheet with all the facts I provided for this [document type].
Here is the draft produced. Identify any factual assertions in the draft that
don't appear in my fact sheet. Flag each addition so I can verify or remove
it.
```

Style Prompts

Style calibration

```
Write in [tone description]. Keep sentences under [word limit] where possible.
Use [language preferences]. Be [advocacy stance]. Follow [citation format].
Structure [organizational preferences]. Here is a sample of our firm's
preferred style for this document type: [paste example].
```

Customize these prompts by adding context about your practice area, jurisdiction, client preferences, or organizational conventions. Save your most effective prompts in a personal prompt library. Over time, your library becomes a collection of proven approaches for your specific drafting needs.

Key Takeaways

1. Effective drafting workflow sequences AI assistance across four stages. Instructions define purpose, context, constraints, structure, and output specifications, with instruction quality heavily influencing draft quality. First draft generation treats AI output as a starting point for revision, not a final product. Revision addresses substance, accuracy, completeness, and style, transforming AI output into work product reflecting professional judgment. Verification confirms the document accomplishes its purpose and contains no errors. Time allocation should weight revision heavily, recognizing that revision is where you add the professional value that makes the document yours.
2. Contract clause work benefits from AI assistance across generation, comparison, and risk identification. Clause generation produces language options you evaluate rather than adopt uncritically, with detailed instructions specifying what each clause must accomplish. Source-bounded generation grounded in verified precedent improves output quality. Clause comparison identifies differences between alternative formulations and explains their implications, though AI comparison doesn't replace reading both versions yourself. Risk identification surfaces potential problems in proposed language, with AI-identified risks requiring prioritization based on your transaction knowledge. Interaction effects between clauses may create risks that neither clause creates alone.
3. Redlining assistance accelerates negotiation analysis while preserving strategic judgment. AI can catalog counterparty changes and explain their effects, helping you focus attention where it matters. Response generation produces options for addressing unacceptable changes. Fallback positions prepare you for impasses. Issue trading identifies opportunities for mutual gain through package deals. Post-negotiation documentation captures agreements and open

items. Throughout, AI accelerates analysis while you retain control over strategic decisions that shape negotiation outcomes.

4. Litigation drafting differs from transactional drafting because facts must trace to the record and authorities must be verified. AI doesn't know your record and cannot satisfy the fact constraint. Source-bounded drafting provides AI with verified facts and authorities as inputs, confining AI to organization and presentation rather than substance generation. Every factual assertion must cite to record support. Every legal assertion must build on verified authorities. Rule 11 certification applies regardless of AI assistance.

5. Fact discipline prevents AI from inventing or embellishing facts. AI generates plausible-seeming factual assertions from training patterns, not from your record. Structured fact inputs, negative instructions, and gap flagging reduce invention risk but don't eliminate it. Verification techniques include line-by-line fact review, comparison against your fact sheet, and assertion-source matrices that document verification. Temporal, quantitative, and characterization assertions require particular scrutiny. Accept incomplete narratives rather than filling gaps with fabrications.

6. Style systems maintain consistency when AI assists drafting across multiple drafters, documents, or matters. Style guides provide the foundation; style prompts translate guides into AI instructions. Document-type conventions recognize that different documents call for different styles. Multi-drafter coordination requires shared style references. Training ensures people can apply the system; iteration improves it over time. Style verification confirms that documents meet standards before finalization.

Chapter Checkpoint

Before moving on, make sure you can:

- ☐ Execute the four-stage drafting workflow (Instructions → First Draft → Revise → Verify), providing comprehensive instructions and treating AI output as raw material for professional revision
- ☐ Generate contract clauses using detailed instructions that specify what each clause must accomplish, then critically evaluate AI output against your client's interests
- ☐ Compare alternative clause formulations to identify differences and their implications, using AI analysis as a starting point while reading both versions yourself
- ☐ Identify risks in proposed contract language and prioritize AI-identified risks based on your knowledge of the transaction context
- ☐ Analyze counterparty redlines systematically, categorizing changes by significance and developing response options that advance your negotiating position
- ☐ Apply source-bounded drafting to litigation documents, providing verified facts with record citations and legal authorities from your research as AI inputs
- ☐ Implement fact discipline techniques that prevent AI from inventing or embellishing facts, including structured inputs, negative instructions, and line-by-line verification
- ☐ Apply style systems that maintain consistency across documents, drafters, and matters through style guides, style prompts, and style verification
- ☐ Document your AI-assisted drafting process in a manner that demonstrates responsible use and supports supervision

What's Next

Chapters 7 and 8 addressed the core work product workflows: research that produces understanding and drafting that creates documents. Chapter 9 turns to litigation and discovery workflows that apply these capabilities to the adversarial context.

Litigation involves not just producing your own work product but engaging with opposing parties, managing large document volumes, preparing for depositions, and building cases that will persuade courts and juries. The litigation context presents distinct challenges for AI assistance. Case theory development requires synthesizing facts, law, and strategy into a coherent narrative. Motion practice involves anticipating and responding to opposing arguments in real time. Discovery generates massive document volumes that must be reviewed, categorized, and produced efficiently. Deposition preparation requires organizing facts, identifying impeachment opportunities, and preparing witnesses to testify effectively. Investigations demand timeline construction, entity relationship mapping, and gap analysis that surfaces what you don't yet know.

Chapter 9 shows how AI can assist across these litigation workflows. You'll learn to use AI for case theory development without letting AI's framing substitute for your strategic judgment. You'll see how AI can help with motion strategy by identifying issues, anticipating responses, and organizing arguments. You'll address discovery planning where AI excels at processing volume but requires careful oversight for privilege and relevance determinations. You'll develop approaches for deposition preparation that leverage AI's organizational capabilities while maintaining the human judgment that effective examination requires. And you'll learn investigation techniques that use AI to build timelines and identify gaps while preserving the analytical rigor that serious investigations demand. The litigation workflows build on everything you've learned while adapting to the specific demands of adversarial practice.

Key Terms

Drafting instructions — Guidance to AI defining what a document must accomplish, what context applies, what constraints limit the draft, how it should be structured, and what output format is expected.

Source-bounded drafting — Providing AI with specific source materials (precedents, facts, legal standards) that constrain the draft to verified content rather than free generation. Maintains quality control while leveraging AI drafting capabilities.

Drafting verification — The final review process confirming that a document accomplishes its intended purpose and that all factual and legal assertions are accurate. Focuses on whether the document works, not just whether citations exist.

Clause generation — Using AI to produce draft contract language for a specific provision. Generated clauses are starting points for evaluation and revision, not final language to adopt.

Interaction effects — How multiple contract provisions work together to create outcomes that neither provision creates alone.

Redline analysis — Systematic review of counterparty markup to identify, categorize, and assess proposed changes. AI can accelerate cataloging; evaluating acceptability requires lawyer judgment.

Issue trading — Exchanging concessions across different contract provisions, where each party gives ground on issues less important to them in exchange for gains on issues more important.

Fact constraint — The requirement that every factual assertion in litigation documents must trace to a verifiable source: the evidentiary record, good-faith allegations with expected support, or matters subject to judicial notice.

Record citation — A specific reference to evidentiary materials in the case file that supports a factual assertion. In motion practice, statements of fact must be supported by record citations.

Rule 11 certification — The implicit certification made when a lawyer signs a court filing, representing that the filing has legal and factual support and is not filed for improper purpose. AI assistance does not reduce this responsibility.

Fact invention — AI generating factual assertions that do not come from your inputs, based on patterns in training data about what facts typically appear in similar documents.

Assertion-source matrix — A verification tool mapping each factual assertion in a document to its supporting source in the record or investigation materials.

Style system — A coordinated approach to maintaining consistent voice, tone, and formatting conventions when AI assists drafting across multiple drafters, documents, or matters.

Terminology consistency — Using the same words and phrases to describe the same concepts throughout a document, avoiding variations that suggest unintended distinctions or ambiguity.

Style verification — Reviewing documents specifically to confirm they meet organizational style standards.

Risk memo — A document advising a client about legal risks in a transaction, contract language, or course of action. Should clearly identify risks, explain significance, and provide actionable recommendations.

Drafting checklist — A standardized list of steps for AI-assisted document creation across the workflow stages of instructions, first draft, revision, and verification.

Chapter 9: AI for Litigation, Discovery, and Investigations

Litigation is adversarial by design. You don't just produce work product; you engage with opponents who are actively working to defeat you. You don't just analyze facts; you construct narratives that persuade judges and juries while your adversary constructs competing narratives. You don't just find authorities; you anticipate how opposing counsel will distinguish them and prepare responses. This adversarial dimension shapes how AI can help with litigation work. The stakes of error are higher because opponents will exploit your mistakes. The value of preparation is greater because thorough preparation creates advantages that carry through trial. AI that accelerates litigation workflows without accounting for the adversarial context misses what makes litigation different from other legal work.

Discovery adds scale challenges that other practice areas rarely face. A commercial litigation matter might involve millions of documents. An investigation might span years of communications across dozens of custodians. The volume makes comprehensive human review impractical, but the consequences of missing important documents or producing privileged materials are severe. AI has assisted discovery for over a decade through technology-assisted review systems that predate the current generative AI wave. Newer AI capabilities layer onto this foundation, offering more sophisticated analysis, summarization, and pattern recognition. Understanding how to use these tools effectively is now a baseline competence for complex litigation.

This chapter applies the principles from earlier chapters—prompt engineering, confidentiality, verification—to the specific demands of litigation and discovery. The research memo you drafted in Chapter 7 and the motions you drafted in Chapter 8 are components of a larger strategy. This chapter focuses on that larger strategy: how AI helps you build the case, manage the evidence, and prepare for the interactions that determine outcomes.

What This Chapter Covers

- Using AI to develop case theory and narrative while maintaining strategic control
- Motion strategy workflows that anticipate opposing arguments and identify winning issues
- Discovery planning and management using AI to handle volume and complexity
- Privilege and relevance review strategies that leverage AI classification capabilities
- Deposition preparation techniques that organize facts and identify impeachment opportunities
- Investigation workflows for timeline construction, entity mapping, and gap analysis
- A Mini-Lab creating a deposition preparation packet using provided case materials
- A practice checklist and prompt pack for litigation and discovery tasks

Chapter Summary

Litigation workflows require integrating AI assistance with adversarial strategy. Case theory development benefits from AI's ability to synthesize facts into narratives, but the strategic choice of which narrative to pursue remains yours. Motion strategy uses AI to identify issues and anticipate responses, helping you select arguments that are not just legally sound but strategically effective. Discovery planning and review rely on AI to process volume, prioritizing documents for human review and identifying patterns that manual review might miss. Deposition preparation uses AI to organize facts and identify inconsistencies, equipping you to take effective testimony. Investigations leverage AI to build timelines and map relationships, surfacing the story hidden in the documents. Throughout these workflows, the verification imperative remains: AI suggestions must be tested against the record and the law. Opponents will test your work; you must test it first.

9.1 Case Theory: Turning Facts into Issues and Narratives

A case theory explains why your client should win. It integrates the undisputed facts, your version of the disputed facts, and the applicable law into a coherent narrative that leads to a favorable judgment. Developing a strong case theory is often the difference between winning and losing. AI cannot create a case theory for you because it lacks the judgment to know what will persuade a particular judge or jury. But AI can assist the development process by synthesizing facts, identifying themes, and stress-testing your narrative against alternatives.

Case theory development begins with facts. You have the client's story, the documents you've reviewed, and perhaps some initial witness accounts. AI can help organize these facts into a chronology that reveals cause-and-effect relationships. "Create a timeline of events based on these interview notes and emails. Identify any gaps where we don't know what happened between key events." Seeing the sequence clearly often suggests narrative themes that aren't obvious from a disorganized file.

Narrative brainstorming uses AI to explore different ways to tell the story. The same facts can support multiple narratives. Is this a story about a breach of trust? A story about commercial incompetence? A story about unavoidable market forces? AI can generate options. "Here are the key facts. Suggest three different narrative themes that could explain these events from my client's perspective. For each theme, identify which facts emphasize it and which facts might undermine it."

This brainstorming helps you avoid anchoring on the first theory that comes to mind. By seeing alternatives, you can make a deliberate choice about which narrative best fits the evidence and the law. You might combine elements from different AI suggestions into a theory stronger than any single option.

Theme testing evaluates your chosen theory against the evidence. A good theory accounts for the bad facts as well as the good ones. AI can help you identify the bad facts your theory needs to explain. "My working theory is that the defendant failed to deliver because they prioritized other more profitable customers. Based on the fact summary I've provided, what are the biggest factual obstacles to this theory? What evidence suggests alternative explanations?" This adversarial testing prepares you to address weaknesses rather than ignoring them.

> **Risk Note** Do not let AI sanitize the bad facts. AI models trained to be helpful may minimize contradictory evidence to support the narrative you asked for. Explicitly prompt AI to identify facts that do not fit your theory so you can address them. A theory that ignores bad facts will collapse under adversarial scrutiny.

Legal alignment ensures your narrative satisfies the legal elements required for relief. A compelling story that doesn't prove a claim is useless in court. AI can check your narrative against the legal framework. "Here is my narrative summary and the elements of the breach of contract claim. Does this narrative address every required element? Are there elements where the narrative is weak or relies on assumptions?" This check confirms that your story does the legal work it needs to do.

Perspective taking helps you anticipate the other side's theory. You can't win if you don't understand the story the opponent will tell. "Based on these facts, what is the most persuasive narrative the opposing party could construct? How would they explain the events I've interpreted as breach?"

Seeing the strongest version of the opponent's case helps you refine your own theory to pre-empt or rebut their themes.

> **Practice Point** Use AI to role-play the opposing counsel opening statement. "Write an opening statement for the defendant based on these facts." Reading the story from their perspective often reveals vulnerabilities in your own theory that you can address before you file your initial pleading.

Iterative refinement improves the theory as discovery proceeds. Case theory isn't static; it evolves as you learn more. When new facts emerge, update your inputs and ask AI to reassess the narrative. "We just learned X from a witness. How does this new fact affect our 'breach of trust' theory? Does it support an alternative theory more effectively?" AI helps you maintain a dynamic theory that matches the current record.

> **Key Point** AI assists case theory by organizing facts, brainstorming narrative themes, identifying weaknesses, and aligning stories with legal elements. It helps you see alternatives and anticipate opposing theories. Your judgment determines which theory is most persuasive and strategically sound. Use AI to expand your options and stress-test your choices, not to make the strategic decision for you.

The human judgment required for case theory involves assessing plausibility and emotional resonance. AI can tell you if a narrative is logical; it can't tell you if it will feel true to a jury. It can't assess whether a particular witness will come across as credible or evasive. It can't weigh the reputational factors that might make a client prefer one narrative over another. You must filter AI's suggestions through your understanding of human nature and the specific decision-makers who will hear the case.

> **Practice Area Note: Criminal Defense** Criminal defense case theory operates differently from civil theory. You don't need to prove what happened — you need to show the prosecution hasn't proven what they claim happened. When using AI for defense theory development, orient prompts toward weakness identification rather than narrative construction: "Based on these facts, what are the biggest gaps in the prosecution's theory that the defendant acted with intent? Where is the circumstantial evidence weakest? What alternative explanations for the physical evidence has the prosecution failed to eliminate?" This reframing prevents AI from constructing an affirmative defense narrative you may not want while surfacing the reasonable doubt arguments that may persuade the factfinder.

9.2 Motion Strategy: Framing Issues and Anticipating Responses

Motions are the mechanism for advancing your case theory before trial. Every motion should serve the larger strategy, whether by resolving claims, limiting evidence, or forcing the opponent to commit to a position. AI can assist with motion strategy by helping you identify which motions to bring, how to frame the arguments, and how to anticipate the opposition's response.

Issue identification prompts ask AI to spot potential motions based on the record. "Review this complaint and the fact summary. What grounds for a motion to dismiss might exist? Are there jurisdictional issues, pleading deficiencies, or affirmative defenses that appear on the face of the complaint?" AI acts as a checklist, surfacing potential procedural and substantive attacks you might consider.

Argument selection helps you choose the strongest grounds. A kitchen-sink motion that argues everything often persuades no one. AI can help you prioritize. "I've identified five potential arguments for dismissal. Based on the authorities I've provided, which two arguments appear strongest? Which arguments might distract from the stronger points?" This evaluation helps you focus your briefing on winning issues.

Framing the question presented shapes how the court views the issue. The way you state the question often determines the answer. AI can generate options for framing. "I need to frame the question presented for a motion to compel arbitration. The key issue is whether the arbitration clause covers this tort claim. Suggest three ways to frame this question: one neutral, one framing it as a broad delegation of authority, and one framing it as a narrow contract interpretation issue." Seeing options helps you choose the frame that best positions your argument.

Anticipating the opposition is critical for motion success. You need to write your opening brief with the opposition's response in mind. AI can help predict that response. "I plan to argue X. What is the strongest response the defendant can make? What authorities might they cite? How would they distinguish my key cases?" This predictive work allows you to "inoculate" your brief against their best arguments, addressing them proactively rather than waiting for the reply brief.

> **Practice Point** Before finalizing your motion strategy, ask AI to "pre-mortem" the motion. "Assume the judge denies this motion. What is the most likely reason for the denial based on the arguments I've outlined?" This exercise highlights the weakest link in your chain of reasoning, allowing you to strengthen it or adjust your strategy.

Reply brief strategy benefits from AI analysis of the opposition. When you receive the opposition's brief, use AI to analyze it. "Summarize the three main arguments in this opposition brief. For each argument, identify the key authorities relied upon and the factual premises assumed." This summary helps you structure your reply to address the actual arguments made, not just the ones you expected.

Oral argument preparation uses AI to simulate the judge's questioning. "I'm arguing this motion to dismiss. My main point is X. Act as a skeptical judge and ask me three hard questions about this argument. Challenge my reading of the key case." This simulation forces you to articulate answers and identifies where your explanation needs clarity.

Strategic timing considerations are often outside AI's capability. Whether to file a motion now or after discovery, whether to combine motions or separate them, whether a motion will educate the opponent more than it helps you—these are judgments requiring knowledge of the specific judge, the opponent's tendencies, and the settlement dynamics. AI can analyze the legal merits; you must analyze the strategic context.

> **Risk Note** Do not let AI convince you to file a frivolous motion. AI will generate arguments for almost any position if asked. The existence of an argument does not make it non-frivolous. You must evaluate whether the argument has a good-faith basis in law and fact. Rule 11 of the Federal Rules of Civil Procedure (Fed. R. Civ. P. 11) applies to the decision to file, not just the content of the filing.

AI supports motion strategy by identifying potential motions, prioritizing arguments, framing questions presented, and anticipating opposition responses. It helps you prepare for oral argument by simulating judicial questioning. Strategic decisions about timing and advisability require your judgment about the specific litigation context. Use AI to ensure you've considered all angles, but rely on your experience to decide which angles to pursue.

9.3 Discovery Planning: Requests, Objections, Proportionality

Discovery is the engine of litigation, but it's often an inefficient engine. Drafting requests, responding to them, and managing the scope of discovery consumes vast resources. AI can streamline discovery planning and execution, helping you target what you need and protect what you shouldn't have to produce.

Drafting discovery requests requires translating your case theory into specific inquiries. You need documents and testimony that prove your claims and disprove the opponent's defenses. AI can help generate comprehensive requests. "Based on the elements of our breach of contract claim and the factual issues regarding the delivery delay, draft a set of document requests targeting internal communications about the delay schedule." AI generates a list of requests that you can refine, ensuring you don't miss obvious categories.

Definitions and instructions are critical for avoiding loopholes. AI can help draft robust definitions. "Draft a definition of 'Meeting Records' that captures not just formal minutes but also informal notes, calendar entries, recordings, and chat logs related to meetings." Clear definitions prevent the opponent from defining their way out of production.

Responding to requests involves identifying valid objections and determining what must be produced. AI can help analyze incoming requests. "Review these document requests and identify which ones are potentially objectionable as overbroad, vague, or seeking privileged information. Suggest specific objections for each." This initial pass helps you focus on the problematic requests while quickly handling the routine ones.

Proportionality analysis is central to modern discovery. Courts limit discovery that is disproportionate to the needs of the case. AI can help you articulate proportionality arguments. "We are objecting to this request for five years of email from ten custodians as disproportionate. Draft an argument explaining why the burden of this production outweighs the likely benefit, given that the contract was negotiated and breached within a six-month period."

Use AI to estimate the burden of discovery requests. "If we have to search these terms across this volume of data, what are the likely steps involved? Help me describe the technical burden of restoring backup tapes for this request." A detailed explanation of burden is more persuasive than a conclusory objection.

Deposition preparation starts with identifying who to depose and what to ask. AI can help identify key witnesses based on the documents. "Based on this chronology of emails, who are the key people involved in the decision to delay the shipment? What specific knowledge does each appear to have?" This analysis helps you build your witness list.

Third-party discovery often requires subpoenas. AI can help draft subpoenas that comply with the specific rules for non-parties. "Draft a Rule 45 subpoena (Fed. R. Civ. P. 45) for the shipping company seeking records of the delivery. Include the required text regarding protection of persons subject to subpoenas."

Discovery dispute resolution often involves meet-and-confer letters and motions to compel. AI can help draft these communications, summarizing the history of the dispute and the specific deficiencies in the opponent's production. "Draft a meet-and-confer letter addressing the defendant's failure to produce the internal Slack messages we requested. Summarize why these messages are relevant and why their 'undue burden' objection is unfounded based on the case law regarding modern communication platforms."

Discovery disputes often turn on specific procedural history. AI doesn't know what happened in your phone calls with opposing counsel unless you tell it. Ensure that any correspondence accurately reflects the meet-and-confer history. Misstating the history damages your credibility with the court.

Verification in discovery drafting means ensuring that your requests actually cover what you need and that your objections are legally sustainable. An AI-generated objection that "the request is overbroad" is useless if you don't have a specific reason why. You must ensure that your discovery positions are defensible.

AI streamlines discovery planning by generating comprehensive requests, identifying valid objections, and articulating proportionality arguments. It helps identify key witnesses and draft third-party subpoenas. Use AI to translate your case theory into specific discovery goals, but verify that requests are targeted and objections are defensible.

9.4 Review Triage and Privilege Spotting

Reviewing documents for relevance and privilege is the most labor-intensive part of litigation. AI has transformed this process more than any other. Technology-Assisted Review (TAR) using predictive coding has been standard for years. Generative AI brings new capabilities for summarization, complex relevance analysis, and privilege log generation.

Relevance review with generative AI goes beyond simple keyword searching or predictive scoring. You can ask the AI to evaluate documents against specific relevance criteria. "Review this batch of emails. Identify any document that discusses 'project delays' or 'supply shortages.' For each relevant document, explain in one sentence why it meets the criteria." This semantic understanding allows for more nuanced review than keyword hits.

Privilege spotting is a high-stakes task. Producing a privileged document can waive privilege for the subject matter. AI can help identify potentially privileged communications. "Flag any document involving an attorney (list of names provided) or discussing legal advice. Also flag documents where employees discuss 'checking with legal' or 'risk assessment.'" AI acts as a safety net, identifying documents that human reviewers might miss.

> **Risk Note** AI cannot make the final privilege determination. Privilege depends on the specific purpose of the communication, legal advice versus business advice. AI can flag candidates, but a lawyer must make the final call on whether privilege applies. Never let AI auto-withhold or auto-produce without human verification on privilege.

Privilege log generation is tedious but essential. Courts require specific descriptions of withheld documents. AI can generate these descriptions efficiently. "For each document flagged as privileged, draft a privilege log description that identifies the sender, recipient, date, and a generic description of the subject matter (e.g., 'Email seeking legal advice regarding contract termination') without revealing the substance of the advice." This automation saves hours of drafting while maintaining consistency.

Key document identification helps you find the "hot docs" early. In a sea of relevant documents, only a few really matter for the case narrative. AI can help surface these. "Identify documents where the tone indicates conflict, urgency, or concern. Flag emails where the project manager admits to a mistake." Finding these documents early shapes your case strategy and deposition preparation.

Review quality control uses AI to check human reviewers. "Review this sample of documents marked 'not relevant' by the review team. Did they miss any documents that discuss 'pricing strategy'?" This second-pass review helps ensure consistency and catches reviewer fatigue errors.

Data privacy review (PII redaction) is increasingly important. Litigation documents often contain personal identifiable information (PII) that must be redacted. AI tools can automatically identify and redact names, social security numbers, and financial details. Verification is still needed, but the automated pass handles the bulk of the work.

When setting up a review workflow, define the "responsive" criteria clearly for the AI. Vague criteria lead to inconsistent results. "Documents about the contract" is vague. "Documents discussing the negotiation, drafting, or execution of the 2023 Supply Agreement" is clear. Test your criteria on a small sample before running the full review.

AI transforms document review from a linear slog into a prioritized, intelligent process. Use AI for relevance classification, privilege spotting, and key document identification. Automate privilege log drafting and PII redaction. Maintain human oversight for final privilege determinations and quality control. Courts have consistently approved technology-assisted review when parties demonstrate reasonable processes for training, validation, and quality control — see the Authority Note in Chapter 6 discussing *Rio Tinto PLC v. Vale S.A.*, 306 F.R.D. 125 (S.D.N.Y. 2015). The goal is to spend lawyer time analyzing the documents that matter, not sifting through the ones that don't.

9.5 Depositions: Outlines, Exhibits, and Impeachment Planning

Depositions are where cases are often won or lost. Effective depositions require deep command of the facts, strategic organization of topics, and readiness to impeach witnesses who deviate from the record. AI can assist with the preparation that makes successful depositions possible.

Deposition outlines structure your questioning. A good outline ensures you cover all necessary topics without becoming a script that prevents you from following up. AI can help generate the outline structure. "Create a deposition outline for the defendant's project manager. The goal is to establish that she knew about the delays two months before notifying us. Topics should include: reporting structure, weekly status meetings, the March 15th email chain, and the decision to divert supplies."

Detailed topic modules can be fleshed out by AI. "Draft a line of questioning to establish the authenticity and receipt of these three emails. Include questions to preclude"I didn't see it" or "it went to spam" defenses." This preparation ensures you have the technical questions ready so you can focus on the substance.

Exhibit selection and organization is crucial. You need to show the witness the right documents at the right time. AI can help organize exhibits. "Review the key documents we've identified. Suggest an order for presenting them to the witness that builds the narrative of knowledge and concealment. Which documents should be grouped together?"

Impeachment planning prepares you for the witness's evasions. "If the witness denies knowing about the delay in March, what documents contradict that? List the specific emails with dates and senders that show she was informed." Having this index ready allows you to impeach effectively in the moment.

Witness preparation (for your own witnesses) benefits from AI role-play. "I'm preparing our client for deposition. The opposing counsel will likely ask about the missing meeting notes. Act as opposing counsel and ask aggressive follow-up questions on this topic so I can help the client practice remaining calm and truthful." This simulation helps identifying where the witness needs more preparation.

Transcript summarization after the deposition helps you use the testimony. "Summarize the key admissions made by the project manager in this transcript. Identify any new names mentioned that we need to investigate. Flag points where her testimony contradicts the documents." This rapid synthesis allows you to adjust your strategy for subsequent depositions.

Don't let the outline become a crutch. Use AI to build a comprehensive roadmap, but be ready to go off-road when the witness gives an unexpected answer. The best questions are often "Why?" and "Tell me more about that," which AI can't script for you.

AI assists deposition preparation by structuring outlines, organizing exhibits, and identifying impeachment materials. It helps prepare your witnesses through role-play simulations. After the deposition, it accelerates transcript analysis. The goal is to walk into the deposition room with a command of the record that allows you to focus on the witness.

9.6 Investigations: Timelines, Entity Charts, and Gap Analysis

Internal investigations and factual inquiries require assembling scattered information into a coherent picture of what happened. You start with fragments—emails, interview notes, financial records—and must construct a unified timeline and narrative. AI excels at this synthesis work, helping you see connections and identify what's missing.

Timeline construction is the backbone of investigation. AI can process mixed inputs to build a master chronology. "Create a chronological timeline of events based on these 50 emails and 3 interview summaries. For each entry, list the date, time, key actors, and source document." This timeline reveals the sequence of events and often highlights inconsistencies between accounts. An investigation timeline template is available on the companion site.

Entity and relationship mapping helps you understand who is involved. "Based on these documents, identify all the individuals and entities mentioned. Create a chart showing their roles and relationships to each other. Who reports to whom? Who was copied on communications between the primary actors?" This mapping is essential in fraud or conspiracy investigations where relationships are central.

Gap analysis identifies what you don't know. "Review the timeline and identify periods of silence or missing information. Are there key meetings referenced for which we have no minutes? Are there email chains that end abruptly?" Identifying these gaps directs your further data collection and witness interviews.

> **Risk Note** Investigations often involve highly sensitive internal matters. Ensure that you are using AI tools with appropriate confidentiality protections, Tier 3 or Tier 4, and that you are not inadvertently waiving privilege by sharing information with unsecured third-party systems.

Pattern recognition can surface misconduct. "Analyze these expense reports and identify any patterns of round numbers, duplicate amounts, or weekend charges. Flag transactions that deviate from the typical spending pattern for this department." AI can spot anomalies in structured data that human review might glaze over.

Interview preparation relies on the synthesized record. "I'm about to interview the CFO. Based on the timeline and documents, what specific questions should I ask her about the approval of the Q3 payments? What documents should I show her to refresh her recollection or challenge her account?"

Report drafting synthesizes the investigation findings. AI can help draft the factual background section of the investigation report, weaving the timeline and documents into a narrative. "Draft a factual summary of the events leading up to the board meeting, citing the relevant interviews and documents." As always, verification is essential—the report must accurately reflect the evidence.

> Practice Area Note: Immigration Asylum credible-fear investigations are one of the most natural AI-assisted workflows in immigration practice. Timeline construction from this section applies directly — country conditions documentation, travel chronology, and persecution event sequencing all benefit from AI synthesis of scattered records. Gap analysis is critical: "Based on this asylum timeline, what periods are undocumented? What corroborating evidence would strengthen the claim for each persecution event?" Immigration practitioners handling asylum cases often work with limited budgets and fragmented records, making AI-assisted investigation organization particularly valuable for building coherent narratives from incomplete documentation.

AI accelerates investigations by turning raw data into structured timelines and relationship maps. It helps identify gaps in the record and patterns of potential misconduct. It supports interview preparation and report drafting. The investigator's role is to direct the inquiry, assess credibility, and make findings; AI handles the data organization that makes those judgments possible.

9.7 Mini-Lab: Deposition Prep Packet

Mini-Lab: Deposition Prep Packet

Objective: Use AI to create a comprehensive deposition preparation packet for an adverse witness, including an outline, exhibit list, and impeachment cheat sheet.

A deposition prep packet template is available on the companion site for download.

Time Required: 60–75 minutes

What You'll Need

- The case scenario from the Chapter 7 Mini-Lab (Greenfield v. Digital Solutions)
- The research memo you produced in Chapter 7
- Access to an AI tool

Step 1: Define the Deposition Goal

You are deposing the owner of Digital Solutions, the tenant who abandoned the property. Your goal is to establish that they abandoned the premises, received the notice (or evaded it), and that the computers left behind belonged to them, not third parties.

Step 2: Generate the Outline Structure

Prompt AI to create a deposition outline structure.

> *"Create a deposition outline for the owner of a commercial tenant in an abandoned property case. The goals are to establish abandonment, notice receipt, and ownership of left-behind items. The outline should cover: lease signing, rent payment history, decision to vacate, steps taken to move out, receipt of landlord notices, and ownership of the specific items left on site."*

Step 3: Develop Specific Modules

Select one key topic (e.g., Notice Receipt) and ask AI to flesh it out.

> *"Draft detailed questions for the 'Notice Receipt' section. The witness will likely deny receiving the mail. Include questions to establish their standard practice for checking mail, confirming the address was correct, and checking if they have any record of the notice. Include questions to lock them into a story before showing the certified mail receipt."*

Step 4: Create an Exhibit List

Ask AI to suggest exhibits based on the scenario.

> *"Based on the facts of the Greenfield case, what specific documents should I use as exhibits in this deposition? List them in the order they should be introduced and explain the purpose of each."*

Step 5: Build an Impeachment Cheat Sheet

Ask AI to anticipate evasive answers and prepare responses.

> *"The witness is likely to claim the computers belonged to customers and they had no right to move them. How should I question them to test this claim? What evidence would disprove it? Create a line of questioning to pin down exactly whose property it was."*

Step 6: Review and Refine

Review the AI-generated materials. Are the questions admissible? Is the tone appropriate? Do the exhibits match the record? Refine the packet to be a practical tool you could take into the deposition room.

Deliverable

A deposition prep packet containing:

1. A strategic outline structured by topic
2. Detailed question modules for key issues
3. An organized exhibit list
4. An impeachment/follow-up guide for anticipated defenses

A good outline isn't a script; it's a safety net. It ensures you don't forget to ask the crucial questions while leaving you free to listen and follow up. Use the AI-generated questions as a starting point, but trust your ears in the room.

Scenario Analysis Guide

After completing your deposition prep packet, use this guide to check your work.

Outline Coverage Check: Your outline should address at minimum: lease formation and payment history (establishing the contractual relationship), the decision to vacate and steps taken (establishing abandonment), all communications regarding the property after vacancy (establishing notice), and specific identification of each item left behind (establishing ownership). If your outline misses any of these areas, the deposition will have gaps the witness can exploit.

Impeachment Readiness Check: For the notice receipt issue, your packet should include questions that lock the witness into a story about their mail practices BEFORE revealing the certified mail receipt. If you show the receipt first, the witness adapts their story. The sequence matters. For the customer-computers claim, your packet should include questions about business records, customer intake forms, and repair logs that would exist if the computers genuinely belonged to third parties. Absence of these records undermines the claim.

Exhibit Sequence Check: Exhibits should build chronologically toward the key confrontation documents. The lease comes first (establishing the relationship), then payment records (showing default), then the notice documents (showing compliance attempts), then photos of the abandoned premises (showing the condition). Save the certified mail receipt for impeachment during the notice-receipt questioning.

Bonus Challenge

Prepare for the other side of the deposition table. Draft a witness preparation memo for Greenfield's property manager, who will be deposed by Digital Solutions' counsel. The memo should identify the likely areas of questioning (Was the notice properly served? Did Greenfield enter the premises lawfully? Were any items damaged during the property manager's inspection?), prepare the witness with truthful responses that don't volunteer harmful information, and flag documents the opposing counsel will likely use as exhibits. This exercise practices the witness preparation skills that complement the adverse deposition skills from the main lab

9.8 Practice Checklist and Prompt Pack

Litigation Workflow Checklist

A printable version of this checklist is available on the companion site for matter files.

Case Theory & Strategy

- ☐ Used AI to create a master timeline from facts/documents
- ☐ Brainstormed narrative themes and counter-narratives
- ☐ Stress-tested theory against "bad facts" identified by AI
- ☐ Used AI to anticipate opposing counsel's case theory Motion Practice
- ☐ Used AI to identify potential grounds for dismissal/judgment
- ☐ Framed Questions Presented with AI assistance
- ☐ Analyzed opposition brief with AI to identify weak points
- ☐ Simulated oral argument Q&A with AI persona Discovery
- ☐ Drafted document requests based on claims/defenses
- ☐ Used AI to review incoming requests for objections
- ☐ Drafted proportionality arguments for discovery disputes
- ☐ Set up AI-assisted review workflow (relevance/privilege)
- ☐ Verified privilege log descriptions generated by AI Deposition & Trial Prep
- ☐ Generated deposition outlines and exhibit orders
- ☐ Created impeachment cheat sheets for key witnesses
- ☐ Summarized deposition transcripts for motion/trial use
- ☐ Used AI to draft potential opening statement themes Prompt Pack for Litigation

Prompt Pack for Litigation

All prompts in this section are available as copy-paste templates on the companion site.

Case Theory Prompts

Document chronology

> Create a chronology of events from these documents. Highlight any inconsistencies between the email record and the witness interview notes.

Narrative themes

> Suggest three narrative themes for the plaintiff in this breach of contract case. For each theme, list the key facts that support it and the facts that undermine it.

Motion Strategy Prompts

Counterargument anticipation

> I am moving to dismiss on the grounds of lack of personal jurisdiction. What are the strongest counterarguments the plaintiff will likely raise? How can I pre-empt them in my opening brief?

Opposition brief summary

> Read this opposition brief. Summarize the three main arguments. For each, identify the factual premise and the primary legal authority cited.

Discovery Prompts

Interrogatory drafting

> Draft a set of interrogatories directed at the defendant's internal decision-making process regarding the product recall. Focus on who knew about the defect and when.

Objection drafting

> Review this request for production. Draft an objection based on undue burden, explaining that the request covers 10 years of data for a claim involving a single transaction.

Deposition Prompts

Question outline

> I am deposing the HR manager in a discrimination case. Create a question outline covering: the hiring process, the decision to fire the plaintiff, and the company's internal investigation. Suggest specific questions to establish pretext.

Transcript analysis

> Summarize this deposition transcript. Identify every instance where the witness admitted to communicating with the competitor.

Key Takeaways

1. Litigation is adversarial. AI helps you prepare for the opponent by anticipating their moves, stress-testing your theories, and identifying your vulnerabilities.
2. Case theory requires narrative construction. AI can synthesize facts into timelines and themes, helping you choose the most persuasive story.
3. Motion strategy benefits from predictive analysis. Use AI to frame issues, anticipate opposition arguments, and simulate judicial scrutiny.
4. Discovery relies on AI for scale. Use AI to draft requests, identify objections, and process large document volumes for relevance and privilege.
5. Deposition prep is about organization. AI helps structure outlines, organize exhibits, and plan for impeachment, freeing you to focus on the witness.
6. Investigations use AI to find the story. Timeline construction, entity mapping, and gap analysis transform scattered data into a coherent investigative record.
7. Verification is the constant. Whether it's a fact in a brief, a privilege call in discovery, or a question in a deposition, you must verify AI's suggestions against the record and the law.

Chapter Checkpoint

Before moving on, make sure you can:

☐ Use AI to generate a case timeline and identify narrative themes from a fact pattern

☐ Develop a motion strategy that anticipates and addresses opposing arguments

☐ Draft discovery requests and objections using AI assistance

☐ Describe the role of AI in relevance and privilege review and the limits of its authority

☐ Create a deposition outline and exhibit plan using AI

☐ Execute an investigation workflow to build a timeline and identify factual gaps

What's Next

We've covered the core legal work: research, drafting, and litigation. Chapter 10 turns to the business side of practice: client communications, intake, and business development. You'll learn how to use AI to explain complex legal concepts to clients, manage intake efficiently, and produce marketing content that builds your reputation. The principles of accuracy and professional responsibility remain central, but the audience shifts from courts and counterparties to the people you serve.

Key Terms

Case theory — The coherent narrative explaining why your client should prevail, integrating facts and law into a persuasive story. Must account for all undisputed facts, explain favorable facts, and neutralize unfavorable ones.

Motion strategy — Deciding which motions to file, when to file them, and how to frame the arguments to advance the overall case theory. Balances likelihood of success against cost and the strategic value of the ruling.

Discovery planning — Determining what evidence you need, what evidence the opponent needs, and how to obtain or protect that evidence efficiently. Includes drafting requests, anticipating disputes, and establishing protocols.

Relevance review — Determining which documents in a collection are responsive to discovery requests and relevant to the claims or defenses. AI accelerates this by classifying documents based on criteria you provide.

Impeachment preparation — Identifying specific prior statements or documents that contradict the witness's likely testimony. AI can map contradictions, creating a "cheat sheet" of page-and-line citations.

Chapter 10: Client Communications, Intake, and Business Development

Legal work exists to serve clients. The research, drafting, and litigation skills covered in previous chapters produce value only when they help real people and organizations navigate legal challenges. Client communications are where that value is delivered: explaining options so clients can make informed decisions, providing updates so clients understand where matters stand, and building the trust that effective representation requires. AI can enhance these communications significantly, helping you translate complex legal concepts into accessible language while maintaining the precision that accuracy demands. The client relationship, however, remains fundamentally human. AI assists the communication; you maintain the connection.

Client-facing work differs from the internal work product covered in earlier chapters. A research memo serves lawyers who share your vocabulary and analytical frameworks. A client letter serves people who may have no legal background and whose primary concern is understanding their situation, not appreciating legal nuance. This difference in audience changes everything about how you communicate. Clarity matters more than comprehensiveness. Answering the client's actual question matters more than demonstrating thoroughness. Tone matters because clients are often stressed, confused, or worried, and your communication either alleviates or compounds those feelings.

Intake processes shape everything that follows in a client relationship. The information gathered at intake determines whether you can identify the client's legal issues, assess conflicts, and provide competent representation. Incomplete intake leads to incomplete understanding, which leads to incomplete advice. AI can help extract and organize information from client narratives, spot issues that warrant further investigation, and ensure that intake captures what you need. But intake is also where clients form their first impressions of working with you. Efficiency that feels impersonal may win the engagement but lose the relationship.

Business development connects your capabilities to people who need them. Marketing content, thought leadership, and client development activities all benefit from AI assistance in ways that parallel the drafting workflows you've learned. The constraints differ: advertising rules impose requirements that don't apply to briefs, and authenticity matters because clients choose lawyers based on perceived fit, not just credentials. AI-generated content that sounds generic undermines the differentiation that effective business development requires.

What This Chapter Covers

- How to summarize client narratives and spot legal issues during intake, capturing essential information while building rapport with new clients
- Techniques for translating complex legal concepts into plain English without losing precision or creating misunderstanding
- Drafting client updates that balance completeness with accessibility, addressing what clients need to know without overwhelming them
- Creating marketing content and thought leadership that reflects your actual expertise while complying with advertising rules
- Considerations for chatbots and intake automation, including when automation serves clients and when it creates barriers
- A Mini-Lab producing a client update letter with appropriate verification, applying the chapter's techniques to a realistic scenario

- A practice checklist and prompt pack providing ready-to-use resources for client communication workflows

Chapter Summary

Intake is where client relationships begin and where the information foundation for representation is established. AI can help summarize client narratives by extracting key facts, dates, parties, and claims from often disorganized initial descriptions. Issue spotting during intake identifies potential legal claims, defenses, or complications that the client's narrative suggests, even when the client hasn't recognized these issues themselves. The balance between efficiency and relationship-building requires attention: clients want to feel heard, and intake processes that feel rushed or impersonal undermine trust even when they capture necessary information.

Plain-English communication bridges the gap between legal complexity and client understanding. Clients need to understand their situations well enough to make informed decisions, but they don't need to understand every legal nuance. AI can help translate legal concepts into accessible language, generate analogies and examples that clarify abstract principles, and identify jargon that should be replaced with everyday terms. The precision imperative remains: simplification that creates misunderstanding serves no one. Verification of AI-generated explanations ensures that accessibility hasn't sacrificed accuracy.

Client updates keep clients informed about matter progress, upcoming events, and decisions they need to make. Effective updates answer the questions clients actually have rather than the questions lawyers think they should ask. AI can help structure updates to address status, next steps, timeline, and action items in formats clients can quickly understand. Updates should calibrate detail to client preferences: some clients want comprehensive information, while others want bottom-line summaries with detail available on request. Learning client preferences and adapting communications accordingly strengthens relationships over time.

Marketing content and thought leadership present your expertise to potential clients. AI can accelerate content creation by drafting articles, blog posts, presentations, and other materials that you then refine to reflect your voice and actual knowledge. Advertising rules constrain what you can say and how you can say it, and compliance verification must occur before publication. Authenticity matters because clients choose lawyers partly based on perceived fit, and generic AI-generated content undermines the differentiation that attracts clients to you specifically.

Chatbots and intake automation can improve accessibility and efficiency when implemented thoughtfully. Automated systems can gather preliminary information, answer frequently asked questions, and route inquiries appropriately. They can also frustrate potential clients who want human interaction or whose situations don't fit automated workflows. The decision to automate should consider what clients actually want, not just what's efficient for your practice. When automation is appropriate, implementation requires attention to accuracy, confidentiality, and the limitations of automated responses to legal questions.

Throughout client-facing work, the verification imperative applies with particular force. Inaccurate communication to clients can create liability, damage relationships, and undermine the trust that effective representation requires. AI-assisted client communications require the same review and verification that any work product requires, adapted to the particular risks of client-facing content. The audience may be less sophisticated than a court, but the stakes of accuracy are just as high.

10.1 Intake: Summarizing Client Narratives and Spotting Issues

Intake is where representation begins. Before you can help a client, you must understand their situation: what happened, who's involved, what they want, and what legal issues their circumstances raise. This understanding comes primarily from what clients tell you, which is often disorganized, incomplete, emotionally charged, and focused on what matters to them rather than what matters legally. Your job during intake is to extract the information you need while making the client feel heard and building the trust that effective representation requires. AI can help with the extraction; the relationship-building remains yours.

Client narratives rarely arrive in legally useful form. A client describing a business dispute might spend twenty minutes on the personal betrayal they feel and thirty seconds on the contract terms that will determine the outcome. A client describing an injury might focus on their pain and disruption while omitting details about when they first noticed symptoms or what warnings they received. Clients organize their stories around emotional significance, not legal relevance. They emphasize what bothers them, not what matters for their claims or defenses.

The intake challenge is extracting legally relevant information without making clients feel interrogated or dismissed. Interrupting a client's narrative to ask about statute of limitations may be legally necessary but relationally damaging. Letting the narrative run without structure may build rapport but miss essential information. Skilled intake balances these concerns, creating space for clients to tell their stories while ensuring you learn what you need to know.

AI can assist intake by processing client narratives after the conversation to extract and organize key information. Rather than taking notes that divide your attention during the client meeting, you can focus on the client and then use AI to summarize what was discussed. A recorded intake session, with appropriate client consent, can be transcribed and analyzed to produce structured summaries that organize the narrative into legally useful categories.

Post-conversation summarization prompts ask AI to extract specific categories of information. "Based on the following intake transcript, extract and organize: key facts in chronological order, all parties mentioned with their roles, all dates mentioned, potential legal claims suggested by the facts, documents the client mentioned or should be asked to provide, and follow-up questions to clarify gaps in the narrative." This extraction produces a structured summary from an unstructured conversation, capturing what you need while you focus on the client during the meeting itself.

If you record intake sessions for later AI processing, ensure you have clear client consent and that your confidentiality obligations are satisfied. Recordings should be handled with the same care as any privileged material. Delete recordings after processing if retention isn't necessary.

Real-time intake assistance is possible but requires careful implementation. AI tools that listen to conversations and provide real-time suggestions or identify gaps can help ensure comprehensive intake. However, visible attention to a screen during a client meeting may signal that you're not fully present. If you use real-time AI assistance, position it so you can glance at suggestions without breaking engagement with the client. Better yet, reserve AI assistance for post-conversation processing and rely on your own attention during the meeting.

Intake questionnaires can gather preliminary information before the initial conversation. Online forms that collect basic facts, documents, and narrative descriptions give you a starting point before you meet the client. AI can process questionnaire responses to identify issues and suggest areas for deeper exploration. The initial conversation then builds on this foundation rather than starting from zero, making meetings more efficient while preserving time for relationship-building.

AI processing of intake questionnaires might identify issues the client hasn't recognized. "The client describes a fall at a store two years ago but mentions they 'never got around to doing anything about it.' Flag potential statute of limitations concern for immediate discussion." This issue spotting ensures that time-sensitive matters receive prompt attention even when clients don't appreciate their urgency.

Statute of limitations issues that arise during intake require immediate attention. Malpractice claims frequently arise from lawyers who took matters without recognizing that limitations had run or was about to run. AI-assisted intake should specifically flag any timing issues that questionnaire responses or initial narratives suggest.

Issue spotting during intake goes beyond limitations concerns. A client describing a contract dispute might mention facts that suggest fraud, breach of fiduciary duty, or other claims beyond the contract itself. A client describing an employment situation might mention details that suggest retaliation, discrimination, or wage violations they haven't identified. Comprehensive issue spotting ensures you identify all potential claims and defenses, not just the ones the client came in asking about.

AI can assist issue spotting by analyzing client narratives against patterns of legal claims. "Based on this intake narrative involving an employment termination, what potential legal claims do the facts suggest? Consider wrongful termination, discrimination, retaliation, breach of contract, defamation, and any other claims the facts might support. For each potential claim, identify the facts that support it and the facts that would need to be established through further investigation."

Issue spotting must be followed by verification through your own legal analysis. AI may identify claims that don't actually apply in your jurisdiction or that require elements the facts don't support. AI may miss claims that require specialized knowledge to recognize. The AI-generated issue list is a starting point for your own analysis, not a substitute for it. Review the suggested issues against your knowledge of applicable law and the specific facts presented.

Conflict checking information must be captured during intake before you learn confidential details that could create conflicts with existing clients. AI can help ensure intake captures all information needed for thorough conflict checking: all parties mentioned in the narrative, related entities, opposing parties, witnesses, and anyone else whose involvement might create a conflict. Prompting for this extraction ensures you don't miss parties mentioned in passing who might later prove to be adverse to existing clients.

Practice Area Note: Personal Injury PI intake has specific data-gathering challenges beyond the general framework described here. You need medical authorization forms signed immediately to preserve access to treatment records. You need insurance information for all parties — the client's health insurance, the defendant's liability coverage, and any applicable UM/UIM coverage. You need accident details with specificity (date, time, location, weather, witnesses) that clients may not think to provide. And you need to document the client's condition at intake as a baseline, because memories of pre-injury function fade quickly. Practice-specific intake templates that capture these PI essentials prevent gaps that weaken claims later. Client document collection should begin during intake when possible. Clients often have relevant documents: contracts, correspondence, photographs, medical records, or other materials that illuminate their situations. Requesting these documents during intake, and setting expectations about what you'll need, starts the process of building a complete factual record. AI can help identify what documents a client's narrative suggests they should have, generating document request lists tailored to the specific situation.

Organizing intake information into usable form serves both immediate and long-term purposes. Immediately, organized intake information supports your initial assessment, engagement decisions, and early case planning. Long-term, organized intake creates a record you can return to as the matter develops. AI can help structure intake information into templates that match how your practice manages matters, ensuring that intake feeds smoothly into ongoing matter management.

Develop intake templates tailored to your practice areas. A personal injury intake captures different information than a business formation intake. Practice-specific templates ensure comprehensive information gathering and allow AI processing to focus on what matters for each matter type.

Preliminary legal assessment based on intake information helps you evaluate whether to accept the matter and on what terms. AI can assist this assessment by analyzing the facts against potential claims, estimating the strength of the client's position, and identifying factors that affect case value or risk. This assessment is preliminary: it's based on the client's narrative without verification and without the depth of research that will follow engagement. But preliminary assessment informs the engagement decision and helps you discuss the matter realistically with the client.

Risk identification during intake surfaces factors that might affect your willingness to take the matter or how you structure the engagement. Red flags might include limitations issues, prior lawyers who withdrew, client expectations that seem unrealistic, or facts suggesting the client's story may not be entirely accurate. AI can help identify these risk factors from intake information, but evaluating what they mean and whether they should affect engagement requires your judgment.

> **Risk Note** Pay attention to why the client is seeking new counsel. If they have had prior lawyers who withdrew or were terminated, understand why before accepting the engagement. Prior counsel problems may reflect difficult client behaviors, unrealistic expectations, or matter complexities that will affect your representation too.

Communication preferences should be established during intake. How does the client prefer to receive updates: email, phone, text, letter? How frequently do they want to hear from you? How much detail do they want in communications? Establishing these preferences early allows you to tailor communications to client needs rather than your defaults. AI can help you track and implement communication preferences across your client base.

Engagement letter preparation flows from intake. The information gathered during intake determines the scope of representation, the fee arrangement, and the terms that should be addressed in the engagement letter. AI can assist engagement letter drafting by incorporating intake-specific details into templates, but engagement letters require careful review to ensure they accurately reflect what you're agreeing to do and on what terms.

Fee discussions benefit from the information gathered during intake. Understanding the matter's complexity, likely duration, and probable outcomes helps you quote fees that are fair to both you and the client. AI can help estimate matter scope based on intake facts, comparing to similar matters you've handled. These estimates inform fee discussions without committing to specific outcomes that depend on facts not yet known.

Fee disputes are a leading source of bar complaints. Clear fee arrangements established during intake, documented in engagement letters, and reinforced through ongoing communication prevent most disputes. Never assume the client understands fee arrangements; explain them clearly and confirm understanding.

Client expectations should be managed from the start. Clients often arrive with unrealistic expectations about timelines, outcomes, or what legal representation involves. Intake is the time to address these expectations: explaining how long matters typically take, what outcomes are realistic, and what the client's role and responsibilities will be. AI can help identify common misconceptions based on the matter type and generate explanations that address them.

Follow-up after intake ensures you have complete information before proceeding. Intake conversations rarely capture everything you need. Follow-up questions, document requests, and clarification of ambiguities should be organized and communicated promptly after intake. AI can generate follow-up lists based on gaps identified in the intake narrative, ensuring nothing falls through the cracks.

Documentation of intake creates a record that may be important later. What the client told you, when they told you, and what advice you provided are all potentially relevant if questions arise about what you knew and when. Intake notes, questionnaire responses, and any recordings should be preserved according to your document retention policies.

Dictate or write detailed intake notes immediately after client meetings while your memory is fresh. Note not just what the client said but your impressions, concerns, and preliminary assessment. These contemporaneous notes are more reliable than later recollections and create a valuable record.

The human element of intake cannot be automated away. Clients choosing a lawyer are making a significant decision based partly on how they feel about you. The intake process is where they form impressions that affect whether they engage you, trust you, and follow your advice. Efficient information extraction serves representation quality, but relationship-building serves representation effectiveness. Both matter. AI helps with extraction; you provide the human connection that clients need.

Balancing efficiency and rapport requires conscious attention to how intake feels from the client's perspective. A client who provides extensive information through a questionnaire, then sits through an interview that merely verifies that information, may feel their time was wasted. A client who receives immediate follow-up and document requests may feel they've hired an efficient lawyer or may feel overwhelmed. Design intake processes that serve both information needs and relationship needs, recognizing that these goals sometimes tension with each other.

Intake establishes the information foundation for representation while building the client relationship. AI can help extract and organize key facts from client narratives, identify potential legal issues the client may not have recognized, flag risk factors and timing concerns, and generate follow-up questions and document requests. Issue spotting must be followed by your own legal analysis to verify suggested claims apply in your jurisdiction. Conflict checking information must be captured before learning confidential details. Client communication preferences and expectations should be established early. The human element of intake, the relationship-building that creates trust, cannot be automated and must receive attention alongside information extraction. Design intake processes that serve both efficiency and connection.

10.2 Plain-English Explanations Without Losing Precision

Clients need to understand their legal situations well enough to make informed decisions. They don't need to understand every nuance of doctrine or procedure, but they need to grasp what's happening, what their options are, and what the consequences of different choices might be. Translating complex legal concepts into language clients can understand is a core lawyering skill, one that many lawyers struggle with because legal training emphasizes precision over accessibility. AI can help bridge this gap, generating explanations that communicate clearly while you ensure those explanations remain accurate.

The translation challenge is real. Legal concepts often don't have everyday equivalents. "Jurisdiction" doesn't mean the same thing to lawyers and non-lawyers. "Discovery" sounds like finding something new, not formal information exchange. "Motion" sounds like physical movement, not a request to a court. Even when words have ordinary meanings, their legal meanings may differ. "Consideration" in contract law isn't the same as being considerate. "Party" isn't a celebration. The vocabulary gap between lawyers and clients creates communication barriers that must be actively overcome.

Beyond vocabulary, legal reasoning follows patterns unfamiliar to most clients. Lawyers think in elements, burdens, and standards. Lawyers distinguish between holdings and dicta, between mandatory and persuasive authority, between questions of law and questions of fact. Clients think in terms of right and wrong, fair and unfair, what happened and what should happen. Effective communication bridges these different frameworks, helping clients understand legal analysis without requiring them to think like lawyers.

AI excels at translation tasks, making it well-suited for generating plain-English explanations. Provide AI with a legal concept and ask for an explanation suitable for a non-lawyer audience. "Explain the concept of summary judgment in terms a client with no legal background would understand. Use everyday analogies where helpful. Avoid legal jargon or explain any legal terms that must be used." AI generates explanations you can review, refine, and use.

Audience calibration improves AI explanations. Different clients have different backgrounds and different capacities for complexity. A business executive may understand more sophisticated explanations than someone encountering the legal system for the first time. A client in a technical field may appreciate precise terminology while a client in a creative field may prefer narrative explanations. Specifying your audience helps AI calibrate appropriately. "Explain what it means that the contract has an integration clause. The client is a small business owner who has signed contracts before but has no legal training. Keep the explanation under 100 words and use a practical example if helpful."

Before sending any AI-generated explanation to a client, read it as if you were the client. Does it actually answer the question the client would have? Does it make sense without background knowledge? Does it create any new confusions? This perspective-taking catches problems that technically accurate explanations may still have.

Analogies and examples make abstract concepts concrete. A client may not understand "joint and several liability" in the abstract, but they understand that if three people share responsibility for a debt, the creditor can collect the whole amount from any one of them and leave that person to sort out shares with the others. AI can generate analogies, though the analogies require review to ensure they fit the specific legal context and don't create misleading impressions.

Analogy generation prompts might ask for multiple options. "Generate three different analogies for explaining the burden of proof in a civil case versus a criminal case. The client understands they have a civil case but keeps asking why the other side wouldn't 'go to jail' for what they did." Multiple analogies give you options to choose from or combine, and they may surface different aspects of the concept that a single analogy might miss.

The precision imperative limits how far simplification can go. Some legal concepts are genuinely complex, and oversimplification creates misunderstanding. A client who understands "you'll probably win" when the actual situation is "you have a strong argument but significant uncertainty remains" has been misled even if the communication was clear. Plain English must remain accurate English. Simplification that sacrifices accuracy serves no one.

Watch for AI explanations that overstate certainty or oversimplify in ways that could create client misunderstanding. AI tends toward confident, clear statements even when the law is uncertain or the situation is complex. Review explanations to ensure they accurately convey the level of certainty the situation actually supports.

Hedging and qualification require particular attention in plain-English communication. Lawyers instinctively hedge because law is full of uncertainty. "It depends," "typically," "generally," "in most cases," and similar qualifications reflect legal reality. But excessive hedging in client communications can be paralyzing. Clients need guidance, not endless contingencies. The skill is hedging appropriately: communicating uncertainty where it genuinely exists while providing clear guidance where the law is settled.

AI can help calibrate hedging by asking for explanations at different confidence levels. "Explain our position on the contract interpretation issue. First, give me a confident version that emphasizes the strength of our argument. Second, give me a hedged version that emphasizes the uncertainties. Third, give me a balanced version that conveys our strong position while acknowledging realistic risks." Comparing these versions helps you find the right calibration for your specific client and situation.

Lawyers who overstate certainty to clients may win engagements but lose trust when predictions don't pan out. Lawyers who hedge everything may seem uncommitted or unknowledgeable. The balance matters for credibility and for helping clients make informed decisions. Communicate confidence where it's warranted and uncertainty where it exists.

Explaining procedures helps clients know what to expect. Clients often have no framework for understanding how legal matters progress. Explaining what discovery is, how motions work, what happens at hearings, and how trials proceed helps clients understand status updates and feel oriented in an unfamiliar process. AI can generate procedural explanations at varying levels of detail, from high-level overviews to step-by-step guides. "Explain to a client what will happen during their deposition. They've never been deposed before and are nervous. Include what they should expect, how to prepare mentally, and what their role is. Keep it reassuring without minimizing the importance of the event."

Visual aids supplement written explanations for some clients. Timelines, flowcharts, and diagrams can communicate processes and relationships more effectively than prose for visually-oriented clients. AI can help generate text descriptions of visual aids that you or a designer can then create. "Describe a simple flowchart showing the stages of a civil lawsuit from filing through trial, with typical timeframes for each stage." The description guides visual aid creation even when AI can't create the visual itself.

Explaining options and tradeoffs is essential for informed client decision-making. Clients often must choose between paths that each have advantages and disadvantages: settle or litigate, accept terms or negotiate further, pursue a claim or let it go. Explaining these tradeoffs clearly, without pushing clients toward your preferred outcome, respects client autonomy and protects you from later claims that you didn't present all options.

Options presentation prompts can generate balanced explanations. "Explain to the client the pros and cons of accepting the current settlement offer versus continuing to litigate. Present both options fairly, including: likely outcomes, risks, costs, timeline, and emotional considerations. Let the client see the full picture without telling them what to do."

Recommendation framing when you do offer a recommendation should be clear about what's your professional judgment versus what's the client's decision. "Based on my assessment of the risks and likely outcomes, I recommend accepting the settlement. But this is your decision. Here are the factors I think you should consider..." This framing provides guidance while respecting that certain decisions belong to the client.

Document that you presented options and that the client made the ultimate decision on significant strategic choices. Client decisions to reject settlement offers, to pursue or not pursue claims, and to accept or reject proposed terms should be confirmed in writing. This documentation protects both you and the client.

Cultural and language considerations affect communication with some clients. Clients from different cultural backgrounds may have different expectations about lawyer-client relationships, different comfort with directness, and different frameworks for understanding legal systems. Clients whose first language isn't English may need explanations calibrated to their language proficiency. AI can help adapt communications to these considerations when you provide appropriate guidance.

Emotional context shapes how explanations are received. A client receiving bad news processes information differently than a client receiving good news. A client who is frightened, angry, or grieving may not absorb detailed explanations. Meeting clients where they are emotionally, acknowledging their feelings before diving into legal analysis, helps ensure your explanations are actually received.

AI can help draft communications that acknowledge emotional dimensions while providing necessary legal information. "Draft an email explaining to the client that their motion was denied. The client was very hopeful about this motion and will be disappointed. Acknowledge the disappointment, explain what the denial means for the case, and explain the path forward. Keep the tone supportive without minimizing the setback."

Communications that ignore the emotional dimensions of legal matters often fail to achieve their purpose. A client who feels unheard won't process the information you're providing. Acknowledging emotions before providing analysis is not wasted words; it's essential for effective communication.

Verifying client understanding confirms that your explanation worked. After providing an explanation, check that the client actually understood. "Does that make sense?" is a weak check because clients often say yes to avoid seeming uninformed. Better approaches ask clients to explain back what they understood or to identify what questions remain. "What questions do you have about how discovery works?" assumes questions exist rather than asking if they do.

Following up complex explanations with written summaries reinforces understanding and creates a record. After a phone call or meeting where you explained something complex, send a follow-up email summarizing the key points. This summary gives clients something to refer back to and confirms in writing what you communicated. AI can help draft these follow-up summaries based on your notes from the conversation.

> Practice Point After explaining complex concepts verbally, send a brief written summary capturing the key points. This reinforcement improves client understanding and creates documentation of what you communicated. The summary need not be elaborate: "As we discussed today, here are the key points about the discovery process..."

Technical terms you must use should be defined when first introduced. Some concepts don't have good plain-English substitutes, or using the proper term helps clients understand documents and proceedings they'll encounter. When you must use technical terms, define them clearly. "The other side filed a motion for summary judgment. That's a request asking the judge to decide the case, or part of it, without a trial, based on their argument that the facts aren't really in dispute."

Avoiding condescension while simplifying requires attention to tone. Explaining concepts simply should not feel like talking down to clients. Adults who aren't lawyers are still intelligent adults. Frame explanations as clarifying an unfamiliar domain, not as accommodating limited capacity. "Legal terminology can be confusing if you haven't encountered it before, so let me explain what this means" respects the client while acknowledging the vocabulary gap.

Written explanation style differs from how you might explain verbally. Written communications can be scanned, skipped around in, and referenced later. Structure written explanations with clear organization, short paragraphs, and perhaps headings or bullets for longer communications. Frontload the most important information since clients may not read to the end. AI-generated explanations may need restructuring to match how clients actually read.

Iterating explanations when the first attempt doesn't work requires patience and flexibility. If a client doesn't understand your explanation, try a different approach rather than repeating the same explanation louder and slower. Different analogies, different framings, different levels of detail may succeed where the first attempt failed. AI can quickly generate alternative explanations when you need a fresh approach. "I tried to explain the statute of limitations to the client using the 'deadline to file' framing, but they keep asking why they can't sue someone for something that clearly happened. Generate an alternative explanation that emphasizes why the law imposes time limits and what purposes those limits serve."

Building client legal literacy over time benefits the representation. Clients who develop some understanding of legal concepts become better clients: they understand status updates more easily, they can participate more meaningfully in strategic decisions, and they have more realistic expectations. Ongoing communication should build on prior explanations rather than starting from zero each time.

Translating legal concepts into plain English serves client understanding and informed decision-making. AI can generate explanations calibrated to specific audiences, provide analogies that make abstract concepts concrete, and offer hedging variations that help you find the right confidence level. The precision imperative limits simplification: plain English must remain accurate English. Explaining procedures, options, and tradeoffs helps clients navigate legal matters and make decisions. Verify client understanding through techniques beyond "does that make sense?" Follow up verbal explanations with written summaries. Attend to emotional context and cultural considerations. Avoid condescension while acknowledging that legal vocabulary is unfamiliar. Iterate when explanations don't succeed on the first attempt.

10.3 Client Updates: Status, Next Steps, and Risk Framing

Clients want to know what's happening with their matters. This desire for information is fundamental, yet lawyers frequently underperform at keeping clients informed.

Busy lawyers let updates slide. Lawyers who don't have news don't reach out. Lawyers assume clients understand silence means things are proceeding normally. Meanwhile, clients wonder what's happening, whether their lawyer is actually working on their case, and whether they should call to check in. Proactive, regular client updates prevent this dynamic, building trust and preventing the anxiety that uninformed clients experience. AI can help you maintain update discipline even when your attention is pulled in many directions.

The update paradox is that clients most need communication when there's least to report. During active periods with hearings, negotiations, or significant developments, updates flow naturally because there's news to share. During dormant periods when the matter is waiting for a court ruling, waiting for the other side to respond, or simply proceeding through routine stages, updates require deliberate effort. These dormant periods are precisely when clients worry most: Is anything happening? Did my lawyer forget about me? Should I be doing something? A brief update explaining that you're waiting for X and expect to hear by Y provides enormous reassurance even when the substantive content is minimal.

Regular update schedules prevent updates from falling through the cracks. Rather than updating when you have news, update on a regular schedule regardless of whether significant developments have occurred. Monthly updates for routine matters, weekly updates during active phases, and immediate updates for significant developments creates a rhythm that clients can rely on. AI can help you maintain this schedule by drafting routine updates that you can quickly review and send.

The Client Communication Trust Model

What AI can touch and what must remain lawyer-to-client

NEVER AI-GENERATED

Legal advice. Strategic recommendations. Decisions about client matters. Anything that commits the lawyer's professional judgment. These are **non-delegable.**

AI-ASSISTED, LAWYER-REVIEWED

Client status updates. Explanations of legal concepts. Drafts of substantive correspondence. AI drafts; **lawyer reviews** and **takes responsibility** before sending.

AI-APPROPRIATE

Scheduling. Administrative updates. Document requests. Meeting confirmations. Routine acknowledgments. Low-sensitivity communication that **does not require legal judgment.**

The closer you get to legal judgment, the less AI belongs in the communication.

Clients trust their lawyer, not their lawyer's tools.

Figure 10.1 — The Client Communication Trust Model

Structured updates address the questions clients actually have. Most clients want to know: What's the current status? What happens next? When will things happen? What do I need to do? Structuring updates around these questions ensures you address client concerns rather than providing information you find interesting but clients find irrelevant. AI can help draft updates that follow this structure consistently.

A status-next-timeline-action framework organizes update content effectively: - Status: Where things stand right now. "The defendant has filed their answer, and we're now in the discovery phase." - Next steps: What will happen next and who does it. "We'll be sending written discovery requests to the defendant this week. They'll have 30 days to respond." - Timeline: When things are expected to occur. "We expect their responses in early April. The court has set a discovery cutoff of July 15 and a trial date of October 3." - Action items: What the client needs to do. "Please gather any emails from January through March regarding the contract negotiations and send them to me by Friday." Adopt a consistent update format that clients learn to expect. When clients know that updates always cover status, next steps, timeline, and action items, they can quickly find the information they need. Consistency also makes updates faster for you to produce because you're working from a familiar structure.

Action items deserve particular emphasis because client inaction can stall matters. When you need something from the client, make the request unmistakably clear. State what you need, why you need it, and when you need it by. Don't bury action items in lengthy paragraphs. Consider highlighting them visually or listing them separately. AI can help ensure updates clearly identify client action items and explain their importance.

Calibrating detail to client preferences serves clients better than one-size-fits-all communication. Some clients want comprehensive information about everything happening in their matter. Others want bottom-line summaries and prefer to ask questions if they want more detail. Learning client preferences early, through direct inquiry or observation, allows you to tailor updates appropriately. AI can help generate updates at different detail levels from the same underlying information.

Communication failures are a leading source of client complaints to bar associations. Most communication complaints aren't about substantive failures but about clients feeling uninformed and ignored. Regular, proactive updates prevent most communication complaints regardless of case outcome.

Risk framing in updates requires careful calibration. Clients need to understand risks to their matters, but how you present those risks affects their ability to process the information and make good decisions. Gratuitous alarm about risks that are unlikely or that clients can't affect creates anxiety without serving any purpose. Downplaying significant risks leaves clients unprepared for adverse developments. Balanced risk communication acknowledges realistic risks while maintaining appropriate perspective.

When reporting adverse developments, context matters. A motion loss might be devastating or merely a procedural setback depending on circumstances. Provide context that helps clients understand what the development actually means. "The court denied our motion to dismiss, which means the case will proceed to discovery. This is disappointing but not unexpected; these motions have a low success rate, and we preserved the arguments for appeal if needed. The merits of your claims remain strong, and discovery will allow us to develop the evidence to support them."

> **Practice Point** Never delay communicating adverse developments. Clients who learn bad news late, especially if they learn it from someone other than you, lose trust that is hard to rebuild. Deliver bad news promptly, clearly, and with context that helps clients understand what it means.

Good news deserves context too. A motion victory, a favorable settlement offer, or a strong deposition performance are worth celebrating, but clients should understand what good developments do and don't mean for the ultimate outcome. "The defendant made a settlement offer of $75,000. This is a meaningful offer that suggests they're concerned about their exposure. It's below what I believe the case is worth, but it's a serious opening. Let's discuss whether you want to counter or continue toward trial."

Proactive problem identification builds trust more than hiding problems until they become unavoidable. If you see a potential issue developing, tell the client early. "I want to flag a concern I've identified. The contract has an arbitration clause that might require us to arbitrate rather than litigate. I'm researching whether arguments exist to avoid it, and I'll update you on what I find. If we end up in arbitration, there are both advantages and disadvantages I should explain."

Explaining delay maintains client patience when matters take longer than expected. Legal matters frequently take longer than anyone anticipates. Courts are slow. Opponents delay. Complexity emerges. When timelines slip, explain why rather than letting clients wonder what happened to expected events. "The defendant requested and received an extension to respond to our discovery requests. This is routine and we didn't object because it preserves our ability to request extensions if we need them. Their responses are now due April 15 instead of March 30."

Cost updates keep clients informed about financial dimensions. If billing approach allows, include billing status in regular updates. If significant expenses are anticipated, provide advance notice so clients can plan. Surprises about fees create conflict even when the fees are reasonable. "Through February, fees total $12,500, which is within the estimate I provided at engagement. Discovery will increase activity significantly; I expect March through May to be the most intensive billing period, with fees likely in the $6,000-8,000 per month range."

> **Risk Note** Never surprise clients with invoices that exceed what they expected. If a matter is costing more than estimated, communicate that before the invoice arrives. Clients who are warned can adjust expectations; clients who are surprised feel ambushed.

Response to client inquiries should be prompt even when you don't have complete information. A client who emails asking about status deserves acknowledgment even if you need to research before providing a full answer. "Got your email. I need to check on the court's ruling and will get back to you tomorrow with a full update." Prompt acknowledgment shows responsiveness; full answers can follow when you have them.

Communication channel preferences affect how clients receive updates. Some clients check email constantly; others rarely. Some clients prefer phone calls; others find them intrusive. Some clients want text messages for time-sensitive matters. Match your communication to client preferences, and confirm those preferences rather than assuming. "What's the best way to reach you for routine updates versus time-sensitive matters?"

Younger clients often prefer text and messaging apps over email and phone. Older clients often prefer traditional channels. Don't assume; ask. And remember that certain communications, like advice requiring documentation, may warrant more formal channels regardless of preference.

Documentation of client communications creates records that protect both you and the client. What you communicated, when, and how the client responded may be relevant if questions later arise. Document significant updates and client instructions in writing even when the initial communication was verbal. "Per our call this morning, you've decided to reject the settlement offer and proceed to trial. I'll prepare for the next steps and be in touch about scheduling your deposition preparation."

Multiple contact updates serve clients with multiple stakeholders. Corporate clients often have several people who need to know about matter status: general counsel, the business person responsible for the relationship, finance contacts for billing, and others. Understanding who needs what information and ensuring appropriate distribution serves these clients well. AI can help adapt updates for different stakeholders from a single underlying update.

Client portals and technology solutions can streamline update delivery for practices with many matters. Portals where clients can check status, view documents, and access billing information provide 24/7 access without requiring lawyer time for routine inquiries. These systems complement rather than replace personalized updates, handling routine information access while you focus on substantive communication.

> **Key Point** Regular client updates prevent the anxiety that uninformed clients experience and build trust that sustains the relationship. Structured updates address what clients actually want to know: status, next steps, timeline, and action items. Calibrate detail level to client preferences. Frame risks appropriately, neither alarming unnecessarily nor downplaying genuine concerns. Communicate adverse developments promptly with context. Provide cost updates before bills surprise clients. Respond promptly to client inquiries even when full answers require research. Document significant communications. Close matters with appropriate final communications. Update templates accelerate routine communications while ensuring completeness. AI can help draft updates consistently, but the relationship judgment about what clients need remains yours.

If you use client portals or automated status tools, ensure clients know how to access them and actually do so. Technology that clients don't use provides no benefit. Some clients will embrace portals; others will ignore them and call you anyway. Know which type each client is.

Closing communications when matters conclude shouldn't be overlooked. Final updates should summarize what happened, explain any ongoing obligations or monitoring needs, discuss document retention, and invite future contact. Matters that dribble to conclusion without clear closing leave clients uncertain about status. "With the settlement finalized and payment received, this matter is concluded. I'll retain the file for [retention period] per firm policy. Please don't hesitate to reach out if questions arise or if I can assist with other matters."

Update templates accelerate routine communications while ensuring completeness. Develop templates for common update scenarios: monthly status updates, motion filing notifications, settlement offer communications, adverse development reports. AI can help create these templates and then populate them with matter-specific information for each use. Templates ensure you don't forget important elements while allowing customization for specific circumstances.

10.4 Marketing Content and Thought Leadership

Legal marketing has evolved beyond yellow pages ads and firm brochures. Clients now expect to find information about lawyers online: websites that explain practice areas, articles that demonstrate expertise, social media presence that shows engagement with relevant topics, and profiles that establish credentials. Thought leadership, the creation of content that positions you as a knowledgeable authority, has become a significant business development tool. AI can accelerate content creation dramatically, but the efficiency gains come with risks: generic content that fails to differentiate, compliance issues with advertising rules, and authenticity concerns when AI-generated material doesn't reflect actual expertise.

The business case for content creation is straightforward. Potential clients searching for lawyers often evaluate expertise through available content. An article explaining recent developments in your practice area signals current knowledge. A guide helping potential clients understand their options demonstrates client focus. Regular blog posts show ongoing engagement with the field. This content serves dual purposes: it helps potential clients find you through search engines, and it helps them choose you by demonstrating relevant expertise.

The time barrier has historically limited content creation. Lawyers who could benefit from thought leadership often don't create content because they lack time. The choice between billable work and marketing content usually favors billable work. AI changes this calculus by dramatically reducing the time required to produce content. A blog post that might have taken three hours to draft can start with an AI-generated first draft produced in minutes, with lawyer time focused on refinement rather than initial creation.

AI-assisted content creation follows the drafting workflow from Chapter 8, adapted for marketing purposes. Instructions define what the content should accomplish, who the audience is, and what message you want to convey. First draft generation produces raw material that captures the core content. Revision transforms generic AI output into material that reflects your voice and actual expertise. Verification ensures accuracy and compliance before publication.

Content strategy should precede content creation. Before generating individual pieces, consider what content would best serve your practice development goals. What topics does your target audience care about? What questions do potential clients frequently ask? What developments in your practice area create opportunities to demonstrate expertise? What formats work best for your audience: blog posts, articles, guides, videos, podcasts? AI can help brainstorm content ideas, but the strategic direction should come from your understanding of your practice and market.

Develop a content calendar that plans topics in advance. A calendar prevents the common pattern of sporadic content creation during slow periods followed by silence during busy periods. AI makes content creation efficient enough that you can maintain a consistent schedule even during demanding times.

Topic selection should balance what you want to talk about with what your audience wants to learn. Lawyers often want to write about interesting legal developments that may not interest potential clients. Clients often want practical guidance that lawyers find too basic to discuss. Effective content bridges this gap, addressing topics that both demonstrate your expertise and serve audience needs. "Recent Changes to Employment Discrimination Law" serves lawyers; "What to Do If You Think You've Been Discriminated Against at Work" serves potential clients.

AI can help identify topics by analyzing what questions people are asking in your practice area. "What are the most common questions individuals have about personal injury claims? What information do they search for online? What misconceptions do they commonly hold?" This analysis surfaces topics that serve audience needs, which you can then develop in ways that demonstrate your expertise.

Marketing content should be accurate, but it's marketing, not legal advice. Draw clear lines between general educational information and advice specific to any reader's situation. Include appropriate disclaimers. Content that creates implied attorney-client relationships or provides advice without proper engagement creates risk.

Prompt design for marketing content should specify audience, purpose, and voice. Generic prompts produce generic content. Specific prompts produce content closer to what you need. "Write a 600-word blog post explaining the recent Supreme Court decision in [case name] and what it means for employers. The audience is HR professionals and business owners, not lawyers. Focus on practical implications rather than legal nuance. Use clear, accessible language. The tone should be helpful and informative, not alarmist or promotional. Include a brief mention that employers should consult counsel for situation-specific guidance."

Voice preservation matters because authentic content requires your voice, not generic AI voice. AI-generated content often sounds similar across different authors, which undermines the differentiation that thought leadership should provide. Revision must inject your perspective, examples from your experience, and the particular way you explain concepts to clients. Providing AI with samples of your previous writing helps it match your voice.

Keep a folder of your best previous writing to use as AI voice calibration samples. Articles you're proud of, client explanations that worked well, and presentations that received positive feedback all serve as models for AI to learn from.

Authenticity concerns arise when AI-generated content presents expertise you don't actually have. If AI drafts an article about complex tax issues and your practice doesn't include tax work, publishing that article misrepresents your capabilities. Content should reflect your actual expertise and experience. AI can help you communicate that expertise efficiently, but it shouldn't create the appearance of expertise you lack.

Adding genuine value means including perspectives AI cannot generate: insights from your actual experience, examples from real matters you've handled (appropriately anonymized), and opinions developed through years of practice. Generic AI content provides generic information anyone could provide. Value-adding content provides what only you can provide based on your specific expertise and experience. "The AI draft explains the general process for business formation. Add a paragraph about the mistakes I commonly see clients make based on my actual experience, and another paragraph about the questions I recommend clients ask themselves before choosing an entity type."

Advertising rules constrain legal marketing in ways that don't apply to most businesses. Rules vary by jurisdiction but commonly address false or misleading statements, claims of specialization without proper certification, inappropriate comparisons to other lawyers, promises of results, and required disclaimers. AI-generated content may violate these rules unless reviewed for compliance before publication.

Rule review must occur before any content is published. Review AI-generated drafts against your jurisdiction's advertising rules. Common issues include superlative claims ("best," "most experienced," "leading"), specialty claims that require certification you don't have, result predictions that imply guarantees, and failure to include required disclaimers. When in doubt, consult your state bar's advertising guidelines or ethics hotline.

Testimonials and endorsements have specific rules in most jurisdictions. If your content incorporates client testimonials or peer endorsements, ensure compliance with applicable rules about substantiation, disclaimer requirements, and prohibited claims. AI might suggest adding testimonials without understanding the regulatory constraints involved.

AI-generated marketing content frequently includes superlative claims and implied promises that violate advertising rules. Review every piece of content for claims like "best," "top," "guaranteed," "proven," and similar language that may create compliance issues.

> **Authority Note** Bar disciplinary authorities monitor lawyer advertising, including online content, under the Model Rules governing lawyer advertising. Violations can result in discipline ranging from private reprimands to suspension. The efficiency of AI-generated content does not reduce your responsibility for compliance. Review everything before publication.

Disclaimers and disclosures should be included as appropriate for the content type. General educational content typically needs a disclaimer that it's not legal advice and doesn't create an attorney-client relationship. Content addressing specific jurisdictions should note those limitations. Contact information and, where required, principal office address should be included. AI can include standard disclaimers, but verify they're appropriate for your jurisdiction and content type.

Platform-specific considerations affect how content should be crafted. A LinkedIn article serves a professional audience and can assume more sophistication than a consumer-facing blog post. A Twitter thread requires concision that a long-form article doesn't. A YouTube script should be conversational in ways that written content shouldn't. AI can adapt content for different platforms, but you should specify the platform and its conventions.

SEO optimization helps content reach its intended audience. Content that potential clients can't find doesn't serve business development purposes. AI can help optimize content for search by incorporating relevant keywords naturally, using appropriate headings and structure, and addressing the questions that people actually search for. Balance SEO optimization against readability: keyword-stuffed content may rank well but read poorly.

Research what terms your potential clients actually search for, and incorporate those terms naturally into your content. "Wrongful termination lawyer" may get more searches than "employment litigation attorney." Meeting clients where they search helps them find you.

Content repurposing extends the value of content creation efforts. A single substantial piece can become multiple content items: a blog post can become a LinkedIn article, a series of social media posts, a section of a client guide, and talking points for a presentation. AI can help repurpose content across formats, extracting core information and adapting it to different contexts and lengths. "Take this 1,500-word article on commercial lease negotiations and create: a 200-word LinkedIn summary with key takeaways, five individual social media posts highlighting specific points, and an outline for a 15-minute presentation on the same topic."

Frequency and consistency matter more than volume. Regular posting on a sustainable schedule builds audience expectations and engagement more effectively than sporadic bursts of activity. A weekly blog post maintained for a year outperforms a dozen posts in one month followed by silence. AI makes consistency achievable by reducing the time required for each piece.

Measuring effectiveness helps you refine your content strategy. Track which content generates engagement, inquiries, and ultimately clients. Analytics tools show what people read, share, and respond to. Use this data to do more of what works and less of what doesn't. AI can help analyze your content performance and suggest adjustments based on what's resonating with your audience.

Competitive differentiation requires content that stands out from what others produce. As AI makes content creation easier, more lawyers will produce more content. Generic AI-generated material will become increasingly common and increasingly ignored. Differentiation comes from unique perspective, specific expertise, genuine voice, and authentic experience that generic content lacks. Use AI for efficiency, but invest the time saved in adding the value that differentiates your content.

Ethical considerations about AI disclosure are evolving. Some clients and readers may want to know whether content was AI-assisted. Some may not care. Current rules generally don't require disclosure of AI assistance in marketing content, but transparency about your process may be valued by some audiences. Consider your audience and whether disclosure serves your relationship with them.

Collaboration with marketing professionals may enhance your content strategy. If your firm has marketing staff or you work with outside marketing consultants, AI-assisted content creation can integrate with their efforts. They may have insights about topics, formats, and distribution that complement your substantive expertise. Clear workflows that leverage both legal expertise and marketing expertise produce better results than either alone.

Don't publish AI-generated content without reading it carefully and editing it to reflect your actual views and voice. Content with your name on it represents you. If you wouldn't say it yourself, don't publish it under your name regardless of how efficiently AI produced it.

Client development beyond content includes relationship nurturing, referral cultivation, and business development activities that content supports but doesn't replace. Content creates awareness and demonstrates expertise; personal relationships convert that awareness into engagements. AI can help you maintain relationships by drafting check-in communications, tracking contacts, and preparing for meetings, but the relationship itself remains human.

AI can accelerate thought leadership content creation dramatically, but efficiency gains come with risks requiring attention. Content strategy should precede content creation, identifying topics that serve both your practice goals and audience needs. Prompt design should specify audience, purpose, and voice. Voice preservation ensures content sounds like you, not like generic AI. Authenticity requires that content reflect expertise you actually have. Advertising rule compliance must be verified before publication, with particular attention to superlative claims, specialty assertions, and required disclaimers. Repurpose content across platforms to extend value. Maintain consistency over time rather than sporadic bursts. Differentiate through unique perspective and genuine experience that generic AI content lacks. Content creates awareness; personal relationships convert awareness into engagements.

10.5 Chatbots and Intake Automation

Automation promises efficiency: answering common questions without lawyer involvement, gathering preliminary information before consultations, and routing inquiries appropriately without human triage. These efficiencies are real and can benefit both law practices and the people they serve. But automation also creates risks: frustrating potential clients who want human interaction, providing inaccurate or misleading information, creating unauthorized practice issues, and building barriers that exclude people whose needs don't fit automated workflows. Implementing automation thoughtfully means understanding both what it can accomplish and what it shouldn't attempt.

The case for automation rests on scale and accessibility. A solo practitioner cannot personally respond to every website inquiry within minutes, but a chatbot can. A firm cannot have staff available around the clock, but automated systems can provide basic information at 3 AM when a worried potential client is searching for help. Automation can handle high volumes of routine inquiries, freeing lawyer and staff time for matters requiring professional judgment. These benefits are genuine and explain why legal chatbots and intake automation have proliferated.

The case against automation rests on relationship and judgment. Clients seeking lawyers are often in difficult situations: facing legal problems they don't understand, feeling vulnerable, and looking for someone they can trust. Automated interactions may feel impersonal, frustrating, or dismissive to people who want human connection. Automated systems cannot exercise judgment about complex or unusual situations, and attempts to handle complexity through automation often fail in ways that harm both the potential client and the practice's reputation.

Chatbot capabilities and limitations should shape implementation decisions. Current chatbots can answer frequently asked questions from predefined responses, collect structured information through guided conversations, route inquiries based on responses to screening questions, provide general information about legal topics, and schedule consultations. Chatbots cannot provide legal advice tailored to specific situations, exercise judgment about complex or unusual circumstances, build the trust that comes from human interaction, or reliably handle questions outside their programmed scope.

Appropriate chatbot uses play to these strengths while avoiding these weaknesses. A chatbot that answers "What are your office hours?" serves users well. A chatbot that attempts to assess whether someone has a valid legal claim ventures into territory that requires professional judgment. A chatbot that collects contact information and basic facts for human follow-up improves efficiency. A chatbot that makes users navigate complex decision trees to reach a human creates frustration.

Before implementing any automation, map the client journey from first contact through engagement. Identify which steps benefit from automation and which require human interaction. Automation should smooth the journey, not create obstacles to reaching human help.

Information accuracy in automated systems requires careful attention. Chatbots respond based on their programming or, for AI-powered chatbots, based on patterns that may produce inaccurate responses. A chatbot that provides incorrect information about statutes of limitations, required filings, or legal rights creates risks for both the potential client and the practice. Automated responses must be accurate, current, and appropriately limited.

Static chatbots with predefined responses can be controlled precisely: they say exactly what you program them to say. The limitation is that they can only address anticipated questions. Unanticipated questions produce frustrating responses like "I don't understand" or inappropriate responses that don't match what was asked. AI-powered chatbots can handle a broader range of questions but with less control over responses. An AI chatbot might answer questions you didn't anticipate, which seems helpful until it provides inaccurate information you never approved. The flexibility that makes AI chatbots useful also makes them dangerous for legal applications where accuracy is essential.

> **Risk Note** AI-powered chatbots for law firm websites require careful testing and monitoring. The same hallucination risks that affect legal research apply when chatbots answer questions about law. A chatbot that invents legal requirements or mischaracterizes procedures creates liability exposure. If you use AI chatbots, constrain them to redirect legal questions to human consultation rather than attempting to answer.

Unauthorized practice concerns arise when automated systems provide information that constitutes legal advice. General legal information, such as explaining what a statute of limitations is, differs from legal advice, such as telling someone whether a statute of limitations bars their specific claim. The line between information and advice is not always clear, and chatbots may cross it without anyone intending them to.

Safe automation stays clearly on the information side of this line. "Personal injury claims in Texas generally must be filed within two years of the injury" is information. "Based on what you've told me, you have six months to file your claim" is advice. Chatbots should provide the former and direct users to consultation for the latter. Clear disclaimers that automated responses are general information, not advice, help maintain the distinction but don't eliminate the risk.

Bar authorities have begun addressing chatbot and automation issues, though guidance remains limited in most jurisdictions. The unauthorized practice rules that apply to human non-lawyers also apply to automated systems. Supervise automated systems as you would supervise non-lawyer staff providing information to the public.

Intake forms and questionnaires represent a different automation category than chatbots. Forms collect information without the conversational interaction that chatbots simulate. Well-designed intake forms gather comprehensive information efficiently, allowing potential clients to provide details at their own pace and ensuring consistent data collection across inquiries. Forms lack chatbots' interactivity but also lack their risks of inappropriate responses.

Effective intake forms balance comprehensiveness with usability. A form that asks for every conceivable piece of information may be thorough but will frustrate users and produce abandonment. A form that asks too little may not gather enough information to be useful. Design forms that collect essential information upfront and allow optional additional detail for users willing to provide it.

Conditional logic improves form usability by showing relevant questions based on prior answers. A user indicating a car accident sees different follow-up questions than a user indicating a slip and fall. This tailoring makes forms feel more relevant and reduces the burden of answering inapplicable questions. Most form builders support conditional logic; use it to create intake experiences that feel customized rather than generic.

Data collection and privacy considerations apply to any automated intake system. Information collected through chatbots and forms must be handled in compliance with privacy obligations. Users should know what information is being collected, how it will be used, and who will access it. Data should be stored securely with appropriate access controls. If your intake systems collect information about potential clients before engagement, confidentiality obligations may already attach even though no formal relationship exists.

Clear privacy policies and terms of use should accompany automated intake systems. Users should understand that providing information doesn't create an attorney-client relationship (unless it does under your jurisdiction's rules), that the information will be reviewed by your practice, and how you handle information from people who don't become clients. Disclaimers should be prominent, not buried in fine print that no one reads.

Treat information from intake automation as potentially confidential even before engagement. A prospective client who shares sensitive facts through your website form may have reasonable expectations of confidentiality. Handle pre-engagement information with care appropriate to its sensitive nature.

Accessibility requirements apply to automated systems under disability law. Websites and digital tools must be accessible to people with disabilities, including those using screen readers, those with motor impairments affecting mouse use, and those with cognitive disabilities affecting comprehension. Automated intake that isn't accessible excludes potential clients who may need legal help. Accessibility is both a legal requirement and an ethical imperative.

Testing automated systems for accessibility should be part of implementation. Use accessibility checking tools, test with screen readers, and consider how your systems work for users with various disabilities. Provide alternative pathways for users who cannot use automated systems, such as phone intake for users who cannot navigate web forms.

Human fallback must be available when automation can't serve user needs. No automated system handles every situation appropriately. Users who don't fit automated workflows, whose questions aren't anticipated, or who simply prefer human interaction need a clear path to reach a person. Automation that traps users in loops without human escape routes frustrates rather than serves.

Easy escalation to human assistance should be prominently available throughout automated interactions. "Would you like to speak with someone?" should be an option at every stage, not a last resort after multiple failed automated attempts. Users who choose automated paths initially should be able to exit to human help when automation isn't working for them.

Pro bono and access to justice considerations affect automation decisions. Automation that makes legal help more accessible, by providing around-the-clock information or handling high volumes of inquiries efficiently, can expand access. Automation that creates digital barriers, requiring technological sophistication that some populations lack, can restrict access. Consider who your automation serves and who it excludes.

Monitoring and improvement should be ongoing after implementation. Track how users interact with automated systems. Where do users drop off? What questions does the chatbot fail to answer?

What intake form fields cause confusion? This data reveals where automation is failing and where improvement is needed. Regular review and refinement keeps automated systems effective as user needs and your practice evolve.

User feedback provides insights that analytics alone cannot capture. Ask users how their automated experience was. What worked well? What frustrated them? This feedback identifies issues that usage data might not reveal and demonstrates client-focus that builds goodwill. Act on feedback by improving systems based on what users report.

Schedule quarterly reviews of your automated intake systems. Analyze usage data, review user feedback, test all pathways yourself, and update content for accuracy. Automated systems that aren't maintained become outdated and ineffective, reflecting poorly on your practice.

Integration with practice management ensures that information collected through automation flows into your matter management systems without manual rekeying. When a potential client completes an intake form, that information should populate your conflict checking system, your contact database, and your matter management platform. Integration eliminates double entry, reduces errors, and makes automation genuinely efficient rather than just front-end convenient.

Vendor selection for automation tools requires the same diligence discussed in Chapter 6. How does the vendor handle data you collect? Where is data stored? What security measures are in place? What happens to data if you stop using the service? Chatbot and form vendors that handle potential client information must meet your confidentiality obligations, and vendor terms should be reviewed before implementation.

Cost-benefit analysis should inform automation decisions. Automation has costs: vendor fees, implementation time, ongoing maintenance, and the risks discussed throughout this section. Benefits include efficiency, accessibility, and consistency. Analyze whether the benefits justify the costs for your practice. A high-volume practice with many routine inquiries may benefit substantially; a practice with few inquiries and complex matters may find automation costs exceed benefits.

> **Practice Area Note: Immigration** Immigration firms are among the heaviest users of intake automation because volume demands it — a busy immigration practice may field dozens of inquiries daily about visa eligibility, status renewals, and asylum procedures. Automated eligibility screeners that ask structured questions about visa category, employment status, and timeline can efficiently route inquiries to the right attorney or identify matters outside the firm's practice. But immigration intake automation carries heightened sensitivity: clients may fear that information provided through automated systems could be shared with enforcement authorities. Clear privacy disclosures and explicit statements about confidentiality are essential for immigration intake automation to serve rather than deter potential clients. Professional responsibility for automated systems rests with you. If your chatbot provides inaccurate information, you're responsible for the consequences. If your intake form fails to collect information needed for conflict checking and a conflict results, you bear the professional responsibility implications. Automation is a tool you deploy, and supervision obligations apply. You must ensure automated systems perform appropriately, which requires testing, monitoring, and maintenance that many practices neglect.

Client expectations about automation vary. Some clients prefer the efficiency and control of automated interaction. They want to schedule consultations online at 11 PM without waiting for office hours. They want to provide information through forms at their own pace rather than answering questions in real time. Other clients interpret automation as impersonal or dismissive. They want to talk to a person, and chatbots feel like barriers rather than assistance. Understanding your client base and their preferences helps you design automation that serves rather than alienates them.

Hybrid approaches combine automation benefits with human relationship-building. Automated systems handle initial data collection, and humans follow up personally once basic information is gathered. Or humans have initial conversations, and automation handles scheduling and document collection. These hybrid approaches capture efficiency where automation works well while preserving human connection where it matters.

Don't set and forget automated systems. Chatbots and forms that worked when implemented may become outdated, inaccurate, or broken without ongoing attention. Schedule regular reviews and test all automated pathways periodically to ensure they still function appropriately.

Automation can improve efficiency and accessibility when implemented thoughtfully, but creates risks requiring careful management. Chatbot capabilities are limited; appropriate uses stay within those limits while routing complex matters to humans. Information accuracy in automated responses requires ongoing attention, with particular caution for AI-powered chatbots that may hallucinate. Unauthorized practice concerns require keeping automated responses clearly on the information side of the information-advice line. Intake forms offer automation benefits with less risk than conversational chatbots. Data privacy, accessibility, and human fallback must be addressed in any implementation. Monitor and improve systems continuously. Cost-benefit analysis should confirm that automation serves your practice and clients before implementation. Professional responsibility for automated systems rests with you.

10.6 Mini-Lab: Client Update Letter with Verification

Objective: Apply the client communication techniques from this chapter to draft a comprehensive client update letter. Practice translating legal developments into plain English, structuring updates around client concerns, framing risks appropriately, and verifying AI-assisted communications before sending.

The complete Mini-Lab packet, including the Jennifer Walsh scenario, case status details, and analysis guide, is available on the companion site for download.

Time Required: 60 to 75 minutes

What You'll Need

- The plain-English techniques from Section 10.2
- The update structure from Section 10.3
- Access to an AI tool for drafting assistance
- A method for documenting your process

Step 1: Review the Scenario and Case Status

Read the following scenario describing the current state of a client's matter. You will draft an update letter communicating recent developments and next steps.

Scenario

You represent Jennifer Walsh, an individual client who was terminated from her position as a regional sales manager at Midwest Manufacturing, Inc. after twelve years of employment. Jennifer believes she was terminated in retaliation for reporting safety violations at the company's main production facility to OSHA. She is 58 years old and believes her age may also have been a factor, as the company has been replacing older managers with younger employees.

You filed suit three months ago alleging retaliation under the whistleblower protection provisions of the Occupational Safety and Health Act and age discrimination under the Age Discrimination in Employment Act. The case is in federal court.

Jennifer is not legally sophisticated. She has never been involved in litigation before and finds the process confusing and stressful. She is currently unemployed and actively seeking new employment. She checks in frequently and has expressed frustration that things seem to be moving slowly. In your initial communications, you learned she prefers detailed updates that help her understand what's happening, rather than brief summaries.

Current Case Status

- Discovery Update: You served written discovery requests on Midwest Manufacturing six weeks ago. Their responses were due last week. They requested and received a two-week extension, so responses are now due in one week. This is routine and you did not object to the extension.
- Motion Practice: Midwest Manufacturing filed a motion to dismiss the OSHA retaliation claim, arguing that Jennifer failed to exhaust administrative remedies before filing suit. You have reviewed the motion and believe it lacks merit because Jennifer did file a complaint with OSHA, which investigated and issued findings before she filed suit. Your opposition brief is due in two weeks.
- Deposition Scheduling: The parties have agreed to begin depositions next month. You anticipate deposing Jennifer's former supervisor, the HR director, and the company's safety officer. Midwest's counsel has indicated they will depose Jennifer. You need to schedule her deposition preparation session.
- Settlement: Midwest's counsel made an informal inquiry about whether Jennifer would be interested in discussing settlement. No specific offer was made. You want to discuss with Jennifer whether she wants to explore settlement discussions or focus on litigation.
- Document Request: You need Jennifer to locate and provide any documents she retained from her employment, particularly any communications about the safety issues she reported and any performance reviews or commendations from her twelve years at the company.

Background Facts for Reference

Jennifer reported safety concerns about inadequate machine guarding to her supervisor in January. When nothing was done, she filed an OSHA complaint in March. OSHA investigated and cited the company for violations in May. Jennifer was terminated in June, officially for "reorganization," though her position was filled by a 34-year-old external hire two weeks later. Jennifer filed an EEOC charge for age discrimination and received a right-to-sue letter. OSHA also found reasonable cause to believe retaliation occurred.

Step 2: Identify Communication Objectives

Before drafting, clarify what this update needs to accomplish. Based on the scenario and Jennifer's characteristics, identify:

- Information objectives: What does Jennifer need to know about case developments?
- Relationship objectives: How can this communication address her frustration and maintain trust?
- Action objectives: What do you need Jennifer to do?
- Tone objectives: What tone is appropriate given the developments and Jennifer's emotional state?

Write out your objectives before proceeding. These objectives will guide your drafting decisions.

Step 3: Outline the Update Structure

Using the status-next-timeline-action framework from Section 10.3, outline what content should appear in each section of your update:

- Status: What is the current state of the case across each area (discovery, motion, depositions, settlement)?
- Next Steps: What will happen next in each area, and what is your role versus the court's versus opposing counsel's?
- Timeline: What are the key dates Jennifer should know about?
- Action Items: What do you need from Jennifer, with clear deadlines?

Step 4: Draft the Update Letter

Using AI assistance, draft a comprehensive update letter to Jennifer. Your prompt should incorporate the communication objectives and structure you've identified, along with guidance about Jennifer's preferences and characteristics.

Sample prompt structure:

Draft a client update letter with the following parameters:

CLIENT: *Jennifer Walsh, individual plaintiff in employment litigation. Not legally sophisticated, first-time litigant, prefers detailed explanations, currently stressed about the slow pace of litigation and her unemployment.*

DEVELOPMENTS TO COMMUNICATE: *[List the developments from the scenario]*

STRUCTURE: *Use status-next-timeline-action framework. For each development area, explain current status, what happens next, relevant timeline, and any client action needed.*

TONE: *Warm, supportive, patient. Acknowledge that the process is slow and stressful. Provide reassurance where appropriate without overpromising outcomes. Use plain English and explain legal concepts when they must be mentioned.*

ACTION ITEMS NEEDED: *(1) Schedule deposition preparation session; (2) Gather and provide employment documents; (3) Consider position on settlement discussions*

LENGTH: *Approximately 2 pages, detailed enough to satisfy a client who wants to understand what's happening. Include appropriate professional closing and invitation for questions.*

Review the AI-generated draft critically before proceeding to verification.

Step 5: Verify the Draft

Apply verification appropriate for client communications. Review the draft for:

- Accuracy verification: Are all factual statements about the case status accurate? Are legal characterizations accurate?
- Tone verification: Does the tone match what's appropriate for this client? Does it acknowledge her concerns without being dismissive or overly alarming?
- Plain-English verification: Would Jennifer understand this letter without legal training? Are legal terms explained when used?

- Completeness verification: Does the letter address all developments? Are all action items clearly stated with deadlines?
- Expectation management verification: Does the letter appropriately set expectations without overpromising?
- Risk framing verification: Is the motion to dismiss discussed in a way that informs without alarming?

Read the draft aloud as if you were Jennifer receiving it. Does it answer the questions she would have? Does it make her feel informed and supported?

Step 6: Revise Based on Verification

Based on your verification review, revise the draft to address any issues identified. Common revisions include:

- Simplifying language that remains too technical
- Adding explanations for legal concepts that were mentioned but not explained
- Adjusting tone where it's too formal, too casual, or doesn't match client needs
- Clarifying action items that might be overlooked
- Adding reassurance where appropriate
- Restructuring for better flow and readability

Document what changes you made and why.

Step 7: Document Your Process

Create a brief record of your AI-assisted drafting process:

- What AI tool did you use?
- What instructions did you provide?
- What verification steps did you complete?
- What revisions did you make between AI draft and final version?
- How long did the process take compared to drafting without AI assistance?

This documentation reinforces the workflow discipline that responsible AI use requires.

Scenario Analysis Guide

After completing your own draft, use the following guide to evaluate your work.

Communication Objectives Analysis

Information objectives should include:

- Explaining the discovery extension and why it's routine
- Describing the motion to dismiss and your assessment of its merit
- Informing about deposition scheduling and what that means for Jennifer
- Communicating the settlement inquiry without pressuring a decision

Relationship objectives should include:

- Acknowledging Jennifer's frustration with the pace of litigation
- Demonstrating that you're actively working on her case

- Building confidence that you have matters under control
- Showing that you understand her situation and care about her wellbeing

Action objectives should include:

- Scheduling the deposition preparation session
- Obtaining employment documents Jennifer may have
- Getting Jennifer's input on settlement discussions

Tone objectives should include:

- Warm and supportive given Jennifer's stress
- Patient and explanatory given her lack of legal sophistication
- Confident without overpromising given case uncertainties
- Professional while remaining accessible

Client communications that ignore the emotional dimensions of litigation often fail. An update that provides information without acknowledging Jennifer's realities misses the relationship dimension that effective client communication requires.

Content Analysis by Section

Discovery Section should explain:

- That you sent discovery requests seeking information from Midwest
- That they asked for extra time to respond, which is normal
- That responses are now expected in one week
- What you'll do with the responses when received
- That this is progress even though it feels slow

Motion Section should explain:

- What a motion to dismiss is in plain English (a request asking the judge to throw out part of the case before trial)
- What Midwest is arguing (that Jennifer didn't follow required procedures)
- Why you believe the motion lacks merit (she did file with OSHA and they investigated)
- What happens next (you'll file a response, the judge will decide)
- That motions always create some uncertainty, but you're confident

Deposition Section should explain:

- What depositions are in plain English (formal interviews under oath)
- Who will be deposed and why those people matter
- That Jennifer will also be deposed by the other side
- Why deposition preparation is important
- That you need to schedule her preparation session

Settlement Section should explain:

- That opposing counsel asked if Jennifer would discuss settlement
- That no specific offer was made

- That exploring discussions doesn't commit her to anything
- That you want her input before responding
- That you'll support whatever direction she wants to go

Action Items should clearly state:

- Please call the office to schedule your deposition preparation session
- Please gather any documents from your employment, particularly safety communications and performance reviews
- Please think about whether you'd like to explore settlement discussions and let me know

Action items buried in lengthy paragraphs often get missed. Make action items visually distinct and unmistakably clear.

Plain-English Analysis

Legal terms that need explanation if used:

- **Discovery:** Formal process of getting information from the other side
- **Motion to dismiss:** Request asking the judge to end part of the case without a trial
- **Exhaust administrative remedies:** Complete required agency procedures before suing
- **Deposition:** Formal interview under oath where lawyers ask questions and a court reporter records answers
- **Opposition brief:** Our written response arguing why the motion should be denied

Sample Update Structure

A well-structured update for this scenario might be organized as follows:

- **Opening (1–2 paragraphs):** Warm greeting, acknowledgment that it's been a month, recognition that litigation moves slowly, assurance that the case is progressing.
- **Discovery Update (1–2 paragraphs):** Explanation of what discovery is, what you requested, the routine extension, when you expect responses, and what you'll do with them.
- **Motion Update (2–3 paragraphs):** Explanation of the motion in plain English, what they're arguing, why you believe it lacks merit, what happens next, timeline for resolution, acknowledgment of uncertainty while expressing confidence.
- **Deposition Update (2–3 paragraphs):** Explanation of what depositions are, who will be deposed and why, that Jennifer will be deposed, importance of preparation, need to schedule preparation session.
- **Settlement Discussion (2–3 paragraphs):** Communication of the inquiry, explanation that no offer was made, framing options without pressure, invitation to discuss by phone.
- **Action Items (clearly formatted list):** Three items with deadlines and brief explanation of why each matters.
- **Closing (1 paragraph):** Invitation to call with questions, assurance you're here to help, warm sign-off.

Deliverable

By the end of this Mini-Lab, you should have:

1. Written communication objectives (information, relationship, action, tone)

2. A content outline following the status-next-timeline-action framework
3. A complete client update letter of approximately 2 pages
4. Documentation of verification steps completed
5. Notes on revisions made between AI draft and final version
6. Documentation of your AI-assisted process

This deliverable demonstrates client communication that combines AI efficiency with relationship sensitivity and verification discipline.

Bonus Challenge

First, adapt the update for a different client profile. Imagine Jennifer is instead a sophisticated business executive who prefers brief, bottom-line communications. Redraft the update for this different client profile in one page or less. Compare the two versions to see how client characteristics shape communication approach.

Second, draft a follow-up communication. Assume Jennifer calls and says she'd like to explore settlement discussions but is worried it might be seen as weakness. Draft a response email that addresses her concerns and reassures her that exploring discussions doesn't commit her to anything.

Third, prepare talking points for the deposition preparation scheduling call. Jennifer will likely have questions. Prepare brief talking points covering: what deposition preparation involves, how long it takes, what she should do to prepare, and what to expect during her actual deposition.

Settlement discussions require client consent and client decision-making. The decision whether to explore settlement, and ultimately whether to accept any offer, belongs to her.

10.7 Practice Checklist and Prompt Pack for Client Communications

This section provides ready-to-use resources for implementing AI-assisted client communications in your practice. The checklists consolidate the workflow requirements from this chapter into references for intake, updates, marketing, and automation. The prompt pack provides templates for common client communication tasks that you can adapt to your specific needs. These resources translate the chapter's concepts into practical tools you can use immediately.

Client Communication Workflow Checklist

Use this checklist for any significant client communication. For routine communications, the checklist serves as a mental framework. For important communications, complete the checklist to ensure quality and appropriateness. A printable version of this checklist is available on the companion site.

Pre-Communication Planning

- ☐ Identified communication purpose (inform, request action, manage expectations, respond to inquiry)
- ☐ Considered client characteristics (legal sophistication, preferences, emotional state)
- ☐ Determined appropriate level of detail based on client preferences
- ☐ Identified key information that must be communicated
- ☐ Identified action items that require clear statement with deadlines
- ☐ Considered tone appropriate to content and client circumstances

- ☐ Selected appropriate communication channel (email, letter, phone, in-person) Drafting with AI Assistance
- ☐ Provided AI with client context and characteristics
- ☐ Specified tone, detail level, and format requirements
- ☐ Included all substantive content that must be communicated
- ☐ Requested plain-English explanations for legal concepts
- ☐ Asked for appropriate structure matching communication type Verification
- ☐ Accuracy verification: All factual statements are correct
- ☐ Plain-English verification: Client will understand without legal training
- ☐ Tone verification: Tone matches client needs and communication content
- ☐ Completeness verification: All necessary information is included
- ☐ Action item verification: Client actions are clearly stated with deadlines
- ☐ Expectation verification: Appropriate expectations are set without overpromising
- ☐ Risk framing verification: Uncertainties communicated appropriately
- ☐ Confidentiality verification: No inadvertent disclosures Finalization
- ☐ Revised based on verification findings
- ☐ Proofread for errors and clarity
- ☐ Confirmed appropriate signature and contact information
- ☐ Documented communication in matter file Intake Checklist Use this checklist when conducting intake for new clients or new matters. A printable intake checklist is available on the companion site.

Pre-Intake Preparation

- ☐ Reviewed any preliminary information client provided (questionnaire, initial inquiry)
- ☐ Prepared intake questions appropriate to matter type
- ☐ Ensured conflict check can be completed before learning confidential details
- ☐ Prepared engagement materials if intake leads to engagement Intake Conversation
- ☐ Allowed client to tell their story without excessive interruption
- ☐ Gathered key facts: parties, dates, events, documents
- ☐ Identified client's goals and expectations
- ☐ Assessed client's emotional state and concerns
- ☐ Captured information needed for conflict checking
- ☐ Noted potential statute of limitations or other timing issues
- ☐ Established communication preferences
- ☐ Discussed fee arrangements in general terms Post-Intake Processing

- ☐ Completed conflict check promptly
- ☐ Used AI to extract and organize key information from notes or recording
- ☐ Identified potential legal issues from client narrative
- ☐ Verified AI-identified issues against your own legal analysis
- ☐ Prepared follow-up questions list for information gaps
- ☐ Requested documents client mentioned or should have
- ☐ Documented intake conversation while memory is fresh
- ☐ Prepared engagement letter if proceeding Create practice-area-specific intake checklists. Personal injury intake captures different information than estate planning intake. Tailored checklists ensure you gather what each matter type requires.

Client Update Checklist

Use this checklist when preparing client status updates. Content Planning

- ☐ Identified all developments since last update
- ☐ Determined which developments client needs to know about
- ☐ Identified upcoming events and deadlines
- ☐ Identified client action items
- ☐ Considered appropriate level of detail for this client
- ☐ Planned risk framing for any adverse or uncertain developments Structure and Drafting
- ☐ Organized using status-next-timeline-action framework
- ☐ Led with most important information
- ☐ Explained legal concepts in plain English
- ☐ Made action items visually distinct and unmistakably clear
- ☐ Included specific deadlines for client actions
- ☐ Provided appropriate context for developments
- ☐ Acknowledged client concerns where relevant
- ☐ Set realistic expectations about timeline and outcomes Verification and Delivery
- ☐ Verified all factual statements for accuracy
- ☐ Confirmed tone matches client needs and content
- ☐ Ensured client will understand without follow-up questions
- ☐ Checked that action items are clear and complete
- ☐ Reviewed risk framing for appropriateness
- ☐ Documented update in matter file
- ☐ Delivered through client's preferred channel

- ☐ Followed up if action items not completed by deadline Communication failures generate more bar complaints than almost any other issue. Regular, proactive updates prevent most complaints regardless of case outcome. Build update discipline into your practice management.

Marketing Content Checklist

Use this checklist when creating marketing content or thought leadership materials.

Strategy and Planning

- ☐ Identified target audience for content
- ☐ Confirmed topic serves audience needs, not just your interests
- ☐ Determined appropriate format (blog post, article, guide, video script)
- ☐ Confirmed topic aligns with expertise you actually have
- ☐ Reviewed content calendar for consistency with overall strategy Drafting with AI Assistance
- ☐ Provided AI with audience description
- ☐ Specified voice and tone matching your authentic style
- ☐ Included sample previous content for voice calibration if available
- ☐ Requested appropriate length and format
- ☐ Asked for SEO-friendly structure where appropriate Content Verification
- ☐ Verified all legal information is accurate and current
- ☐ Confirmed content reflects expertise you actually have
- ☐ Reviewed for authentic voice versus generic AI voice
- ☐ Added unique perspective from your actual experience
- ☐ Checked for appropriate hedging and qualifications Compliance Verification
- ☐ Reviewed against jurisdiction's advertising rules
- ☐ Removed superlative claims (best, top, leading) unless substantiated
- ☐ Verified any specialty claims comply with certification rules
- ☐ Confirmed no implied guarantees of results
- ☐ Included required disclaimers (not legal advice, no attorney-client relationship)
- ☐ Added appropriate jurisdiction limitations
- ☐ Included required contact information Publication
- ☐ Final proofread for errors
- ☐ Optimized for platform (LinkedIn, website, etc.)
- ☐ Scheduled for consistent publication rhythm
- ☐ Planned promotion and distribution

- ☐ Set reminder to monitor engagement and respond to comments AI-generated marketing content frequently includes claims that violate advertising rules. Review every piece for superlatives, implied guarantees, and unsupported specialty claims before publication.

Automation Implementation Checklist

Use this checklist when implementing or reviewing chatbots and intake automation.
Planning and Design

- ☐ Identified specific goals automation should accomplish
- ☐ Mapped client journey to identify appropriate automation points
- ☐ Determined what information automation should collect
- ☐ Designed human escalation pathways at all stages
- ☐ Confirmed automation won't create unauthorized practice issues
- ☐ Planned for users who prefer human interaction Content and Accuracy
- ☐ Drafted all automated responses for accuracy review
- ☐ Verified legal information is accurate and current
- ☐ Confirmed responses stay on information side of advice line
- ☐ Included appropriate disclaimers
- ☐ Tested all conversation pathways for accuracy
- ☐ For AI-powered chatbots, constrained responses to prevent hallucination Compliance and Privacy
- ☐ Included privacy policy and terms of use
- ☐ Disclosed what information is collected and how it's used
- ☐ Confirmed data handling meets confidentiality obligations
- ☐ Verified vendor meets security requirements (per Chapter 6)
- ☐ Included disclaimer that no attorney-client relationship is created
- ☐ Tested accessibility for users with disabilities Implementation and Monitoring
- ☐ Tested all pathways before deployment
- ☐ Trained staff on how automation integrates with their work
- ☐ Established integration with practice management systems
- ☐ Set up analytics to track user behavior
- ☐ Scheduled regular accuracy reviews
- ☐ Created feedback mechanism for users
- ☐ Planned quarterly comprehensive review Prompt Pack for Client Communications

Prompt Pack for Client Communications

The following prompts are templates you can adapt for your specific client communication needs. Replace bracketed placeholders with your actual information. All prompts in this section are available as copy-paste templates on the companion site.

Intake Processing Prompts

Narrative summarization

```
Based on the following intake notes from a client meeting, extract and
organize the key information:

INTAKE NOTES: [Your notes from the client conversation]
Matter type: [practice area]

Please organize into:
  • Key facts in chronological order
  • All parties mentioned with their roles
  • All dates mentioned with associated events
  • Potential legal claims suggested by the facts
  • Documents the client mentioned or should be asked to provide
  • Gaps in information requiring follow-up questions
```

Issue spotting

```
Based on this intake narrative involving [matter type], what potential legal
claims or defenses do the facts suggest?

NARRATIVE: [Client's story as captured in intake]
Jurisdiction: [state/federal]

For each potential claim or defense:
  • Identify the legal theory
  • Note facts that support it
  • Note facts that would need to be established
  • Flag any timing concerns (statutes of limitations, filing deadlines)
  • Indicate confidence level (strong support, possible, speculative)
```

Follow-up questions

```
Based on the following intake summary, what follow-up questions should I ask
to complete my understanding of the client's situation?

INTAKE SUMMARY: [Summary of what you know]
MATTER TYPE: [practice area]
POTENTIAL CLAIMS: [claims you've identified]

Generate follow-up questions organized by:
  • Factual gaps that affect claim viability
  • Information needed for damages assessment
  • Documents that would support the claims
  • Potential defenses the client should address
```

Customize intake prompts for your practice areas. A personal injury intake extraction differs from a corporate formation intake extraction. Practice-specific prompts produce more useful results.

Plain-English Explanation Prompts

Concept explanation

```
Explain [legal concept] in plain English for a client who has no legal
training.

CLIENT CONTEXT: [brief description of client and their situation]

Guidelines: Use everyday language, not legal jargon. If legal terms must be
used, define them. Include a practical example or analogy if helpful. Keep the
explanation under [word count] words. The client should understand this
without follow-up questions.
```

Procedure explanation

```
Explain the [legal procedure, e.g., deposition, discovery, motion practice] to
a client who has never been through litigation before.

CLIENT CONTEXT: [client's situation and emotional state]

Include: what it is in plain terms, what happens during the process, what the
client's role is, what to expect emotionally, how long it typically takes, and
why this step matters for their case. Tone should be [reassuring/matter-of-
fact/encouraging].
```

Options explanation

```
Explain to a client the pros and cons of [Option A] versus [Option B].

CONTEXT: [describe the decision the client faces]

For each option, address: likely outcomes, risks involved, costs (financial
and otherwise), timeline implications, and emotional considerations. Present
both options fairly without pushing toward either. The client should feel
informed to make their own decision.
```

Clients who don't understand their options can't give informed consent to strategy decisions. Plain-English explanations aren't just good service; they're essential to ethical practice.

Client Update Prompts

Comprehensive update

```
Draft a client update letter with the following parameters:

CLIENT: [name and key characteristics: sophistication level, preferences,
emotional state]
MATTER: [brief case description]
DEVELOPMENTS TO COMMUNICATE: [List all developments since last update]
UPCOMING EVENTS: [List upcoming deadlines and events]
ACTION ITEMS NEEDED FROM CLIENT: [List what you need the client to do with
deadlines]

STRUCTURE: Use status-next-timeline-action framework
TONE: [warm/professional/matter-of-fact] given [client circumstances]
LENGTH: [target length based on client preference]

Include invitation to call with questions.
```

Adverse development communication

```
Draft a communication to a client about an adverse development in their case.

CLIENT: [name and characteristics]
ADVERSE DEVELOPMENT: [what happened]
CONTEXT: [why this happened and what it means for the case]
PATH FORWARD: [what happens next, what options exist]

The communication should: deliver the news clearly and promptly; provide
context that helps the client understand what it means; avoid both minimizing
and catastrophizing; explain the path forward; maintain appropriate confidence
without overpromising; acknowledge any emotional impact; invite client contact
for questions.

TONE: [supportive but realistic]
```

Action request

```
Draft a communication requesting specific actions from a client.

CLIENT: [name and characteristics]
ACTIONS NEEDED: [List each action with deadline and why it matters]

The communication should: make each action item unmistakably clear; explain
why each item matters for the case; provide specific deadlines; offer
assistance if the client has difficulty; be formatted so action items cannot
be overlooked. Keep [brief/moderate length] based on client preference.
```

Marketing Content Prompts

Blog post draft

```
Write a [word count]-word blog post about [topic].

AUDIENCE: [who will read this: potential clients, referral sources, industry
professionals]
PURPOSE: [educate, demonstrate expertise, answer common questions]
KEY POINTS TO COVER: [List main points]

GUIDELINES: Use accessible language appropriate for [audience]. Include
practical takeaways readers can use. Avoid legal jargon or explain it when
necessary. Do not make claims that would violate advertising rules. Include
disclaimer that this is general information, not legal advice. End with
appropriate call to action.
```

Content repurposing

```
Take the following [article/blog post/guide] and create:

ORIGINAL CONTENT: [Paste original content]

Please create:
  • a [word count]-word LinkedIn summary with key takeaways
  • [Number] individual social media posts highlighting specific points (each
under 280 characters)
  • an outline for a [length]-minute presentation on the same topic
  • a brief email newsletter version ([word count] words)
```

```
Maintain the same key messages across all formats while adapting to each
platform's conventions.
```

Topic brainstorm

```
Suggest [number] blog post or article topics for a [practice area] lawyer.

TARGET AUDIENCE: [who you want to reach]
GOALS: [demonstrate expertise, answer client questions, improve search
visibility]

For each topic suggestion: provide a working title, brief description of what
the post would cover, why this topic serves the target audience, and suggested
keywords for SEO. Focus on topics that address questions potential clients
actually have, not just topics lawyers find interesting.
```

Build a content calendar with topics planned months in advance. AI makes content creation efficient enough to maintain consistent publication. Consistency matters more than volume.

Automation Content Prompts

Chatbot response drafting

```
Draft responses for a law firm website chatbot for the following frequently
asked questions.

PRACTICE AREA: [your practice area]
JURISDICTION: [where you practice]
QUESTIONS TO ADDRESS: [List common questions]

Guidelines for each response: Keep responses under [word count]. Provide
helpful general information. Do not provide specific legal advice. Include
disclaimer that this is general information. Direct users to schedule a
consultation for specific situations. Maintain [tone] that represents the firm
well. Flag any questions where automated response risks crossing into legal
advice.
```

Intake form content

```
Draft content for an online intake form for [practice area] matters.

Include: welcome message explaining the form's purpose; questions to gather
essential information; explanations for any questions that might confuse
respondents; privacy notice explaining how information will be used;
disclaimer about attorney-client relationship; thank you message with
expectation-setting about follow-up. Keep language accessible to people
without legal training.
```

Chatbot responses should stay clearly on the general information side of the advice line. Draft responses conservatively and have them reviewed for unauthorized practice issues before deployment.

Documentation Template

Use this template to document your AI-assisted client communication process. A fillable version of this template is available on the companion site.

CLIENT COMMUNICATION DOCUMENTATION

Matter	
Client	
Communication Type	
Date	

Communication Purpose

☐ Status update
☐ Response to inquiry
☐ Settlement/strategy discussion
☐ Other: ________________
☐ Action request
☐ Adverse development
☐ Engagement/closing

AI Assistance Used

☐ Initial draft generation
☐ Tone adjustment
☐ None
☐ Plain-English translation
☐ Structure/organization

Client Characteristics Considered

☐ Legal sophistication level
☐ Emotional state
☐ Communication preferences
☐ Detail level preference

Verification Completed

☐ Accuracy of all factual statements
☐ Clarity for lay reader
☐ Clear action items with deadlines
☐ No confidentiality issues
☐ Appropriateness of tone
☐ Completeness of information
☐ Appropriate risk framing

Revisions Made (description of changes from AI draft)

Communication Sent	
Sent via	
Date/time	

Key Takeaways

1. Intake establishes both the information foundation for representation and the relationship foundation for effective client service. AI can help extract and organize key facts from client narratives, identifying parties, dates, events, and potential legal issues from often disorganized initial descriptions. Issue spotting during intake surfaces claims, defenses, and complications the client may not have recognized, including time-sensitive matters like approaching limitations periods. Conflict checking information must be captured before learning confidential details that could create problems with existing clients. The human element of intake cannot be automated: clients choosing lawyers want to feel heard and to build trust with someone they'll rely on during difficult situations. Balance efficiency with rapport, recognizing that these goals sometimes tension with each other.
2. Plain-English communication bridges the gap between legal complexity and client understanding. Clients need to understand their situations well enough to make informed decisions, but they don't need to master legal nuance. AI can help translate legal concepts into accessible language, generate analogies that make abstract principles concrete, and identify jargon that should be replaced with everyday terms. The precision imperative limits how far simplification can go: plain English must remain accurate English, and oversimplification that creates misunderstanding serves no one. Verify that AI-generated explanations actually convey what clients need to know without sacrificing accuracy. Calibrate hedging appropriately, communicating uncertainty where it genuinely exists while providing clear guidance where the law is settled.
3. Client updates prevent the anxiety that uninformed clients experience and build trust that sustains the relationship. The update paradox is that clients most need communication when there's least to report; dormant periods when matters are waiting require deliberate update effort. The status-next-timeline-action framework organizes updates around what clients actually want to know: where things stand, what happens next, when things will happen, and what they need to do. Action items deserve particular emphasis and should be unmistakably clear with specific deadlines. Calibrate detail level to client preferences rather than applying one-size-fits-all communication. Frame risks appropriately, neither alarming unnecessarily nor downplaying genuine concerns. Communicate adverse developments promptly with context that helps clients understand what they mean.
4. Marketing content and thought leadership position you as a knowledgeable authority while requiring attention to authenticity and compliance. AI can accelerate content creation dramatically, but efficiency gains come with risks. Content strategy should precede content creation, identifying topics that serve both your practice goals and audience needs. Voice preservation ensures content sounds like you rather than generic AI. Authenticity requires that content reflect expertise you actually have; AI should help you communicate your knowledge efficiently, not create the appearance of expertise you lack. Advertising rule compliance must be verified before publication, with particular attention to superlative claims, specialty assertions, and required disclaimers. Differentiate through unique perspective and genuine experience that generic AI content lacks.
5. Chatbots and intake automation can improve accessibility and efficiency when implemented thoughtfully, but create risks requiring careful management. Chatbot capabilities are limited:

they can answer frequently asked questions, collect structured information, and route inquiries, but they cannot provide legal advice, exercise judgment about complex situations, or build trust through human connection. Unauthorized practice concerns require keeping automated responses clearly on the general information side of the information-advice line. AI-powered chatbots carry hallucination risks that make them dangerous for legal applications without careful constraints. Human fallback must be available at every stage for users whose needs don't fit automated workflows or who prefer human interaction. Monitor and improve automated systems continuously; systems that aren't maintained become outdated and reflect poorly on your practice.

6. Throughout client-facing work, the verification imperative applies with particular force. Inaccurate communication to clients can create liability, damage relationships, and undermine the trust that effective representation requires. AI-assisted client communications require the same review and verification that any work product requires, adapted to the particular risks of client-facing content. The audience may be less sophisticated than a court, but the stakes of accuracy are just as high.

Chapter Checkpoint

Before moving on, make sure you can:

- ☐ Process intake information using AI to extract key facts, identify potential legal issues, and organize client narratives while maintaining the relationship-building that effective intake requires
- ☐ Translate legal concepts into plain English that clients can understand without legal training, using analogies and examples while maintaining accuracy and appropriate precision
- ☐ Structure client updates using the status-next-timeline-action framework, calibrating detail level to client preferences and making action items unmistakably clear
- ☐ Frame risks appropriately in client communications, acknowledging genuine uncertainties without unnecessary alarm and communicating adverse developments promptly with helpful context
- ☐ Create marketing content that reflects your authentic voice and actual expertise while complying with advertising rules, differentiating through genuine perspective rather than generic AI output
- ☐ Evaluate chatbot and automation implementations for appropriate limitations, unauthorized practice concerns, human fallback availability, and ongoing monitoring requirements
- ☐ Verify AI-assisted client communications for accuracy, appropriate tone, clarity for lay readers, and completeness before sending
- ☐ Document significant client communications and maintain records that protect both you and the client

What's Next

Chapters 7 through 10 addressed the workflows where lawyers produce value: research that creates understanding, drafting that produces documents, litigation that deploys both in adversarial contexts, and client communications that deliver value to the people legal work serves. Chapter 11 shifts focus from individual workflows to practice-level operations. How do you capture what you learn so it benefits future matters? How do you build systems that make your practice more efficient over time? How do you ensure consistency across lawyers and matters? These questions address the infrastructure that supports individual workflows.

Matter playbooks transform experience into reusable guidance. When you handle a matter type repeatedly, you develop knowledge about what works: which clauses to prioritize in certain contracts, which discovery requests yield useful information in particular case types, which issues commonly arise and how to address them. This knowledge often lives only in individual lawyers' heads, lost when they leave or simply unavailable to colleagues who could benefit. Playbooks capture this knowledge in forms that can be shared, refined, and applied to future matters. AI can help create playbooks by synthesizing patterns from past work, but the expertise being captured is yours.

Knowledge management addresses the broader challenge of preserving and reusing work product. Research memos, contract clauses, brief sections, and client explanations all represent intellectual effort that could benefit future matters. Without systems for organizing and accessing this knowledge, lawyers recreate work that's already been done. With effective knowledge management, past work becomes a resource that accelerates future work. AI can help organize, search, and adapt existing work product, but the underlying knowledge management discipline requires human attention to what's worth preserving and how to make it findable.

Time entries and billing narratives present a different operational challenge. Recording time accurately and describing work clearly serves both billing integrity and client relationships. AI can help draft billing narratives that are clear, professional, and appropriately detailed, but the underlying time must be accurately tracked and the descriptions must honestly reflect work performed. Chapter 11 addresses how AI can assist billing workflows while maintaining the professionalism and accuracy that billing requires.

Template systems and training round out practice operations. Templates that capture your preferred approaches, standard language, and quality standards help ensure consistency across matters and lawyers. Training that establishes prompt standards and review expectations helps teams use AI effectively and safely. Chapter 11 addresses these operational dimensions that determine whether AI assistance improves your practice systematically or remains ad hoc and inconsistent.

Key Terms

Intake — Gathering information from new clients or about new matters: client narrative, relevant documents, parties, and potential issues. Effective intake establishes the foundation for competent representation while building the relationship.

Issue spotting — Identifying potential legal claims, defenses, or complications a client's factual situation raises, including issues the client may not have recognized.

Plain-English communication — Explaining legal concepts in everyday language clients can understand without legal training, replacing jargon with accessible terms while maintaining accuracy.

Procedural explanation — Describing to clients what will happen in their legal matter: steps, timeline, and what will be expected of them.

Client comprehension verification — Checking that a client actually understood an explanation, using techniques beyond "does that make sense?"

Update schedule — A predetermined rhythm for communicating with clients about matter status, ensuring regular communication regardless of whether significant developments have occurred.

Risk framing — How potential negative developments are presented to clients, balancing the need to inform against the risks of creating unnecessary anxiety.

Thought leadership — Content positioning you as a knowledgeable authority in your field through articles, guides, presentations, and other materials.

Voice preservation — Ensuring AI-assisted content reflects your authentic voice, perspective, and expertise rather than generic AI style.

Content repurposing — Adapting a single piece of content for use in multiple formats and channels, extending the value of content creation efforts.

Intake automation — Using technology, including chatbots, online forms, and automated workflows, to gather preliminary information from potential clients and route inquiries before human involvement.

Conditional logic (in intake forms) — Showing or hiding questions based on previous answers, creating customized form experiences that ask only relevant questions.

Human fallback — The availability of human assistance when automated systems cannot meet user needs, ensuring automation supplements rather than replaces human interaction.

Status-next-timeline-action framework — A structure for client updates organized around the questions clients actually have: Where do things stand? What happens next? When? What do I need to do?

Risk framing in client communications — Presenting uncertainties in ways that inform without unnecessarily alarming, acknowledging genuine risks while maintaining appropriate confidence.

Client communication checklist — A structured review ensuring communications address client needs, maintain appropriate tone, and include necessary information.

Automation implementation checklist — A checklist ensuring chatbots and intake automation are accurate, compliant, accessible, and appropriately limited to avoid unauthorized practice issues.

Update prompts — Prompts that specify client characteristics so AI can calibrate tone, detail level, and approach. A sophisticated business client receives different communication than a first-time litigant.

Chapter 11: Practice Management and Knowledge Ops

Individual excellence compounds into organizational capability when systems capture what you learn. A lawyer who handles fifty employment matters develops intuitions about which discovery requests yield useful information, which contract provisions cause problems, and which arguments persuade which judges. Without systems to capture this knowledge, it remains locked in one person's head, unavailable to colleagues, vulnerable to departure, and often forgotten even by the lawyer who learned it. With effective systems, experience becomes infrastructure that accelerates every subsequent matter. AI can help build and leverage these systems, but the underlying discipline of capturing, organizing, and reusing knowledge requires deliberate attention that technology alone cannot provide.

Practice management encompasses the operational dimensions that determine whether a law practice runs efficiently or struggles with reinvented wheels and inconsistent quality. Matter playbooks codify how to handle recurring matter types. Knowledge management preserves work product for future use. Billing systems capture time and describe work. Templates ensure consistency across documents and lawyers. Training establishes standards that teams follow. Each dimension represents an opportunity to systematize what works, and each benefits from AI assistance in ways that parallel the workflow applications covered in earlier chapters.

The shift from individual to organizational thinking matters because law practices are more than collections of individual matters. Patterns emerge across matters: the same issues recur, the same documents get drafted, the same questions get asked. Practices that recognize and leverage these patterns operate more efficiently than practices that treat each matter as entirely new. The knowledge that experienced lawyers accumulate represents significant value, but only if that knowledge can be preserved, shared, and applied. AI makes this preservation and application more feasible than ever before.

Knowledge operations, sometimes called "knowledge management" or "KM," have been discussed in law firms for decades with mixed results. Traditional knowledge management initiatives often failed because capturing knowledge required effort that busy lawyers wouldn't invest, and retrieved knowledge was difficult to find or adapt to new contexts. AI changes this calculus in important ways. AI can help extract knowledge from existing work product without requiring lawyers to write separate knowledge documents. AI can help search and retrieve relevant past work. AI can help adapt retrieved work to new contexts. These capabilities make knowledge operations more practical, though they don't eliminate the need for human judgment about what's worth capturing and how to organize it.

What This Chapter Covers

- How to create matter playbooks that transform your experience handling specific matter types into reusable workflows, capturing what you've learned about effective approaches
- Techniques for knowledge management that preserve valuable work product, including research memos, contract clauses, and brief sections, making past work accessible and adaptable for future matters
- Approaches to time entries and billing narratives that maintain professionalism and accuracy while leveraging AI to improve efficiency and clarity
- Template systems that capture your preferred approaches and standard language, ensuring consistency across documents and lawyers while allowing appropriate customization

- Methods for training teams on prompt standards and review expectations, building organizational capability rather than depending on individual skill
- A Mini-Lab producing a matter playbook with integrated prompt pack, demonstrating how to systematize expertise for a matter type you handle regularly

Chapter Summary

Matter playbooks capture what you know about handling specific matter types, translating experience into guidance that benefits future matters. A playbook for a particular matter type might include typical phases and milestones, common issues and how to address them, key documents with preferred approaches, critical deadlines and how to track them, and lessons learned from past matters. AI can help create playbooks by analyzing patterns in your past work, synthesizing common elements, and organizing knowledge into usable formats. Playbooks evolve as you learn more, with each matter potentially contributing refinements. The goal is capturing expertise in forms that can be shared across a practice rather than rediscovered by each lawyer handling similar matters.

Knowledge management addresses preserving and reusing valuable work product beyond individual matters. Research memos represent significant intellectual investment that could benefit future matters involving similar issues. Contract clauses that you've refined through negotiation and experience represent accumulated wisdom about what works. Brief sections that effectively explain recurring legal concepts can be adapted rather than rewritten. AI can help organize this knowledge by extracting key content, tagging for searchability, and identifying connections across documents. AI can also help retrieve and adapt past work, finding relevant precedents and adjusting them for new contexts. Controlled reuse requires attention to ensuring that adapted content actually fits the new context rather than carrying forward assumptions that no longer apply.

Time entries and billing narratives present opportunities for AI assistance that require particular attention to professionalism and accuracy. AI can help draft billing narratives that clearly describe work performed, using professional language and appropriate detail. AI can help identify entries that might raise questions and suggest improvements. But billing integrity requires that time be accurately recorded and that descriptions honestly reflect work performed. AI assistance with billing narratives must not become a tool for inflating time or obscuring what was actually done. The efficiency gains from AI-assisted billing should benefit clients through clearer narratives and reduced billing friction, not through manipulation of billing records.

Template systems ensure consistency across documents and lawyers while capturing preferred approaches. Document templates with standard sections, approved language, and appropriate structure help maintain quality even when multiple lawyers contribute to work product. Prompt templates establish standard approaches for common AI-assisted tasks, ensuring that different lawyers using AI for similar tasks get consistent results. Template libraries require curation to keep templates current and to retire templates that no longer reflect best practices. AI can help maintain template libraries by identifying outdated content and suggesting updates based on recent work.

Training teams on AI use builds organizational capability that doesn't depend on individual skill or initiative. Training should cover prompt standards that establish how your practice approaches common tasks, review expectations that define what verification is required for different work types, and tool-specific guidance for the AI tools your practice uses. Ongoing training addresses new capabilities as tools evolve and new lessons as experience accumulates. Documentation of standards helps new lawyers adopt established practices and helps experienced lawyers maintain consistency. The investment in training pays returns through reduced errors, more consistent quality, and more efficient use of AI capabilities across the practice.

Throughout practice management and knowledge operations, the underlying principle is leveraging AI to build systems rather than just completing individual tasks. Individual AI assistance helps with today's matter. Systematic AI assistance helps with every future matter of the same type. The discipline of moving from individual to systematic thinking determines whether AI transforms your practice or merely provides occasional convenience.

11.1 Matter Playbooks: Turning Experience into Workflows

Experience teaches lessons that shouldn't require relearning.

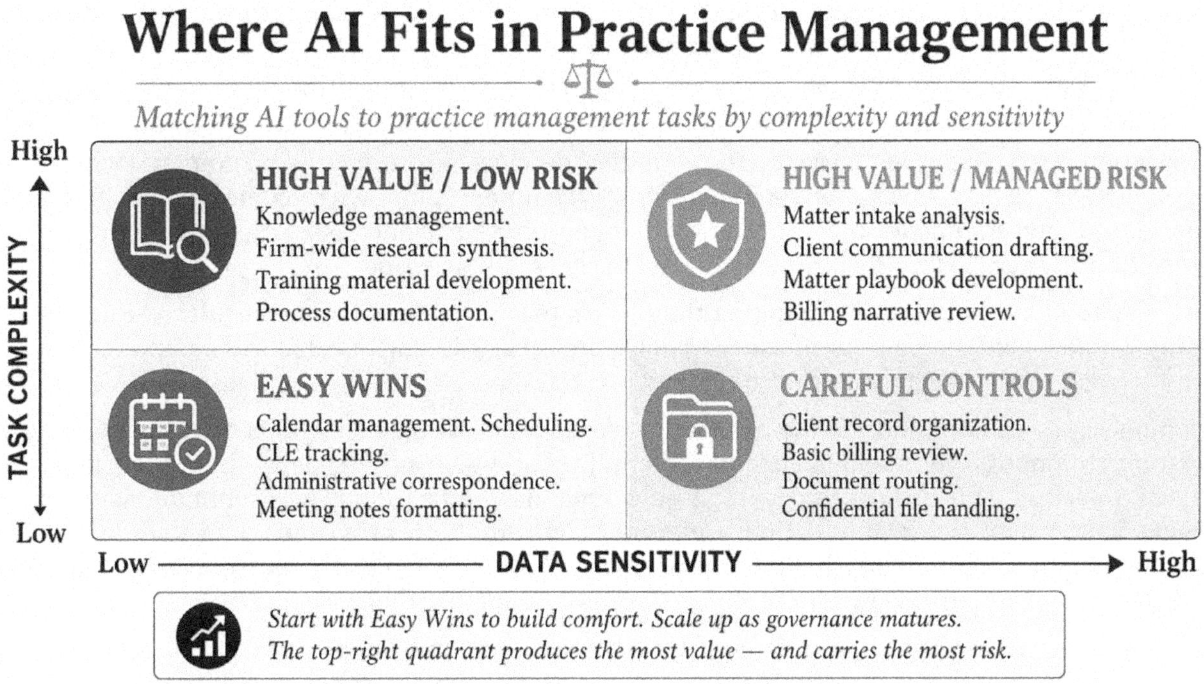

Figure 11.1 — Where AI Fits in Practice Management

After handling your tenth employment discrimination case, you know which discovery requests yield the documents that matter. After negotiating your twentieth software license, you know which provisions cause problems and which concessions you can safely make. After closing your fiftieth real estate transaction, you know which title issues require attention and which are routine. This accumulated knowledge makes you effective, but it often remains implicit, difficult to articulate, and unavailable to colleagues who could benefit from what you've learned. Matter playbooks make this knowledge explicit, capturing experience in forms that guide future matters and can be shared across a practice.

A matter playbook is a structured guide for handling a specific matter type. Unlike general practice guides that explain legal doctrine, playbooks focus on practical workflow: what to do, in what order, with what attention to which issues. A playbook for defending a trade secret misappropriation case might cover initial case assessment, discovery strategy specific to trade secret claims, common defenses and how to develop them, expert witness considerations, and settlement dynamics typical of these cases. Each section captures what experienced practitioners know about handling this matter type effectively.

The value of playbooks increases with matter complexity and repetition. Simple matters that you handle occasionally may not justify playbook investment. Complex matters that you handle regularly benefit enormously from systematic capture of what you've learned. A litigation practice that handles dozens of similar cases annually can dramatically improve efficiency and consistency by documenting their developed approaches. A transactional practice that repeatedly negotiates the same contract types can capture negotiation wisdom that took years to accumulate.

Playbooks transform individual expertise into organizational knowledge. Playbook components typically include several categories of knowledge. Phase structure describes how matters of this type typically progress, identifying milestones, decision points, and transitions between phases. Common issues catalogs the problems that frequently arise and proven approaches for addressing them. Key documents identifies the documents central to this matter type, with guidance on drafting or reviewing them effectively. Deadlines and triggers notes the time-sensitive elements requiring tracking. Lessons learned captures insights from past matters that should inform future ones. Each component serves a different aspect of matter handling.

Phase structure provides the backbone of a playbook. Employment litigation might proceed through initial assessment, pleading stage, written discovery, depositions, dispositive motions, and trial preparation. Each phase has characteristic activities, deliverables, and decision points. Documenting this structure helps lawyers new to a matter type understand what comes when and helps experienced lawyers ensure they haven't overlooked phase-specific tasks. AI can help draft phase descriptions by synthesizing patterns from past matters you describe.

Start playbook development by mapping the phases of matters you handle regularly. Even before adding detailed guidance, a clear phase map helps organize thinking and ensures completeness. Add depth to each phase as experience accumulates.

Common issues sections address the problems that recur across matters of this type. In construction litigation, common issues might include documentation of change orders, delay claim analysis, expert witness selection, and insurance coverage questions. In venture financing, common issues might include option pool sizing, liquidation preferences, anti-dilution provisions, and founder vesting. Documenting these issues and effective approaches for handling them prevents reinventing solutions and helps less experienced lawyers recognize issues they might otherwise miss.

Issue documentation should capture both what the issue is and how to address it. "Clients often underestimate the importance of contemporaneous project documentation" identifies an issue but doesn't help address it. "When clients have weak project documentation, consider these approaches for reconstructing the record: interviewing project participants early while memories are fresh, gathering documentation from subcontractors and suppliers who may have better records, and analyzing electronic files for metadata showing project timeline" provides actionable guidance. AI can help develop this actionable guidance by generating options and considerations for issues you identify.

Key documents sections identify the documents central to the matter type and provide guidance for handling them effectively. A commercial lease playbook might address the lease agreement itself, tenant improvement specifications, subordination agreements, estoppel certificates, and common area maintenance reconciliations. For each document type, the playbook might note critical provisions to focus on, common problems to watch for, and language preferences that reflect your practice's experience. This guidance accelerates document work by focusing attention where it matters.

Judges in specialized courts often develop preferences and patterns that practitioners learn through experience. A playbook for matters in a particular court or before particular judges might capture this institutional knowledge: preferred brief formats, procedural expectations, and approaches that tend to succeed or fail with specific decision-makers.

Document guidance connects naturally to the drafting workflows from Chapter 8 and the prompt techniques from Chapter 3. A playbook might include preferred prompts for drafting key documents, building on the Prompt Blueprint and incorporating matter-specific context. When you develop a prompt that works well for a recurring document type, capturing it in the playbook makes that prompt available for future matters. Over time, the playbook accumulates a prompt pack tailored to this matter type.

Deadline and trigger sections address the time-sensitive elements of matters. Some deadlines are obvious: statutes of limitations, contractual notice periods, court-imposed schedules. Others are less obvious but equally important: optimal timing for certain discovery requests, windows for settlement discussions, deadlines that trigger other obligations. Playbooks should identify these time-sensitive elements so they can be calendared and monitored. Missing a deadline because you didn't know it existed is a preventable failure that playbooks help avoid.

Trigger events require attention beyond simple deadlines. In litigation, receiving certain discovery responses might trigger the need for specific follow-up. In transactions, achieving certain milestones might trigger disclosure obligations or closing conditions. Documenting these triggers ensures they receive appropriate attention when they occur. AI can help identify potential triggers by analyzing the matter type and suggesting events that warrant monitoring.

Lessons learned sections capture insights from specific past matters. After a matter concludes, reflection on what worked and what didn't yields knowledge that should inform future matters. "In the Henderson matter, we learned that requesting electronic access logs early in discovery was crucial because the defendant's IT department had a short retention policy" captures specific knowledge applicable to similar future matters. Systematically capturing these lessons builds institutional memory that improves over time.

Post-matter review practices support lessons learned capture. After significant matters conclude, schedule a brief review session to discuss what went well, what could improve, and what specific insights should be preserved. AI can help facilitate this review by suggesting questions to consider and organizing the resulting insights into playbook format. The investment of an hour after matter conclusion can yield knowledge that saves many hours in future matters.

> **Practice Point** Schedule post-matter reviews within two weeks of conclusion while memories are fresh. Even fifteen minutes of structured reflection captures valuable lessons. Document the insights in your playbook immediately rather than trusting you will remember to add them later.

AI assistance in playbook creation takes several forms. Pattern extraction analyzes your past work to identify common elements. "Based on the following matter files from five employment discrimination cases I've handled, identify common phases, recurring issues, and documents that appeared in all matters." Guidance development generates practical approaches for issues you've identified. "For the common issue of inadequate client documentation in construction cases, suggest practical strategies for reconstructing the project record." Prompt creation develops reusable prompts for recurring tasks. "Create a prompt for drafting the initial case assessment memo in employment discrimination matters, incorporating the factors I typically consider."

Playbook maintenance ensures playbooks remain current as your practice evolves. Laws change, court rules change, best practices change, and your own experience continues to accumulate insights. Stale playbooks that reflect outdated approaches can mislead rather than help. Schedule periodic playbook reviews to update content, incorporate recent lessons learned, and retire guidance that no longer applies. AI can help identify potentially outdated content by flagging references to old rules or approaches that conflict with more recent entries.

Sharing playbooks across a practice multiplies their value. A playbook developed by one experienced lawyer can accelerate the work of every lawyer handling similar matters. Sharing requires attention to ensuring playbooks are accessible, understandable by lawyers who didn't create them, and appropriate for the range of contexts where they might be applied. Playbooks written only for the author's use may require revision before they're useful to others.

> **Risk Note** Playbooks must be updated when underlying law or procedure changes. A playbook reflecting pre-amendment rules or superseded procedures can cause serious errors. Include last-reviewed dates and schedule regular reviews to ensure playbooks remain accurate.

Playbook format should balance comprehensiveness with usability. An exhaustive playbook that covers every possible scenario may be too dense to use effectively. A skeletal playbook that provides only high-level guidance may not offer enough help when specific questions arise. Layered formats work well: a one-page overview for quick orientation, followed by detailed sections that lawyers consult when working on specific phases or issues. AI can help create these layered views by summarizing detailed content into overviews.

Template integration connects playbooks to the template systems discussed later in this chapter. Playbooks might reference specific templates for key documents, ensuring that lawyers using the playbook can quickly access the preferred starting points. This integration creates a comprehensive resource: the playbook provides workflow guidance, and linked templates provide document starting points. Together, they equip lawyers to handle matters efficiently from start to finish.

Starting small makes playbook development manageable. You don't need to create a comprehensive playbook immediately. Start with the matter type you handle most frequently. Begin with phase structure and common issues. Add depth over time as you handle more matters and accumulate more lessons. A partial playbook that covers the most important aspects of a matter type provides immediate value while you continue building it out. Perfectionism about playbook comprehensiveness can prevent you from capturing any knowledge at all. A playbook starter template is available on the companion site to help you begin.

Matter playbooks transform individual expertise into organizational knowledge by documenting practical workflows for specific matter types. Playbook components include phase structure, common issues with actionable guidance, key documents with handling instructions, deadlines and trigger events, and lessons learned from past matters. AI can help extract patterns from past work, generate practical guidance for identified issues, and create reusable prompts for recurring tasks. Post-matter reviews capture lessons while memories are fresh. Playbook maintenance ensures content remains current as law, procedure, and practice evolve. Sharing playbooks across a practice multiplies their value. Start with the matter types you handle most frequently and build depth over time. The discipline of capturing what you learn creates compounding returns as each matter benefits from all previous matters of that type.

11.2 Knowledge Management: Memos, Clause Libraries, Controlled Reuse

Every research memo represents hours of work that could benefit future matters involving similar issues. Every contract clause refined through negotiation represents accumulated wisdom about what language works. Every brief section that effectively explains a recurring legal concept could be adapted rather than rewritten from scratch. Yet in most practices, this valuable work product disappears into closed matter files, effectively lost even though it technically still exists. Knowledge management addresses this waste by creating systems that preserve valuable work product and make it accessible for future use. AI makes effective knowledge management more achievable than ever before, though it doesn't eliminate the need for human judgment about what's worth preserving and how to organize it.

The knowledge management challenge has frustrated law firms for decades. Traditional approaches required lawyers to write separate knowledge documents summarizing their work, a task that busy lawyers rarely prioritized. Retrieval systems were often difficult to search, returning too many irrelevant results or missing relevant documents entirely. Even when relevant past work was found, adapting it to new contexts required significant effort. These frictions meant that despite substantial investment in knowledge management systems, most lawyers continued to start from scratch on matters that could have benefited from past work.

AI changes the knowledge management calculus in three important ways. First, AI can help extract knowledge from existing work product without requiring lawyers to create separate documents. A research memo already contains the knowledge; AI can help identify and tag its key elements for future retrieval. Second, AI can improve retrieval by understanding queries conceptually rather than just matching keywords. A search for "enforceability of non-compete agreements" can surface relevant memos even if they use different terminology. Third, AI can help adapt retrieved work to new contexts, adjusting for different jurisdictions, fact patterns, or client needs. These capabilities make knowledge operations more practical in ways it wasn't before.

Research memo preservation creates a searchable repository of legal analysis. Chapter 7 addressed producing research memos for individual matters. Knowledge management extends this by preserving memos in forms that benefit future matters. A memo analyzing the enforceability of liquidated damages clauses under Texas law represents significant research investment. When a future matter raises similar issues, that memo should be findable and usable rather than forgotten in a closed file.

Memo tagging enables effective retrieval. Tags might identify the legal issues addressed, the jurisdiction, the outcome of the analysis, and any limitations on the memo's conclusions. AI can help generate tags by analyzing memo content and suggesting relevant categories. Consistent tagging across memos builds a searchable index that grows more valuable as more memos are added. The investment in tagging pays returns every time a tagged memo helps a future matter.

When completing a research memo, spend five minutes tagging it for future retrieval before closing the file. Note the key issues, jurisdiction, and any limitations. This small investment makes the memo findable when similar issues arise months or years later.

Clause libraries preserve contract language that has been refined through use and negotiation. A non-compete clause that has survived multiple negotiations represents accumulated knowledge about what language works, what courts enforce, and what counterparties accept. Preserving this clause in a searchable library makes it available for future contracts. Building clause libraries over time creates a resource that accelerates drafting and improves quality by starting from proven language rather than drafting anew.

Clause organization requires structure that matches how lawyers search for clauses. Organization by clause type (indemnification, limitation of liability, confidentiality) allows lawyers to find clauses addressing specific issues. Organization by contract type (software license, employment agreement, asset purchase) allows lawyers to find clauses appropriate for specific transactions. Both organizational schemes have value; effective libraries support multiple access paths. AI can help maintain organization by suggesting categories for new clauses and identifying clauses that might fit multiple categories.

Clause annotation adds value beyond the language itself. Notes explaining why certain language was chosen, what alternatives were considered, and how the clause has performed in negotiations help lawyers use clauses effectively. A clause with a note saying "counterparty accepted this language without negotiation in eight of ten deals" conveys different information than one saying "this language always generates pushback but has been upheld in litigation." AI can help generate initial annotations based on clause content, which lawyers then refine based on actual experience.

Courts interpreting contract language often look to industry practice and common usage. A clause library that documents how language has been interpreted in litigation provides valuable context for drafting. When you learn how a court interpreted particular language, note that interpretation in your clause library.

Brief banks preserve persuasive writing that explains recurring legal concepts. Many legal concepts arise repeatedly across matters: summary judgment standards, contract interpretation principles, damages calculations, evidentiary foundations. Writing that effectively explains these concepts can be adapted for future briefs rather than rewritten. A well-crafted explanation of the standard for piercing the corporate veil, refined through multiple uses, represents intellectual capital worth preserving.

Brief section retrieval benefits from conceptual search. A lawyer writing about fraudulent transfer might benefit from a brief section originally written about alter ego liability, since both involve reaching assets held by nominally separate entities. Keyword search would miss this connection; conceptual search powered by AI can surface relevant material even when terminology differs. This capability expands the useful scope of brief banks beyond exact matches.

Controlled reuse prevents the problems that careless copying creates. Retrieving past work product is only valuable if the retrieved content actually fits the new context. A research memo analyzing Texas law doesn't help a matter governed by California law unless the analysis is updated. A clause designed for a large enterprise transaction may be inappropriate for a small business deal. A brief section written for a conservative court may need adjustment for a different judicial audience. Controlled reuse means verifying that retrieved content fits the current context before incorporating it.

Context verification should be systematic rather than assumed. When retrieving a past memo, check whether the law has changed, whether the jurisdiction matches, and whether any factual distinctions affect the analysis. When retrieving a clause, confirm that it's appropriate for the transaction type, the counterparty sophistication, and the risk profile of the current deal. When retrieving a brief section, verify that the legal standards match the current court and that the rhetorical approach fits the current case. AI can help by identifying potential mismatches, but the verification judgment remains yours.

> **Risk Note** Past work product may reflect law or circumstances that have changed. Before reusing any retrieved content, verify that it remains accurate and appropriate. A memo citing a statute that has been amended or a clause reflecting a regulation that has changed can create serious problems if reused without verification.

Adaptation transforms retrieved content for new contexts. Even when past work is relevant, it rarely fits new contexts perfectly. A memo analyzing an issue under federal law might need adaptation for a state law matter. A clause designed for one industry might need adjustment for another. AI can help with adaptation by adjusting for specified differences: "Adapt this analysis of the federal standard to address how Texas courts apply a similar state law doctrine." Adaptation should be followed by verification to ensure the adapted content is accurate and appropriate.

When adapting past work product, explicitly note what you changed and why. This documentation helps you verify the adaptation was appropriate and helps future users understand the relationship between the original and adapted versions.

Curation keeps knowledge repositories valuable as they grow. A knowledge management system that accumulates outdated, low-quality, or duplicate content becomes harder to use over time. Curation involves reviewing content periodically, retiring outdated material, consolidating duplicates, and elevating high-quality content. Without curation, knowledge repositories degrade. AI can help identify candidates for retirement by flagging content that hasn't been accessed, that contains outdated citations, or that has been superseded by newer versions.

Quality signals help users assess retrieved content quickly. Knowing that a memo was written by a partner with thirty years of experience in the area conveys different reliability than knowing it was written by a summer associate. Knowing that a clause has been used successfully in fifty transactions conveys different confidence than knowing it's never been tested. Capturing and displaying these quality signals helps users make informed decisions about whether and how to use retrieved content.

Contribution incentives encourage lawyers to add to knowledge repositories. If contributing knowledge requires effort but provides no recognition or reward, contributions will be sparse. Practices that want robust knowledge repositories should recognize contributions, perhaps by tracking who contributed high-value content and acknowledging those contributions in reviews or other contexts. Making contribution easy also helps: AI that can extract and tag content from completed work product reduces the burden of contribution.

Judges have access to research databases showing how their decisions are cited and used. Lawyers can similarly benefit from knowing how their work product is reused. Systems that show lawyers when their memos, clauses, or brief sections are accessed create feedback that encourages quality contributions.

Integration with workflows makes knowledge management part of normal work rather than a separate activity. When starting research, the workflow should include checking whether relevant past memos exist. When drafting contracts, the workflow should include reviewing relevant clause libraries. When writing briefs, the workflow should include searching for relevant past sections. Building these checks into standard workflows ensures that knowledge repositories are actually used, which in turn encourages contribution because lawyers see their contributions providing value.

AI-assisted search should be the default retrieval method. Natural language queries like "What do our past memos say about enforcing non-competes against low-level employees?" are more intuitive than constructing Boolean searches. AI can interpret these queries, search across repositories, and present results with relevance explanations. This accessibility encourages use, which encourages contribution, which builds repository value over time.

Security and confidentiality apply to knowledge repositories as they do to all client information. A memo written for one client contains confidential information that shouldn't be accessible when helping a different client. Clause libraries built from client transactions contain information about those clients' businesses. Access controls must ensure that confidential information is only accessible by those authorized to see it. When adapting past work for new clients, confidential information from the original client must be removed.

Knowledge management preserves valuable work product for future use, preventing the waste of recreating work that's already been done. AI improves knowledge management by extracting knowledge from existing work, enabling conceptual search beyond keyword matching, and helping adapt retrieved content for new contexts. Research memos should be tagged for retrieval. Clause libraries should include annotations explaining use and history. Brief banks preserve persuasive writing on recurring concepts. Controlled reuse requires verifying that retrieved content fits the current context, checking for legal changes, jurisdictional differences, and factual distinctions. Curation keeps repositories valuable as they grow. Contribution incentives encourage lawyers to add to repositories. Integration with workflows ensures repositories are actually used. Security controls protect confidential information across clients.

11.3 Time Entries, Billing Narratives, and Professionalism

Time recording is where work becomes revenue, yet many lawyers treat it as an afterthought. Entries get batched at day's end or week's end, memories fade, and descriptions become vague approximations rather than accurate records. "Research and drafting" tells a client nothing about what was accomplished. "Review documents" could mean anything from skimming a few pages to analyzing thousands. These imprecise entries frustrate clients, invite billing disputes, and undermine the professional relationship that effective representation requires. AI can help improve both the efficiency and the quality of time recording, but only if applied with attention to the accuracy and professionalism that billing demands.

The billing narrative serves multiple purposes beyond calculating fees. For clients, narratives explain what work was performed and why it was valuable. For matters, narratives create a record of activity that can inform future work and help reconstruct what happened if questions arise. For lawyers, the discipline of describing work clearly forces reflection on whether the work was necessary and efficiently performed. Good billing narratives serve all these purposes; vague narratives serve none of them.

Contemporaneous time recording produces more accurate entries than reconstructed time. Recording time as you work, or immediately after completing a task, captures duration and description while memory is fresh. Waiting until day's end introduces error: tasks blur together, durations become estimates, and descriptions lose specificity. AI cannot fix the fundamental problem of inaccurate underlying time; it can only help describe time that has been accurately tracked. The discipline of contemporaneous recording must come first.

AI assistance with billing narratives takes several forms. Description improvement transforms vague entries into clear, professional narratives. An entry like "email re: contract" becomes "Reviewed and responded to client's questions regarding indemnification provisions in proposed vendor agreement." Batch processing can improve multiple entries simultaneously, applying consistent style and appropriate detail across a day's or week's entries. Compliance checking can identify entries that might raise questions: unusual durations, vague descriptions, or patterns that could suggest inefficiency.

Description improvement prompts should specify the level of detail appropriate for the client and matter. "Improve this time entry to clearly describe the work performed. The client is a sophisticated corporate client who expects detailed narratives. Entry: 'Research re: liability issue (2.5 hrs)'" might produce: "Researched case law and secondary sources regarding limitation of liability provisions in software licensing agreements; analyzed enforceability under Delaware law; prepared summary of key authorities for use in contract negotiation." The improved entry tells the client what was researched, why, and what it produced.

Develop narrative templates for recurring task types. Research entries should identify the issue researched, the sources consulted, and the output produced. Drafting entries should identify the document, the substantive work performed, and the progress made. Review entries should identify what was reviewed, the purpose, and any issues identified. Consistent templates produce consistent quality. Billing narrative templates for common task types are available on the companion site.

Professional tone in billing narratives matters. Entries should be clear, specific, and businesslike. Avoid both excessive brevity ("Calls") and excessive elaboration that might suggest padding. The goal is communicating what was done in a way that a sophisticated client would find informative and appropriate. AI can help achieve this tone, but review is necessary to ensure entries match your practice's standards and the client's expectations.

Ethical constraints on billing apply regardless of AI assistance. Time entries must accurately reflect work actually performed. AI that improves descriptions must not transform accurate records into inaccurate ones. An entry that takes 0.3 hours of vague work and produces a detailed narrative suggesting 0.3 hours of sophisticated analysis may be misleading if the work wasn't actually sophisticated. The description must match the reality of what was done. AI should clarify and professionalize descriptions, not inflate them.

AI-improved billing narratives must still accurately describe work performed. Using AI to generate detailed descriptions that don't reflect actual work is fraudulent billing regardless of the technology involved. The efficiency gain from AI should improve narrative quality, not enable billing manipulation.

Block billing, where multiple tasks are combined in a single time entry without separate times for each, creates problems that AI cannot solve. An entry reading "Telephone conference with client; review opposing motion; research regarding summary judgment standard (3.2 hrs)" doesn't tell the client how time was allocated across these different tasks. Some clients and courts prohibit or penalize block billing. Even where permitted, block billing invites questions and disputes. Recording tasks separately with individual times produces clearer records that better serve clients and protect lawyers from billing challenges.

Courts reviewing fee petitions regularly reduce awards based on vague or block-billed entries. Entries that don't clearly identify the work performed or that combine multiple tasks without breakdowns are often reduced or disallowed entirely. The billing discipline that avoids these reductions benefits matters where fees may be recoverable.

Task categorization helps clients understand how their money is spent. Categorizing entries as research, drafting, communication, review, court appearances, or other task types allows clients to see how effort is distributed. This transparency builds trust and can surface efficiency opportunities: a client who sees that document review consumed sixty percent of litigation costs might want to discuss review strategies. AI can help categorize entries consistently, applying the same categories across all lawyers and matters.

Billing narrative style should match client expectations and engagement terms. Some clients want maximum detail; others find excessive detail tedious and prefer concise summaries. Some engagements involve fee arrangements where detailed narratives are essential for justifying fees; others involve flat fees where less detail suffices. Understanding what each client expects and calibrating narratives accordingly serves the relationship better than applying uniform style regardless of context.

AI can help identify billing patterns that might raise concerns. Entries with unusual durations, clusters of similar entries suggesting inefficiency, or patterns that might appear suspicious to an auditor can be flagged for review before bills go out. This quality control function helps catch problems before they reach clients, preventing disputes and protecting the practice's reputation. The flagging should prompt human review, not automatic changes.

Before sending bills, review AI-flagged entries carefully. Not every flag indicates a problem, but each deserves consideration. An entry that AI flags as unusually long might reflect complex work that genuinely required the time, but the flag prompts you to verify that the entry accurately reflects what happened.

Client communication about billing should accompany invoices when appropriate. A bill that arrives without context may prompt questions that a brief cover note could prevent. AI can help draft cover notes that contextualize the invoice, highlight significant accomplishments during the period, and set expectations for upcoming work. These communications extend the client update practices from Chapter 10 to the billing context.

Prebill review ensures that bills reflect accurate, appropriate charges before clients see them. Review should verify that entries are accurate, descriptions are clear and professional, charges are appropriate for the work performed, and the bill as a whole will meet client expectations. AI can assist this review by generating summaries of the bill, flagging potential issues, and comparing charges to similar prior periods. Human review must confirm that the bill is one you're comfortable sending.

Write-offs and adjustments sometimes become necessary even with careful time recording. Work that took longer than it should, tasks that didn't produce value, or duplication across team members may warrant adjustment. These decisions require judgment about what's fair to charge and what maintains client relationships. AI might help identify candidates for adjustment by analyzing entries against benchmarks, but the decision to adjust is a professional judgment call.

Efficiency gains from AI should benefit clients, not just the practice. If AI helps you produce a research memo in two hours that previously took four, the ethical approach is billing for two hours, not four. The value to the client is the memo, and if you can produce it more efficiently, the client should share in that efficiency. This principle applies generally to AI assistance: efficiency gains that improve your productivity should translate to better value for clients, which in turn builds relationships and reputation.

Fee discussions should address AI use when relevant. If AI significantly affects how you work on a matter, clients may want to know. Some clients will appreciate the efficiency; others may have questions about quality or approach. Transparency about AI use in billing contexts extends the client communication principles from Chapter 10. You need not detail every AI-assisted task, but clients asking about your use of technology deserve honest answers.

> **Authority Note** Some courts now require disclosure of AI use in fee petitions. Even where not required, transparency about AI assistance may be wise when seeking fee awards. A fee petition that claims full hourly rates for work substantially performed by AI may raise questions about the reasonableness of the charges.

Training on billing practices should cover AI assistance specifically. Team members should understand what AI assistance with billing narratives is appropriate, what review is required, and what ethical constraints apply. Consistent practices across a team produce consistent billing quality and prevent individual lawyers from inadvertently crossing ethical lines. The training on AI use discussed in Section 11.5 should include billing practices as a component.

Documentation of billing practices protects against later questions. Records showing that time was recorded contemporaneously, that entries were reviewed for accuracy, and that prebill review occurred demonstrate billing integrity if questions arise. This documentation parallels the audit-ready habits discussed in Chapter 4: maintaining records that demonstrate appropriate process protects against challenges even when nothing improper occurred.

AI can improve billing narrative quality and efficiency, but must be applied with attention to accuracy and professionalism. Contemporaneous time recording produces accurate entries that AI can then improve; AI cannot fix fundamentally inaccurate underlying time. Description improvement should clarify and professionalize entries without inflating or misrepresenting work performed. Block billing creates problems AI cannot solve; record tasks separately with individual times. Prebill review ensures bills reflect accurate, appropriate charges before reaching clients. Efficiency gains from AI should benefit clients through better value, not just the practice through higher margins. Transparency about AI use in billing contexts extends client communication principles. Training should cover AI assistance with billing specifically. Throughout, the ethical constraint is clear: entries must accurately reflect work actually performed, regardless of how descriptions are generated.

11.4 Template Systems: Standard Sections and Reusable Structures

Consistency matters more than most lawyers acknowledge. A client receiving documents from your practice should encounter a consistent level of quality, a consistent approach to common issues, and a consistent professional presentation regardless of which lawyer prepared the work. Inconsistency signals disorganization and undermines client confidence. Yet achieving consistency is difficult when different lawyers approach the same tasks differently, prefer different language for similar provisions, and format documents according to individual preference rather than practice standards. Template systems address this challenge by establishing standard starting points that embody your practice's preferred approaches, ensuring consistency while still allowing appropriate customization.

Document templates provide starting points for recurring document types. A template for a nondisclosure agreement captures your practice's preferred structure, language choices, and formatting. A template for a motion to compel includes standard sections, preferred framing, and appropriate formatting for the target court. A template for a client engagement letter incorporates required disclosures, standard terms, and professional presentation. Starting from templates rather than blank pages accelerates production while ensuring that baseline quality and consistency are built in from the start.

The value of templates increases with document complexity and frequency. Simple documents that you produce rarely may not justify template investment. Complex documents that you produce regularly benefit enormously from templates that capture accumulated wisdom about structure, content, and approach. A commercial lease template that reflects dozens of prior transactions embeds knowledge that would take years to reacquire. A litigation brief template that incorporates formatting requirements, standard sections, and proven structural approaches prevents reinventing the wheel with each new filing.

Template components typically include structural elements, standard language, placeholder guidance, and formatting specifications. Structural elements define the document's organization: what sections appear, in what order, with what headings. Standard language provides preferred text for recurring content: boilerplate provisions, standard introductions, common explanations. Placeholder guidance indicates where customization is needed and what information should be inserted. Formatting specifications ensure consistent appearance: fonts, spacing, margins, heading styles. Together, these components create documents that look and feel consistent across the practice.

Standard language sections require particular attention because they embed substantive choices. A limitation of liability provision in a contract template reflects decisions about risk allocation that may or may not be appropriate for every transaction. An introduction to a brief that characterizes the standard of review reflects legal positions that must be accurate for the specific court and motion type. Standard language saves time, but users must verify that the standard language fits the current context. Templates should flag sections requiring verification rather than encouraging blind acceptance of standard text.

Placeholder design affects template usability. Placeholders should clearly indicate what information is needed and provide guidance for users who may not immediately know what to insert. "[CLIENT

NAME]" tells users what to insert but not where to find it. "[Insert client's legal name as it appears in the engagement letter]" provides actionable guidance. "[DATE]" is ambiguous; "[Insert agreement effective date, typically the last signature date]" clarifies. Well-designed placeholders make templates usable by lawyers unfamiliar with the document type, not just by the template's creator.

> **Practice Point** Mark template sections that require verification with clear indicators. Bracketed instructions like "[VERIFY: Confirm standard of review applies to this motion type]" or "[CUSTOMIZE: Adjust indemnification scope based on deal terms]" prompt users to think rather than simply accepting template content. Templates should facilitate quality work, not enable shortcuts that create problems.

Formatting consistency through templates serves both professional presentation and practical efficiency. Documents with consistent formatting look professional and signal attention to detail. Consistent formatting also simplifies production: lawyers don't waste time making formatting decisions or fixing inconsistencies. Template formatting should comply with any applicable court rules or client requirements and should reflect your practice's quality standards.

Courts increasingly specify formatting requirements for filed documents: font sizes, margin widths, line spacing, and page limits. Templates for court filings should incorporate these requirements by default, preventing rejection for formatting noncompliance. Include court-specific templates when practices before different courts require different formatting.

Prompt templates extend the template concept to AI-assisted tasks. Just as document templates provide standard starting points for recurring documents, prompt templates provide standard starting points for recurring AI tasks. A prompt template for drafting contract summaries might include standard instructions, output format specifications, and the prompt patterns from Chapter 3. A prompt template for research queries might incorporate the role definitions, context structures, and verification requirements your practice has found effective. Standardizing prompts produces more consistent AI outputs across lawyers and matters.

Prompt template development should follow the prompt engineering principles from Chapter 3, adapted for recurring use. The Prompt Blueprint components, including role, context, task, format, and guardrails, provide the structure. Practice experience identifies what instructions produce best results for specific task types. Capturing this knowledge in prompt templates makes it available to all lawyers in the practice, not just those who developed the effective approaches through trial and error.

Template libraries organize templates for discoverability and use. A disorganized collection of templates that lawyers can't find provides little value. Organization by document type, practice area, jurisdiction, or client allows lawyers to quickly locate relevant templates. Naming conventions that clearly identify template purpose prevent confusion among similar templates. Version control ensures lawyers use current templates rather than outdated versions. The investment in library organization pays returns every time a lawyer quickly finds the right template.

Template curation maintains library quality over time. Templates become outdated as law changes, court rules evolve, and practice preferences shift. Templates may proliferate as different lawyers create variations for specific needs, creating confusion about which template to use. Curation involves reviewing templates periodically, retiring outdated versions, consolidating unnecessary variations, and updating templates to reflect current best practices. Without curation, template libraries degrade into confusing collections that lawyers avoid rather than use.

Outdated templates can cause serious problems. A contract template with provisions superseded by statutory changes, a brief template citing overruled authority, or a court filing template with incorrect formatting for current rules all create risks. Schedule regular template reviews and retire or update templates when they no longer reflect current law and practice.

Template ownership clarifies who's responsible for creating, maintaining, and updating templates. Without clear ownership, templates may be created but never maintained, updated by multiple people creating conflicting versions, or allowed to become outdated because no one feels responsible. Assigning ownership, whether to specific lawyers, practice groups, or administrative staff, ensures accountability for template quality. Owners should have authority to make updates and responsibility for periodic review.

Customization guidelines address when and how templates should be modified. Templates should be customized when circumstances require it; they should not be followed blindly when customization is needed. But excessive customization defeats the consistency purpose of templates. Guidelines should indicate which template elements are fixed, which are routinely customized, and which require approval for modification. These guidelines help lawyers understand how much latitude they have while maintaining appropriate consistency.

Distinguish between template sections that should never change, sections that require customization for each use, and sections that may be modified when circumstances warrant. Clear marking prevents both inappropriate rigidity and excessive deviation from standard approaches.

Integration with document assembly tools enhances template value for high-volume practices. Document assembly systems can populate templates automatically based on matter data, generating draft documents with client names, dates, deal terms, and other information already inserted. This automation accelerates production further while reducing errors from manual placeholder completion. Integration requires templates designed for assembly, with placeholders that map to data sources the assembly system can access.

Template training ensures lawyers know how to use templates effectively. Training should cover where to find templates, how to select the right template for a task, what customization is expected versus optional, and how to provide feedback for template improvement. New lawyers should be introduced to templates during onboarding; experienced lawyers should be updated when significant template changes occur. Training investment produces returns through consistent, efficient template use.

Some courts have their own template forms for common filings. Using court-provided templates, where available, ensures compliance with local requirements and may be expected by the court. Integrate court templates into your template library alongside your own templates, clearly marking which are court-provided versus practice-developed.

Template feedback loops improve templates based on actual use. Lawyers using templates encounter situations the template creator didn't anticipate, find language that doesn't work in practice, or develop improvements worth incorporating. Systems for capturing this feedback, whether formal submission processes or informal communication channels, turn daily use into template improvement. Templates maintained through feedback loops become progressively better; templates without feedback loops stagnate.

AI assistance in template maintenance can identify improvement opportunities. AI can review templates for potentially outdated citations, flag language that might conflict with recent legal developments, and suggest updates based on recent work product that deviated from templates. These suggestions should prompt human review rather than automatic updates, but AI-assisted monitoring reduces the burden of keeping templates current.

Client-specific templates address clients with particular requirements. Large institutional clients may require specific language, formatting, or provisions in documents prepared for them. Creating client-specific template variations ensures these requirements are consistently met. Client templates should be clearly distinguished from general templates to prevent using client-specific provisions for other clients inappropriately.

Template systems ensure consistency across documents and lawyers while accelerating production through standard starting points. Document templates capture preferred structure, standard language, placeholder guidance, and formatting specifications. Prompt templates standardize AI-assisted tasks for consistent outputs across the practice. Template libraries require organization for discoverability and curation for ongoing quality. Template ownership ensures accountability for maintenance. Customization guidelines indicate when modification is appropriate. Integration with document assembly tools enhances value for high-volume practices. Feedback loops improve templates based on actual use. AI can assist template maintenance by identifying potential updates. Throughout, templates should facilitate quality work rather than enable shortcuts: users must verify that template content fits their specific context.

11.5 Training Teams: Prompt Standards and Review Expectations

Individual AI skill doesn't scale. A lawyer who develops sophisticated prompt techniques through months of experimentation has valuable capability, but that capability lives in one person's head. Colleagues who could benefit from those techniques don't know they exist. New lawyers joining the practice start from zero despite the accumulated learning that preceded them. When the skilled lawyer leaves, the capability walks out the door. Training transforms individual skill into organizational capability by establishing standards that everyone follows, documentation that preserves knowledge, and ongoing development that keeps the organization current as AI capabilities evolve.

The training challenge for AI differs from traditional legal training in important ways. Traditional legal skills change slowly; the fundamentals of legal research, writing, and analysis remain relatively stable over a career. AI capabilities change rapidly; tools that didn't exist a year ago may be essential today, and today's best practices may be obsolete tomorrow. Training programs must balance establishing foundational skills with maintaining flexibility to incorporate new developments. The goal is building judgment and adaptability, not just teaching current techniques that may soon change.

Foundational training establishes baseline competence for all lawyers using AI. This foundation should cover how AI works at a conceptual level, drawing on Chapter 2's explanation of what language models can and cannot do. It should address professional responsibility obligations from Chapter 4, ensuring every lawyer understands the competence, supervision, and accuracy requirements that apply to AI-assisted work. It should introduce the prompt engineering fundamentals from Chapter 3, including the Prompt Blueprint and core prompt patterns. This foundation ensures that everyone starts with the same understanding of capabilities, limitations, and responsibilities.

Prompt standards establish how your practice approaches common AI tasks. Rather than each lawyer developing individual approaches through trial and error, prompt standards codify effective approaches for recurring tasks. Standards might specify that research queries always include jurisdiction constraints and verification requirements. Standards might require that drafting prompts use the role-context-task-format-guardrails structure from the Prompt Blueprint. Standards might mandate specific output formats for certain deliverables. These standards ensure consistency while embodying best practices the organization has developed.

Standard development should involve lawyers with AI experience contributing what they've learned. A lawyer who has refined an effective approach to contract review prompts should share that approach for potential adoption as a standard. A lawyer who discovered that certain prompt structures consistently produce better research results should contribute that discovery. Aggregating individual learning into organizational standards captures value that would otherwise remain siloed. The process of developing standards also builds buy-in: lawyers who contributed to standards are more likely to follow them.

Document not just what your prompt standards are but why they exist. Standards that lawyers understand as embodying hard-won lessons are more likely to be followed than standards that seem arbitrary. "Always include jurisdiction in research prompts because AI otherwise defaults to general statements that may not reflect your state's law" explains the reasoning behind the requirement.

Review expectations define what verification is required for different work types. Chapter 2's reliability ladder and Chapter 4's professional responsibility framework establish that different AI outputs require different levels of review. Training should translate these frameworks into concrete expectations: what review is required for AI-assisted research, for AI-assisted drafting, for AI-assisted analysis. These expectations should be specific enough that lawyers know what's required and supervisors can verify compliance.

Review expectations should scale with risk and output type. A quick email summarizing a client call might require only a read-through for accuracy and tone. A research memo might require verification of every cited authority and independent assessment of the analysis. A brief filed with a court requires the comprehensive verification discussed in Chapter 4, including citation checking, quote verification, and accuracy confirmation for every factual statement. Clear expectations prevent both under-review that allows errors and over-review that wastes time on low-risk outputs.

Courts that have sanctioned lawyers for AI-related failures have emphasized the lawyer's responsibility to verify AI outputs before filing. Training that establishes clear review expectations and ensures lawyers understand their verification obligations helps prevent these failures. Document that training occurred as part of your audit trail.

Tool-specific training addresses the particular AI tools your practice uses. General prompt engineering skills transfer across tools, but each tool has specific capabilities, interfaces, and best practices. Training on your practice's tools should cover how to access and use the tool effectively, what the tool does well and where it struggles, how to interpret and verify the tool's outputs, and any tool-specific settings or configurations that affect results. This specific training complements foundational training by addressing the actual tools lawyers will use daily.

Tool training should be updated when tools change significantly. AI tools evolve rapidly, with new features, changed capabilities, and updated interfaces. Training that was accurate six months ago may be outdated today. Assign responsibility for monitoring tool changes and updating training accordingly. When significant changes occur, communicate updates to all users rather than assuming they'll discover changes independently.

Workflow integration training shows how AI fits into matter workflows. Rather than treating AI as a standalone skill, training should demonstrate how AI assistance integrates with the workflows from Chapters 7 through 11. How does AI-assisted research fit into the research workflow? How does AI-assisted drafting connect to the drafting workflow? How do AI-generated billing narratives integrate with time recording practices? This integration focus helps lawyers see AI as part of their normal work rather than a separate activity requiring context-switching.

Hands-on practice develops skill more effectively than passive instruction. Training should include exercises where lawyers actually use AI tools under guidance, make mistakes in safe environments, and receive feedback on their approaches. The Mini-Labs throughout this book model this hands-on approach. Real skill development requires practice, not just explanation. Build practice exercises into training programs and create opportunities for lawyers to develop proficiency before applying AI to client work.

Create a "sandbox" environment where lawyers can practice AI techniques without risk to client matters. Practice exercises using hypothetical facts or sanitized versions of past matters allow experimentation and skill development without the pressure of real client work. Mistakes in practice become learning opportunities rather than problems.

Supervision training addresses how supervisors should oversee AI-assisted work. The supervision obligations discussed in Chapter 4 require that supervising lawyers understand what's being supervised. Training for supervisors should cover how to evaluate AI-assisted work product, what questions to ask about how AI was used, how to spot common AI failure modes in work submitted for review, and how to provide feedback that improves AI use over time. Supervisors who understand AI can supervise it effectively; supervisors who don't understand AI may approve problematic work or reject appropriate work.

Ongoing training maintains competence as AI evolves. Initial training provides foundation, but capabilities change rapidly. Regular updates, whether through formal training sessions, written updates, or informal knowledge sharing, keep lawyers current. Some practices designate AI champions who stay current with developments and disseminate relevant updates to colleagues. Whatever mechanism you use, ongoing learning must be part of how your practice approaches AI, not an afterthought.

AI capabilities change faster than most training programs update. A lawyer trained on AI six months ago may be using outdated techniques or unaware of new capabilities. Build ongoing learning into your training approach rather than treating initial training as sufficient.

Documentation of standards and training serves multiple purposes. Written standards provide reference that lawyers can consult when questions arise. Training records demonstrate that lawyers received appropriate instruction, relevant for professional responsibility purposes if questions arise about supervision. Documentation of what training covered and when allows identification of gaps that need addressing. Maintain this documentation as part of your practice's governance records.

New lawyer onboarding should include AI training as a standard component. Lawyers joining your practice should receive foundational training and tool-specific training as part of their onboarding process. They should be introduced to your prompt standards and review expectations from the start. Early training prevents bad habits from forming and ensures new lawyers can contribute effectively to AI-assisted workflows immediately. Don't assume new lawyers already know what they need; training and experience vary widely.

Bar associations are beginning to require or recommend AI competence for practicing lawyers. Some jurisdictions have issued ethics opinions addressing AI use, and more guidance is expected. Training programs that establish competence and document that training occurred position lawyers to demonstrate compliance with evolving competence expectations.

Feedback mechanisms help training programs improve. Lawyers using AI encounter situations that training didn't address, discover techniques worth sharing, and identify gaps in their preparation. Systems for capturing this feedback, whether formal channels or informal communication, turn daily practice into training improvement. Training that incorporates feedback becomes progressively more useful; training developed in isolation from practice reality stagnates.

Culture matters as much as formal training. A practice culture that encourages experimentation, rewards sharing what works, and treats AI mistakes as learning opportunities will develop AI capability faster than a culture that punishes experimentation or hoards individual knowledge. Leaders should model appropriate AI use, share their own learning publicly, and create psychological safety for lawyers developing new skills. Formal training programs operate within cultural context that significantly affects their effectiveness.

Competency assessment verifies that training achieved its purpose. Assessment might include practical exercises demonstrating skill application, review of actual AI-assisted work product, or observation of AI use in practice. Assessment identifies lawyers who need additional support and validates that training programs are effective. Without assessment, you don't know whether training translated into competence. Build assessment into your training approach, not as punishment for failure but as information for improvement.

> **Key Point** Training transforms individual AI skill into organizational capability by establishing standards everyone follows and documentation that preserves knowledge. Foundational training covers how AI works, professional responsibility obligations, and prompt engineering fundamentals. Prompt standards codify effective approaches for recurring tasks. Review expectations define verification requirements for different work types and risk levels. Tool-specific training addresses the particular tools your practice uses. Workflow integration shows how AI fits into matter workflows. Hands-on practice develops skill more effectively than passive instruction. Supervision training ensures supervisors can effectively oversee AI-assisted work. Ongoing training maintains competence as AI evolves. Documentation serves reference, compliance, and improvement purposes. Culture that encourages experimentation and sharing accelerates capability development. Throughout, the goal is building organizational capability that does not depend on individual skill.

11.6 Mini-Lab: Create a Matter Playbook + Prompt Pack

Objective: Apply the practice management techniques from this chapter to create a matter playbook for a matter type you handle regularly. Practice capturing workflow knowledge, identifying common issues, and developing integrated prompts that accelerate future matters. This exercise demonstrates how to systematize expertise so that each matter benefits from all previous matters of that type.

A matter playbook starter template and the trade secret defense scenario are available on the companion site for download.

Time Required: 90 to 120 minutes

What You'll Need

- The playbook concepts from Section 11.1
- The knowledge management principles from Section 11.2
- The template concepts from Section 11.4
- Access to an AI tool for development assistance
- Notes or recollections from past matters of your chosen type
- A method for documenting your playbook

Step 1: Select Your Matter Type

Choose a matter type you handle with some regularity. The matter type should be specific enough to have consistent patterns but broad enough to occur multiple times per year. Good candidates include:

- A specific transaction type (commercial lease, asset purchase, Series A financing)

- A specific litigation type (employment discrimination defense, contract dispute, personal injury)
- A specific regulatory matter type (licensing application, compliance audit response)
- A specific advisory matter type (entity formation, estate plan, policy review)

If you don't have a matter type from your own practice, use the following scenario for this exercise:

> **Scenario Matter Type: Trade Secret Misappropriation Defense**
>
> You represent companies accused of misappropriating trade secrets, typically when a former employee of another company joins your client and the former employer claims the employee brought confidential information. These matters typically involve temporary restraining order motions at the outset, followed by discovery focused on what information the employee accessed and brought, expert testimony on whether the information qualifies as a trade secret, and either dispositive motions or trial on liability and damages.

Write down your selected matter type and a brief description of what these matters typically involve.

Step 2: Map the Phase Structure

Identify the typical phases these matters progress through. For each phase, note:

- Phase name and description
- Typical duration
- Key activities during this phase
- Deliverables produced
- Decision points or milestones
- Transition triggers to the next phase

Use AI assistance to help structure your phases.

> *"I handle [matter type] matters regularly. Help me map the typical phases these matters progress through from inception to conclusion. For each phase, identify the key activities, typical deliverables, and decision points. I'll refine based on my actual experience."*

Review and refine the AI-generated phases based on your actual experience. The AI output provides a starting framework; your experience provides the accuracy.

Step 3: Identify Common Issues

List the issues that commonly arise in this matter type. For each issue:

- Describe the issue clearly
- Explain why it matters (what goes wrong if it's not addressed)
- Provide actionable guidance for handling it

- Note any variations based on specific circumstances
- Cross-reference to relevant phases where the issue typically arises

Think through past matters you've handled. What problems recurred? What did you wish you'd known earlier? What advice would you give a colleague handling this matter type for the first time?

> *"For [matter type] matters, what issues commonly arise that practitioners should be prepared for? For each issue, explain why it matters and suggest practical approaches for handling it. Focus on practical workflow issues, not just legal doctrine."*

Supplement AI-generated issues with issues from your own experience that the AI may not have identified.

The most valuable playbook content often comes from mistakes you've made or near-misses you've experienced. What went wrong in past matters that you don't want to repeat? What did you discover late that you should have addressed early? These hard-won lessons are exactly what playbooks should capture.

Step 4: Document Key Documents

Identify the key documents in this matter type. For each document:

- Document name and purpose
- When in the matter lifecycle it's typically produced
- Critical elements that require attention
- Common problems to watch for
- Cross-reference to any templates you have or should develop
- Prompts that help with drafting or reviewing this document

Consider both documents you produce and documents you receive and must analyze.

> *"For [matter type] matters, what are the key documents that practitioners produce or analyze? For each document, identify critical elements requiring attention and common problems to watch for."*

Step 5: Identify Deadlines and Triggers

Document the time-sensitive elements of this matter type:

- Statutory or regulatory deadlines (statutes of limitations, filing requirements, notice periods)
- Court-imposed deadlines typical for this matter type
- Contractual deadlines that commonly apply
- Trigger events that activate obligations or require responses
- Internal deadlines for quality work (when to start drafting to allow adequate review time)

For each deadline or trigger: describe what it is, identify how to calculate or identify it, note consequences of missing it, and suggest calendar and tracking approaches.

> **Risk Note** Missed deadlines are a leading source of malpractice claims. Playbooks should clearly identify every deadline and trigger that applies to the matter type, with guidance on tracking and ensuring compliance. When in doubt, include the deadline.

Step 6: Capture Lessons Learned

Document insights from past matters that should inform future ones. These lessons might include:

- What worked well that you want to repeat
- What didn't work that you want to avoid
- Surprises that you want to be prepared for next time
- Efficiency improvements you've discovered
- Relationship or communication insights

For each lesson: describe it clearly, explain the context that generated the insight, provide actionable guidance for applying it, and note any limitations on when it applies.

> *"For trade secret misappropriation defense matters, what lessons do experienced practitioners typically learn that they wish they'd known earlier? Focus on practical insights about workflow, strategy, and common pitfalls."*

Step 7: Develop the Integrated Prompt Pack

Create prompts for recurring tasks in this matter type. For each major phase and key document, develop at least one prompt that would accelerate the work. Your prompt pack should include prompts for:

- Initial matter assessment
- Key research tasks specific to this matter type
- Drafting key documents
- Analysis tasks (document review, deposition preparation, etc.)
- Client communication specific to this matter type

Each prompt should follow the Prompt Blueprint structure from Chapter 3: Role, Context, Task, Format, Guardrails.

Prompts should be specific enough to be useful but flexible enough to adapt to particular matters. A prompt that requires extensive modification for each use provides little value. A prompt that works across matters with only fact-specific customization provides significant value.

> *"I'm creating a prompt pack for [matter type] matters. Develop a prompt for [specific task] that follows this structure: Role, Context, Task, Format, Guardrails. The prompt should be reusable across matters of this type with minimal customization. Include placeholders for matter-specific information."*

Step 8: Organize and Format the Playbook

Compile your work into a structured playbook document. An effective format might include:

- Overview: One-page summary of the matter type and playbook purpose
- Phase Structure: Visual or narrative map of matter progression
- Common Issues: Organized by phase or by issue category, with actionable guidance
- Key Documents: List with critical elements and cross-references to prompts
- Deadlines and Triggers: Comprehensive list with tracking guidance
- Lessons Learned: Organized insights from past experience
- Prompt Pack: All prompts organized by phase or task type
- Templates Reference: Cross-references to any templates used for this matter type
- Resources: External resources, checklists, or references relevant to this matter type

Format the playbook for actual use. A playbook that looks impressive but is difficult to navigate provides less value than a simpler document that lawyers actually consult. Consider both digital use (searchable, hyperlinked) and potential print use (clear organization, readable formatting).

Step 9: Test and Refine

Review your playbook critically:

- Completeness: Does it cover all phases and major issues? Are there gaps a lawyer using this playbook would notice?
- Accuracy: Does the guidance reflect current law and practice? Are there elements that might be outdated or jurisdiction-specific?
- Usability: Can a lawyer unfamiliar with this matter type follow the playbook effectively? Is the organization intuitive?
- Prompt Quality: Do the prompts follow the Prompt Blueprint? Would they produce useful outputs with minimal customization?

If possible, have a colleague review the playbook and provide feedback. Fresh eyes often spot gaps or unclear guidance that the creator misses.

Plan to refine the playbook after using it on actual matters. No playbook is complete on first draft. Build in a process for capturing refinements as you apply the playbook and discover what's missing or what doesn't work as expected.

Step 10: Document Your Process

Create a brief record of how you developed this playbook:

- What matter type did you select and why?
- What sources of knowledge did you draw on (past matters, AI assistance, colleague input)?
- What was most valuable about AI assistance in the development process?
- What elements required your expertise that AI could not provide?
- How will you maintain and update this playbook over time?

This documentation helps you replicate the process for other matter types and contributes to your practice's knowledge about effective playbook development.

Scenario Analysis Guide

If you used the trade secret misappropriation defense scenario, use the following guide to evaluate your playbook. If you used your own matter type, adapt this guide to assess your work against the standards it reflects.

Phase Structure Analysis

A comprehensive phase structure for trade secret defense might include:

- Phase 1: Emergency Response (Days 1–14) — Respond to TRO motion if filed, conduct rapid factual investigation, interview accused employee, secure evidence.
- Phase 2: Initial Case Assessment (Weeks 2–6) — Analyze complaint, identify claimed secrets, assess protectability, evaluate damages exposure.
- Phase 3: Pleading and Early Motion Practice (Weeks 4–12) — Answer, counterclaims, early dispositive motions.
- Phase 4: Discovery (Months 3–12) — Discovery focused on trade secret identification, depositions, expert retention.
- Phase 5: Expert Phase (Months 8–14) — Expert reports, Daubert challenges.
- Phase 6: Dispositive Motions (Months 12–16) — Summary judgment on identification, protectability, or misappropriation.
- Phase 7: Trial Preparation and Trial (Months 14–20) — Trial materials, verdict or settlement.

Common Issues Analysis

Key issues should include:

- **Trade secret identification:** Forcing plaintiff to identify secrets with particularity.
- **Employee knowledge vs. company information:** Distinguishing general skills from specific confidential information.
- **Independent development defense:** Investigating independent creation.
- **Evidence preservation:** Preventing spoliation allegations.
- **Inevitable disclosure doctrine:** Jurisdiction-specific defense strategies.
- **Damages complexity:** Actual damages vs. unjust enrichment vs. royalties.
- **Injunction scope:** Negotiating appropriate limitations.

Trade secret cases often involve parallel employment agreement claims (non-compete, non-solicitation). Ensure your playbook addresses these related claims that frequently accompany trade secret allegations.

Key Documents Analysis

Key documents should include:

- TRO/preliminary injunction opposition
- Answer and counterclaims
- Trade secret identification discovery
- Technical expert report
- Summary judgment motion

Prompt Pack Analysis

Your prompt pack should include prompts for:

- Initial case assessment: Analyzing complaint and assessing risks
- Discovery request drafting: Interrogatories and document requests forcing identification
- Deposition preparation: Outline for deposing plaintiff's witness on trade secret identification

Test your prompts by actually using them with AI. Prompts that seem complete on paper sometimes produce disappointing results in practice. Refine based on actual outputs until prompts consistently produce useful work product.

Deliverable

By the end of this Mini-Lab, you should have:

1. A selected matter type with brief description
2. A documented phase structure with key activities and deliverables
3. A common issues section with actionable guidance
4. A key documents section with handling guidance
5. A deadlines and triggers section
6. A lessons learned section
7. An integrated prompt pack with at least five prompts
8. A formatted playbook document
9. Documentation of your development process

Bonus Challenge

Extend your playbook in three directions:

1. New matter checklist: Create a checklist for opening a new matter of this type.
2. Client communication template set: Develop templates for recurring client communications in this matter type.
3. Post-matter review protocol: Create a structured process for reviewing completed matters to capture lessons learned.

Some courts and practice areas have developed their own checklists or practice guides. Review whether such resources exist for your matter type and incorporate relevant elements into your playbook. Don't duplicate effort when authoritative guidance already exists.

11.7 Practice Checklist

Use this checklist to ensure your practice management systems are comprehensive and effective. A printable version of this checklist is available on the companion site. Playbook Development

- ☐ Selected matter types appropriate for playbooks
- ☐ Mapped phases, deliverables, and decision points
- ☐ Documented common issues and guidance
- ☐ Identified key documents and handling instructions
- ☐ Listed deadlines and triggers
- ☐ Captured lessons learned from past matters

☐ Integrated reusable prompts

☐ Established maintenance schedule Knowledge Management

☐ Tagged research memos for retrieval

☐ Organized clause library with annotations

☐ Maintained brief bank with conceptual search support

☐ Verified context for retrieved content before reuse

☐ Curated repositories to remove outdated material

☐ Established contribution incentives

☐ Integrated KM checks into standard workflows

☐ Secured confidential information within repositories Billing and Time

☐ Implemented contemporaneous time recording

☐ Used AI to improve description quality (not accuracy)

☐ Avoided block billing where possible

☐ Conducted prebill review for accuracy and appropriateness

☐ Ensured efficiency gains benefit clients

☐ Disclosed AI use in billing where appropriate

☐ Trained team on billing ethics and AI assistance Templates

☐ Developed document templates for recurring types

☐ Created prompt templates for AI tasks

☐ Organized template library for discoverability

☐ Assigned ownership for template maintenance

☐ Established customization guidelines

☐ Implemented feedback loops for improvement

☐ Checked for court-specific template requirements Training and Governance

☐ Delivered foundational AI training

☐ Established prompt standards

☐ Defined review expectations by work type

☐ Provided tool-specific training

☐ Integrated AI into workflow training

☐ Supported hands-on practice

☐ Trained supervisors on oversight

☐ Maintained training documentation

Chapter Checkpoint

Before moving on, make sure you can:

- ☐ Create matter playbooks that capture phase structure, common issues, key documents, deadlines, and lessons learned for matter types you handle regularly
- ☐ Develop integrated prompt packs aligned with playbook phases and tasks, providing reusable prompts that accelerate recurring work
- ☐ Implement knowledge management practices that preserve research memos, clause libraries, and brief sections for future retrieval and adaptation
- ☐ Apply controlled reuse principles, verifying that retrieved content fits new contexts before incorporating it
- ☐ Use AI to improve billing narrative quality while maintaining accuracy and professionalism in time entries
- ☐ Develop and maintain document templates and prompt templates that ensure consistency across your practice
- ☐ Establish prompt standards and review expectations that create organizational capability rather than depending on individual skill
- ☐ Design training programs that cover foundational AI skills, tool-specific guidance, and workflow integration with ongoing updates as capabilities evolve
- ☐ Build feedback mechanisms that improve playbooks, templates, and training based on actual use

What's Next

Part III addressed the core workflows where lawyers produce value: research, drafting, litigation, client communications, and practice management. These chapters provided techniques for applying AI to specific tasks and systems for capturing what you learn.

Part IV shifts focus from individual and practice-level concerns to organizational governance and the broader landscape of AI regulation and evolution. Governance determines whether AI use across an organization is consistent, compliant, and effective. Chapter 12 addresses the policies, training programs, audit trails, and incident response procedures that organizations need to manage AI use responsibly. Who decides what AI tools are approved? What policies govern acceptable use and data handling? How do you ensure that lawyers follow established standards? What happens when something goes wrong? These governance questions become increasingly important as AI use scales beyond individual adoption to organization-wide deployment.

Governance models vary based on organization size and structure. A solo practitioner governs their own AI use through personal discipline and documentation. A small firm might rely on informal coordination among partners. A large organization needs formal governance structures: designated decision-makers, written policies, training requirements, and audit mechanisms. Chapter 12 addresses governance across this spectrum, recognizing that appropriate governance scales with organizational complexity.

Policies that work are specific enough to provide guidance but flexible enough to accommodate evolving capabilities and diverse practice needs. Vague policies that say "use AI responsibly" provide no real guidance. Rigid policies that specify exactly which tools and techniques are permitted become outdated as capabilities change. Effective policies establish principles and guardrails while allowing professional judgment within those boundaries. Chapter 12 addresses how to draft policies that actually guide behavior without creating bureaucratic obstacles to beneficial AI use.

Audit trails document how AI was used, enabling review if questions arise and demonstrating appropriate process even when nothing went wrong. Chapter 4 introduced audit-ready habits; Chapter 12 extends these concepts to organizational audit requirements. What should be logged? How long should logs be retained? Who reviews them? These questions require answers that balance documentation value against administrative burden.

Incident response addresses what happens when AI use causes problems. A hallucinated citation in a filed brief, a confidentiality breach through an AI tool, or a client complaint about AI-assisted work all require response. Chapter 12 addresses how to contain incidents, document what happened, make required notifications, and learn from failures to prevent recurrence.

Beyond governance, Chapter 13 addresses the regulatory landscape and the future evolution of AI capabilities. AI regulation is emerging across jurisdictions, with different approaches to transparency requirements, risk classification, and liability allocation. Understanding these regulatory trends helps you anticipate compliance requirements and position your practice for changes ahead. The chapter also addresses emerging capabilities, from AI agents that take actions autonomously to multimodal tools that process images and audio alongside text, helping you understand what's coming and how to prepare. Finally, Chapter 13 identifies the durable human skills that will remain valuable regardless of how AI capabilities evolve, and provides a practical roadmap for continuing your AI integration beyond this book.

Key Terms

Matter playbook — A structured guide documenting the practical workflow for handling a specific matter type, capturing what experienced practitioners know about effective approaches, common issues, and lessons learned.

Trigger events — Occurrences during a matter that require specific responses or activate particular obligations, as distinguished from calendar deadlines that arrive on fixed dates.

Knowledge management — The systematic capture, organization, and reuse of valuable work product and expertise across a practice, preventing the waste of recreating work that has already been done.

Clause library — An organized collection of contract provisions preserved for reuse, including the clause language, annotations explaining its use and history, and metadata for retrieval.

Controlled reuse — Verifying that retrieved work product fits the current context before incorporating it, including checking for legal changes, jurisdictional differences, and factual distinctions that might make past work inapplicable or misleading.

Contemporaneous time recording — Capturing time entries as work is performed or immediately after, rather than reconstructing time at the end of a day or week.

Prebill review — Reviewing invoices before sending them to clients, verifying accuracy, appropriateness, and professionalism of all entries.

Document template — A standardized starting point for a recurring document type, capturing preferred structure, standard language, formatting conventions, and quality standards.

Prompt template — A standardized starting point for recurring AI-assisted tasks, capturing effective instructions, output specifications, and quality requirements.

Foundational AI training — Baseline education that all lawyers using AI should receive, covering how AI works, professional responsibility obligations, and fundamental prompt engineering skills.

Tool-specific training — Education focused on the particular AI tools a practice uses, covering features, capabilities, limitations, and best practices specific to those tools as distinct from general AI skills that transfer across platforms.

Phase structure — Documentation of how matters of a specific type typically progress, identifying stages, milestones, and transitions that help lawyers understand where they are in a matter and what comes next.

Integrated prompt pack — A collection of prompts designed for a specific matter type, organized to align with the matter's phase structure and key tasks, allowing lawyers to access relevant prompts at each stage of the matter.

Chapter 12: Governance: Policies, Training, Audit Trails, Incident Response

Individual competence is necessary but not sufficient. A lawyer who uses AI skillfully produces good work on their matters. An organization where every lawyer uses AI skillfully, following consistent standards, with appropriate oversight and documentation, produces good work across all matters while managing risk at an organizational level. The gap between individual competence and organizational capability is governance: the structures, policies, and processes that ensure AI use across the organization is consistent, compliant, and effective. Governance transforms ad hoc individual practices into systematic organizational approaches that scale as AI use expands.

Governance becomes essential as AI adoption moves from early experimentation to routine use. When a few lawyers experiment with AI, informal coordination may suffice. When AI becomes embedded in how the organization works, informal approaches break down. Which tools are approved? What data can be input? What review is required before AI-assisted work goes out the door? Who decides when new tools can be adopted? Without clear answers to these questions, practices diverge, risks multiply, and the organization loses control over how AI is used in its name.

The governance challenge is designing structures that provide necessary oversight without creating bureaucratic obstacles that prevent beneficial use. Governance that is too restrictive drives AI use into the shadows, creating the very risks it seeks to prevent. Governance that is too loose fails to provide the guidance and protection that lawyers need. Effective governance balances enablement with control, creating a framework where responsible AI use is the path of least resistance.

What This Chapter Covers

- Governance models for different organization sizes, from solo practices to large firms
- Developing policies that provide clear guidance on acceptable use, data handling, and review requirements
- Designing training programs that build organizational competence and ensure policy compliance
- Establishing audit trails that document AI use for accountability and defensibility
- Implementing quality assurance processes that verify compliance with established standards
- Creating incident response protocols for when AI use goes wrong
- Vendor management strategies for overseeing third-party AI providers

Chapter Summary

Governance models determine who makes decisions about AI use and how those decisions are enforced. Models range from centralized committees to distributed practice group leadership, but all require clear authority and accountability. Policies translate governance decisions into rules that guide daily behavior. Effective policies address acceptable use, data classification and handling, required review and verification, and disclosure obligations. Vague policies like "use AI responsibly" fail to provide necessary guidance; specific policies that address real scenarios succeed.

Training is the primary mechanism for implementing governance. Training programs must cover not just tool mechanics but also policy requirements, professional responsibility obligations, and the specific workflows the organization has adopted. Documentation through audit trails creates the record that demonstrates compliance and supports accountability. What to log, how long to retain logs, and who reviews them are governance decisions that balance transparency against administrative burden.

Quality assurance moves beyond individual verification to organizational oversight. Spot checks, peer review, and systematic monitoring ensure that standards are actually being followed. Incident response protocols prepare the organization for inevitable failures, ensuring that when a hallucination is filed or a confidentiality breach occurs, the response is managed, documented, and effective. Vendor management extends governance to the third-party providers on whom the organization relies, ensuring their practices align with the organization's obligations.

Throughout, the goal of governance is not to eliminate risk but to manage it to an acceptable level while enabling the organization to capture the benefits of AI. Governance that prevents all risk also prevents all progress. Governance that enables responsible innovation creates a competitive advantage.

12.1 Governance Models: Who Owns AI Decisions

Governance begins with clarity about decision rights.

Who has the authority to approve a new AI tool? Who decides whether a particular type of client data can be entered into a system? Who determines whether AI-assisted work meets quality standards? Without defined decision rights, organizations drift. IT might approve a tool based on security while the General Counsel prohibits it based on liability concerns. Associates might use tools that partners haven't authorized. Practice groups might adopt incompatible standards.

Governance models vary based on organization size and culture, but three primary models are common: Centralized, Distributed, and Hybrid.

The Centralized Model places authority in a single body, often an AI Committee or Innovation Committee comprising representatives from legal, IT, risk, and practice leadership. This model ensures consistency and rigorous risk management. Decisions are uniform across the organization. The downside is speed; centralized bodies can become bottlenecks that slow adoption and frustrate practitioners who need tools now. Centralized governance works best for foundational decisions—which enterprise platforms to license, what the baseline policies are—but may struggle with practice-specific nuances.

The Distributed Model pushes authority down to practice groups or offices. The litigation group decides what litigation tools to use; the corporate group decides on transactional tools. This model enables speed and ensures that decisions reflect specific practice needs. Litigators understand litigation risks better than a central committee might. The downside is inconsistency; the firm might end up with fragmented tools, incompatible data standards, and varying levels of risk tolerance that create liability exposure. Distributed governance works best for tool selection within approved categories but requires central guardrails.

The Hybrid Model combines central standards with local implementation. A central body establishes the risk framework, approves core enterprise tools, and sets baseline policies. Practice groups then have authority to select specific tools within that framework and to develop practice-specific workflows. This model attempts to balance consistency with agility. It requires clear communication about where central authority ends and local authority begins.

For solo practitioners and small firms, governance is personal discipline documented as policy. You are the committee. Governance means taking the time to write down your rules for yourself and your staff, rather than making ad hoc decisions. "I will only use enterprise-grade tools for client data" is a governance decision. Documenting it transforms it from an intention into a policy.

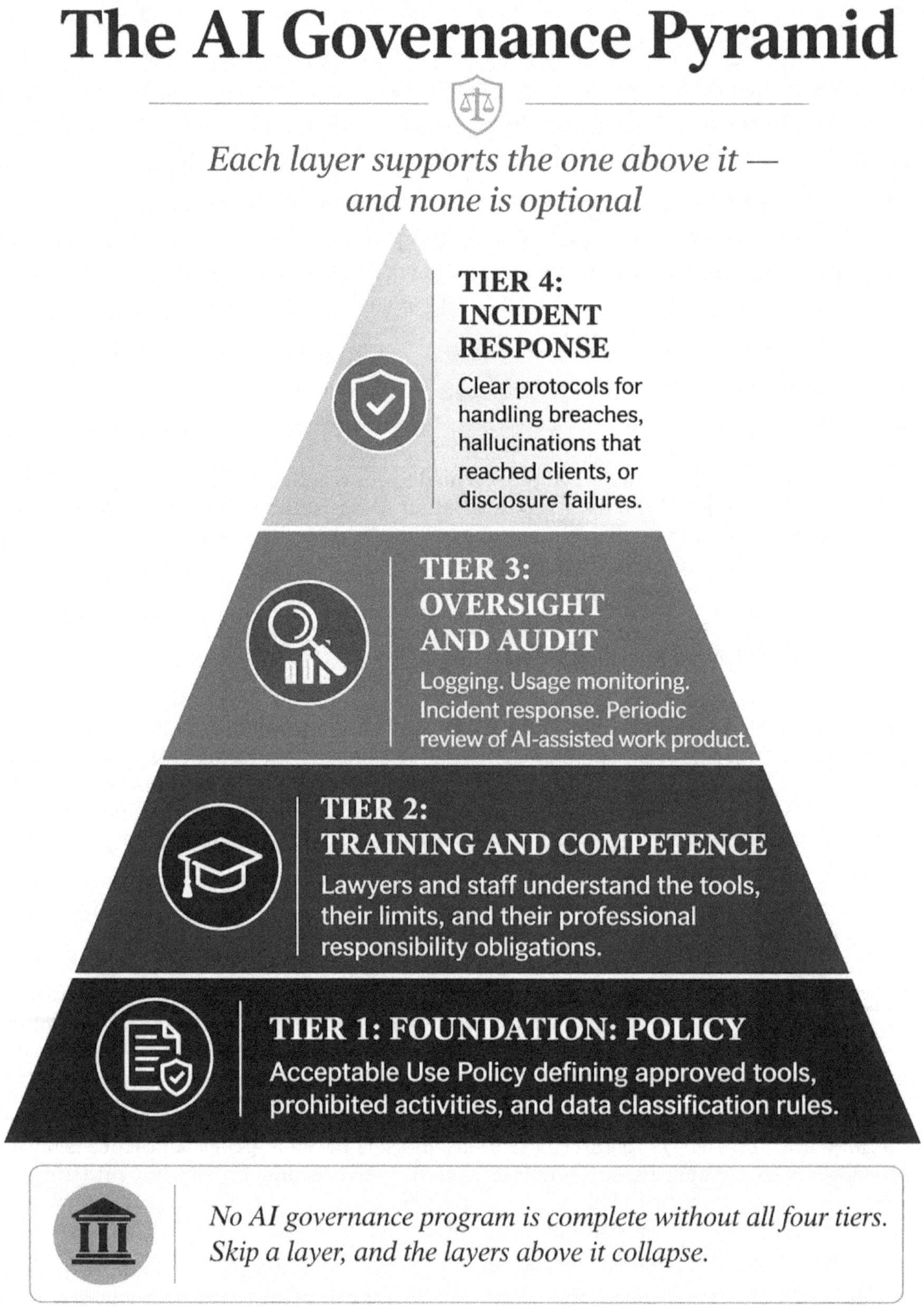

Figure 12.1 — The AI Governance Pyramid

For mid-sized firms, governance often lives with a designated partner or a small committee. The challenge is ensuring that this responsibility is real, not nominal. The designated leader needs time and resources to evaluate tools and monitor usage. "The tech partner handles that" fails if the tech partner is billing 2,000 hours and has no time to actually govern.

For large organizations, governance requires dedicated resources. Knowledge management professionals, innovation teams, and risk counsel play key roles. The challenge is connecting governance to practice. Policies written by committees that don't practice law often fail to address the reality of legal work. Governance bodies must include practicing lawyers to ensure relevance and buy-in.

> **Practice Point** If you are in a small firm without a formal committee, designate an "AI Lead," a lawyer rather than just IT staff, responsible for staying current on AI risks and opportunities. Having one person whose job includes thinking about governance prevents AI issues from falling through the cracks.

Decision frameworks help governance bodies make consistent choices. A framework might categorize decisions by risk level: - Low Risk: Tools using only public data, non-generative tools, or established vendors. Approval can be expedited or delegated. - Medium Risk: Generative tools using internal firm data (not client data), or tools assisting with low-stakes drafting. Approval requires standard review. - High Risk: Tools processing confidential client data, tools generating substantive legal advice, or client-facing AI. Approval requires full committee review and specific risk mitigation plans. This tiered approach prevents the committee from being overwhelmed by minor decisions while ensuring high-stakes decisions receive appropriate scrutiny.

Governance also defines accountability. Who is responsible if an AI tool causes a problem? Is it the individual lawyer who used it? The practice group leader? The committee that approved it? Clarity about accountability aligns incentives. If lawyers know they remain individually responsible for AI-assisted work (as professional ethics require), they are more likely to verify output. If the firm assumes all responsibility, individual vigilance may decline. Governance must reinforce, not dilute, individual professional responsibility.

> **Risk Note** Avoid "Governance Theater," creating committees and policies that exist on paper but do not influence actual behavior. If the committee meets once a year and policies are never read, you have no governance. Effective governance is visible, active, and integrated into daily practice.

Finally, governance must be adaptive. The AI landscape changes monthly. A governance structure that requires six months to approve a tool will be permanently behind. Governance bodies need mechanisms for rapid review and pilot programs that allow testing before full approval. "Provisional approval" for a defined pilot group allows the firm to learn about a tool's risks and benefits without committing the entire organization.

Governance defines who decides. Whether centralized, distributed, or hybrid, an effective model clarifies authority, establishes accountability, and balances risk management with innovation. Solo and small firms need governance just as much as large ones, even if the structure is simpler. Governance must be active and adaptive to keep pace with rapid technological change.

12.2 Policies That Work: Acceptable Use, Data Handling, Review Gates

Policies are the operating instructions for governance. They translate abstract principles ("use AI responsibly") into concrete rules ("do not enter client names into ChatGPT"). Good policies are clear, enforceable, and realistic. Bad policies are vague, ignored, or so restrictive they prevent work from getting done.

An effective AI policy needs to address four core areas: **Acceptable Use, Data Handling, Review Requirements,** and **Disclosure.**

Acceptable Use

Defines which tools are permitted and for what purposes.

- **Approved Tools List:** Explicitly list which tools are authorized. *"Use only the firm-provisioned Enterprise GPT-4 environment."*
- **Prohibited Tools:** Explicitly ban categories of tools if necessary. *"Do not use personal accounts for free-tier consumer AI tools for any work purposes."*
- **Prohibited Tasks:** Identify tasks where AI use is forbidden. *"Do not use AI to generate final legal opinions without human drafting." "Do not use AI to make final privilege determinations."*

Data Handling

Implements the confidentiality framework from Chapter 5.

- **Data Classification:** Define what data can go where. *"Public data may be used with Tool A. Client confidential data may ONLY be used with Tool B (Enterprise)."*
- **Anonymization:** Require specific anonymization steps before data entry. *"Remove client names, deal values, and specific locations before using Tool C."*
- **Retention:** Specify settings. *"Ensure 'history off' or 'no training' settings are active."*

Review Requirements

Operationalizes the supervision and candor obligations from Chapter 4.

- **Verification Mandatory:** All AI-generated citations, facts, and legal assertions must be independently verified against authoritative sources.
- **No "Copy-Paste":** AI output must never be copied directly into court filings without line-by-line review.
- **Supervision:** Associates must inform supervising partners when AI has been used for substantive drafting.

Disclosure

Addresses transparency with clients and courts.

- **Client Notification:** Follow engagement letter terms regarding AI disclosure.
- **Court Rules:** Comply with all standing orders and local rules regarding AI disclosure.

Drafting Policies for Compliance

Policies should be written in plain language, avoiding tech jargon where possible. They should be accessible — stored on the intranet homepage, not buried in an HR manual.

- **Bad Policy:** *"Employees must exercise due care with AI and protect confidential data."* (Too vague.)
- **Good Policy:** *"Do not enter client names, deal terms, or non-public case facts into any web-based AI tool unless it is on the 'Approved for Confidential Data' list."* (Specific and actionable.)

Updating Policies AI policies cannot be static. A policy written two years ago is likely obsolete today. Establish a review cycle—quarterly or bi-annually—to update the Approved Tools list and refine rules based on new capabilities and risks.

Enforcement and Amnesty A policy without enforcement is a suggestion. Firms need mechanisms to detect policy violations (see Audit Trails, below). However, initially, consider an "amnesty" approach or a focus on education rather than punishment. If a lawyer used a consumer tool by mistake, the goal is to get them onto the secure enterprise tool, not to fire them. Reporting mistakes should be encouraged so the firm can assess if any data breach occurred.

> Practice Point Do not ban AI entirely. A policy of "No AI allowed" is often ignored, leading to widespread shadow AI use. It is safer to provide an approved, secure path for AI use than to ban it and drive users to insecure personal accounts.

Client-Specific Policies Some clients impose their own AI guidelines, often stricter than the firm's. Governance must track these "outside counsel guidelines." If Client Bank prohibits all AI use, the firm's policy must ensure that lawyers working on Client Bank matters are aware of and follow that restriction, even if the firm generally allows AI.

> Risk Note Inconsistency creates liability. If your policy says "All AI citations must be verified," and a lawyer fails to verify one that turns out to be fake, the policy violation itself becomes evidence of negligence. Do not write policies you do not intend to enforce or cannot realistically follow.

The "Human in the Loop" Standard Almost all AI policies should articulate a "Human in the Loop" (HITL) standard. This means AI never executes a final action (filing a brief, sending a settlement offer, executing a trade) without human review. This is the ultimate backstop against AI error.

Policies translate governance into rules. Effective policies are specific about tools (Approved/Prohibited), data (Classification/Handling), and behavior (Verification/Disclosure). They must be accessible, realistic, and updated frequently. A "Human in the Loop" requirement is the fundamental safety valve for all professional AI use. Appendix D provides a comprehensive framework of governance questions to resolve with your own counsel before finalizing AI policies.

12.3 Training and Competency: Building Reliable Habits

Policy tells people what to do; training teaches them how to do it. You cannot govern AI use effectively if your lawyers don't know how to use the tools or don't understand the risks. Under Model Rule 1.1 (Competence), training is not just an HR benefit; it's a professional requirement.

Effective AI training moves beyond "tech training" (how to click the buttons) to "process training" (how to work responsibly). It should cover three dimensions: **Tool Fluency, Risk Awareness,** and **Prompt Engineering.**

Tool Fluency: Lawyers need to know how the specific approved tools work. If the firm uses a legal research AI platform, training should cover its specific features, how it differs from traditional search, and how to navigate its interface. If the firm provides a secure GPT wrapper, lawyers need to know how to access it and how to manage conversation histories.

Risk Awareness: This is the compliance layer. Training must cover the "Why" behind the policies.

- **Hallucination:** Show examples of AI fabricating cases. Let lawyers see it happen so they understand the danger is real.

- **Confidentiality:** Explain how training data works and why entering client data into public tools is a breach.
- **Bias:** Discuss how AI can reflect historical biases and why judgment is needed.
- **Verification:** Teach the specific verification protocols (e.g., "Check every citation in Westlaw").

Prompt Engineering: This is the skill layer (see Chapter 3). Training lawyers to write better prompts improves efficiency and quality. A lawyer who knows how to use role, context, and constraints gets better work product than one who types generic queries. Investing in prompt training turns AI from a toy into a tool.

Mandatory AI training is becoming a standard expectation for technological competence. Just as firms mandate harassment training or cybersecurity awareness training, AI competence training ensures a baseline of responsible practice across the organization.

Training Formats

- **On-Demand Basics:** Short videos or modules covering the core policy and risks. Every new hire should take this.
- **Practice-Specific Workshops:** Live sessions tailored to specific groups. "AI for Litigators" covers different tools and risks than "AI for Deal Lawyers."
- **Sandbox Sessions:** Guided hands-on time where lawyers can play with tools in a safe environment using non-confidential data. This builds confidence and reveals limitations.
- **Lunch & Learns:** Regular updates on new features or emerging risks.

Certification

Some firms implement internal certification. To get access to the advanced AI tools, a lawyer must complete the training module and pass a short quiz on the risks and policies. This "license to drive" approach ensures that powerful tools are put in educated hands.

Mentorship and Reverse Mentorship

AI skill doesn't always correlate with seniority. Often, junior associates are more fluent than partners. Governance should encourage "reverse mentorship," where juniors help partners understand tool capabilities, while partners teach juniors about legal judgment and risk. This cross-pollination strengthens the organization.

Training as Governance

Training is also a risk management tool. If an error occurs, the ability to show that the firm provided comprehensive training on proper AI use helps demonstrate that the firm took reasonable measures to ensure compliance (satisfying supervisory duties under Rules 5.1/5.3). It distinguishes a rogue act from a systemic failure.

Include "AI Spotting" in training. Teach lawyers how to recognize AI writing in documents they receive from others — opposing counsel, clients, or vendors. Recognizing the "generic AI voice" or hallucinated citations in incoming work is a new defensive skill.

The "Drift" Problem

Skills decay and tools change. Training cannot be "one and done." Governance requires a schedule for refresher training, especially when major model updates occur (e.g., the shift from GPT-3.5 to GPT-4 changed the risk profile significantly).

Training bridges the gap between policy and practice. It must cover tool fluency, risk awareness (especially hallucinations and confidentiality), and prompt engineering. Use varied formats and consider internal certification to control access. Training documents the firm's commitment to competence and supervision.

12.4 Audit Trails: What to Log and Why

In a digital environment, if you didn't log it, you can't prove it happened (or didn't happen). Audit trails are the evidence of governance. They allow the organization to monitor compliance, investigate incidents, and defend its practices.

What to Log

Privacy concerns balance against oversight needs. You typically don't want to log every keystroke, but you need enough visibility to manage risk.

- **Tool Usage:** Who is using which tool? High usage might indicate a champion; zero usage might indicate a need for training (or use of shadow AI).
- **Prompt Inputs:** Logging the actual text of prompts is sensitive. It captures the user's intent but creates a repository of potentially confidential queries.

    Governance Decision: Many firms log prompts but restrict access to those logs strictly to the General Counsel or compliance officer, accessed only for cause.

- **Outputs:** Logging generated text helps in reconstructing work product if accuracy is challenged later.
- **Verification Steps:** Ideally, workflow tools should allow users to "check a box" or record that they verified the output. "Citation check completed: [Date]."

Why Log?

- **Compliance Monitoring:** Are users pasting large blocks of client contracts into tools not approved for confidential data? Logs can trigger alerts for data leakage patterns (e.g., regex searches for "Confidential" or social security number patterns).
- **Cost Allocation:** If tools charge by usage, logs allow the firm to allocate costs to specific matters or practice groups.
- **Incident Investigation:** If a brief is filed with a fake citation, logs allow the firm to determine: Did the lawyer ask for citations? Did the AI provide the fake one? Or did the lawyer type it in manually? This factual record is crucial for internal discipline and external defense.
- **Usage Analytics:** Understanding how people use AI helps refine training and purchasing. If everyone is using the tool for summarization but no one is using it for drafting, the firm can adjust its strategy.

Retention Policies

How long do you keep AI logs?

- **Billing records:** Keep according to financial retention policies.
- **Substantive prompts/outputs:** Consider keeping them with the matter file. If the AI output informed legal advice, it is part of the file. If it was just brainstorming, it might be transient.
- **General logs:** Often retained for a set period (e.g., 6 months or 1 year) for security auditing, then purged to reduce data liability.

The "Black Box" Problem

Vendor logs vs. firm logs. If you use a consumer tool or a vendor's cloud platform, they have the logs, you might not. Enterprise contracts should specify your right to access or export usage logs. If you can't see who used the tool, you can't govern it.

> **Risk Note** Do not log more than you can protect. A central log of every question every lawyer asked about every case is a high-value target for hackers. Ensure audit logs are encrypted and access-controlled with the highest security.

Transparency with Users Lawyers should know their AI usage is logged. This isn't about spying; it's about compliance. A banner stating "Usage of this tool is monitored for compliance and security" sets the right expectation and deters misuse.

Audit trails validate governance. Log tool usage, and consider logging prompts/outputs with strict access controls. Logs serve compliance, cost allocation, and investigation needs. Ensure you have access to logs from vendors. Align retention with general file retention policies.

12.5 Quality Assurance: Sampling, Checklists, Escalation

Governance requires checking that the work meets the standard. You can't review every AI output, but you can implement systems that ensure quality.

Checklists

The simplest QA tool. Every AI-assisted task should have a corresponding checklist (like the ones in this book).

- "Did you verify citations?"
- "Did you check for data privacy?"
- "Did you review for bias?"

Requiring lawyers to complete a checklist before filing or sending creates a "governance gate" that forces a pause and a review. A sample AI verification checklist is available on the companion site.

Sampling / Spot Checks

Practice group leaders or KM professionals should periodically review a random sample of AI-assisted work. This isn't about re-doing the work, but checking process.

- Did the lawyer save the verification log?
- Does the billing narrative match the AI usage log?
- Are there stylistic signs of unedited AI text (e.g., "In conclusion," "It is important to note")?

Regular spot checks signal that governance is active, not passive.

Peer Review

For high-stakes documents (Rung 4 or 5 on the reliability ladder), require a "human peer review" of the AI components. A second set of eyes is the best defense against hallucination blindness (where the drafter trusts the AI too much because the output looks plausible). The reviewer should be explicitly told: "AI was used for the initial draft of this section; please scrutinize citations and logic."

Automated QA

Some advanced tools can automate parts of QA.

- **Citation Checkers:** Software that scans a brief and verifies every citation against Westlaw/Lexis automatically. This should be mandatory for every filing.

- **Confidentiality Scanners:** Tools that scan outgoing documents or prompts for PII or client names.

Escalation Pathways

What does a lawyer do if the AI produces something weird, unethical, or dangerous? Governance needs an escalation path.

- "If the AI refuses to answer due to 'policy violations,' contact [Risk Officer]."
- "If the AI suggests a legal argument that seems novel or questionable, discuss with [Practice Group Leader]."
- "If you suspect a data breach, contact [IT Security] immediately."

Lawyers need to know who to call so they don't hide problems. Quality Assurance is also about feedback to the vendor. If a tool consistently fails on a certain type of contract, QA data allows the firm to go back to the vendor and demand improvements or switch tools.

12.6 Incident Response: Containment, Documentation, Notifications

Despite policies and training, things will go wrong. An associate will upload a merger agreement to a public chatbot. A partner will file a brief with a fake case. Governance includes planning for failure.

The AI Incident Response Plan

This should be a supplement to the firm's general data breach or crisis management plan. It needs specific protocols for AI-specific failures.

Scenario A: Confidentiality Breach (Data Leak)

- **Containment:** Can we delete the chat? Can we contact the vendor to purge the data? Can we revoke the API key?
- **Assessment:** What exactly was shared? Was it PII? Trade secrets? Strategy?
- **Notification:** Do we need to tell the client? (Usually yes, under Rule 1.4). Do we need to notify regulators (if PII/HIPAA)?
- **Remediation:** How do we prevent recurrence?

Scenario B: Accuracy Failure (The "Mata" Scenario)

- **Immediate Action:** If filed, duty of candor (Rule 3.3) requires immediate correction. Withdraw the filing, inform the court, inform opposing counsel. Do not wait for them to find it.
- **Investigation:** How did it happen? Was it a lack of verification? A rogue employee? A tool failure?
- **Documentation:** Preserve all prompts, outputs, and drafts to prove lack of intent to deceive (though negligence is still a problem).

Scenario C: Shadow AI Discovery

- **Discovery:** IT notices massive traffic to a non-approved AI site.
- **Audit:** Interview the users. What did they send? Why did they use it?
- **Response:** Block access? Discipline? Amnesty if they self-report?

Post-Mortem

After any incident, conduct a "lessons learned" review. Update policies and training. If a tool caused the issue, re-evaluate the vendor. An incident response protocol template is available on the companion site.

> **Risk Note** Silence makes it worse. In almost every AI sanctions case, the court was harder on the lawyers because they tried to hide the error or doubled down when challenged. A governance culture of "report early, fix early" saves careers.

12.7 Vendor Management Basics

You are responsible for the tools you give your lawyers. Vendor management is the governance of your supply chain.

Procurement Gates Don't let credit cards buy AI. All AI purchases must go through a central procurement process that triggers Security, Legal, and IT reviews.

Continuous Monitoring Vendors change terms. A vendor that didn't train on data last year might change its Terms of Service today. Governance requires someone (or some tool) to monitor ToS changes and re-evaluate the risk.

Dependency Management If your entire practice depends on one AI tool, what happens if that tool goes down or the company fails? Governance involves continuity planning. Do you have a backup? Can you revert to manual processes?

> **Key Point** Governance is the difference between a tool and a capability. By establishing clear decision rights, practical policies, training, audit trails, quality assurance, and incident response, you transform AI from a risky experiment into a reliable, professional asset.

12.8 Mini-Lab: Build an AI Governance Packet

Mini-Lab: Build a Firm AI Governance Starter Pack

Objective: Create the core governance documents for a small-to-mid-sized firm (or a specific practice group).

A governance starter pack template is available on the companion site for download.

Time Required: 60 minutes

Instructions

1. **Draft an Acceptable Use Policy (AUP):**
   - **Define Approved Tools** (create a hypothetical list).
   - **Define Prohibited Data** (what data tier stays out?).
   - **Define Verification Requirement** (what must be checked?).

Sample prompt:

> *"Draft a 1-page AI Acceptable Use Policy for a boutique litigation firm. We use [Tool A] and [Tool B]. We prohibit putting client medical records or social security numbers into any AI. We require verification of all case law."*

2. **Create a Training Outline:**
    - **Outline a 1-hour "AI License" course.** What 5 topics are mandatory? (e.g., Intro to LLMs, Confidentiality/Privacy, Prompting 101, Verification Lab, Ethics).
3. **Design a Verification Checklist:**
    - **Create a 5-point checklist** that must be attached to any AI-assisted filing.

Deliverable

A governance "starter pack" containing a policy draft, training outline, and QA checklist.

Scenario Analysis Guide

After completing your governance packet, check your work against these essentials.

- **AUP Coverage Check:** Your Acceptable Use Policy should address at minimum: approved tools by name, prohibited tools or tool categories, data classification rules specifying what information can enter which tools, verification requirements for AI-assisted work product, and disclosure obligations to clients and courts. If your AUP doesn't answer the question "Can I put this client document into this tool?" with a clear yes or no, it needs more specificity.
- **Training Outline Check:** Your five mandatory topics should cover at minimum: how AI works (capabilities and limitations), confidentiality and data handling rules, prompt engineering fundamentals, verification and quality control procedures, and professional responsibility obligations. If a new hire could complete your training and still not know how to verify an AI-generated citation, the training has a gap.
- **Verification Checklist Check:** Your checklist should be completable in under two minutes per document. If it's longer, lawyers will skip it. Every item should be a concrete action ("Verified each citation in Westlaw"), not an aspiration ("Ensured quality").

Bonus Challenge

Draft an incident response protocol for the "Mata scenario" — a lawyer in your firm discovers that a brief filed yesterday contains an AI-generated citation that doesn't exist. Your protocol should address:

- Immediate steps (within one hour)
- Court notification procedures
- Client communication
- Internal investigation to determine how verification failed
- Documentation requirements

This exercise practices the incident response planning from Section 12.6 applied to the most common AI failure mode in legal practice.

Key Takeaways

1. Governance works only when decision rights are clear, active, and tied to real accountability rather than committee formalities.
2. Policies should translate principles into specific rules about acceptable use, data handling, review requirements, and disclosure obligations.
3. Training is the bridge between policy and practice, and it must cover tool fluency, risk awareness, and prompt discipline.

4. Audit trails support compliance, investigations, and defensibility, but organizations should not log more than they can securely protect.
5. Quality assurance requires checklists, sampling, escalation paths, and a culture where problems are surfaced early rather than hidden.
6. Incident response should be planned before something goes wrong, with clear steps for containment, documentation, notification, and remediation.

Chapter Checkpoint

Before moving on, make sure you can:

- ☐ Identify which governance model fits your organization
- ☐ Draft a policy that distinguishes between acceptable and prohibited AI use
- ☐ Explain the three pillars of AI training (Tool Fluency, Risk, Prompting)
- ☐ Describe what should be in an audit log
- ☐ Outline the first three steps of an AI incident response plan

What's Next

We have covered the internal structures of practice. Chapter 13 looks outward at the future: the regulatory landscape, emerging technologies like agents, and how to future-proof your career in an AI-driven profession.

Key Terms

AI Governance — The framework of decision rights, policies, and processes that an organization uses to manage the risks and opportunities of artificial intelligence. It determines who makes decisions, what rules apply, and how compliance is ensured.

Acceptable Use Policy (AUP) for AI — A document outlining the permissible ways employees can use artificial intelligence tools within an organization, specifically addressing approved tools, prohibited activities, and data security requirements.

Audit Trail (in AI governance) — The chronological record of system activities, including user logins, prompts entered, and outputs generated. It provides the necessary evidence to reconstruct events for compliance audits, billing verification, and incident investigations.

Governance model — The organizational structure determining who has authority over AI-related decisions, ranging from centralized committees to distributed practice group leadership to hybrid approaches combining central standards with local implementation.

Shadow AI — Unauthorized use of AI tools that have not been approved by the organization's governance process, often driven by restrictive policies that push users toward unsanctioned alternatives.

Human-in-the-loop (HITL) — The requirement that a human professional review and approve AI outputs before they are finalized, filed, or sent to clients. The fundamental safety mechanism for professional AI use.

Incident response protocol — A predefined plan for responding to AI-related failures, including steps for containment, investigation, notification of affected parties, and remediation to prevent recurrence.

Quality assurance sampling — Periodic review of a random selection of AI-assisted work product to verify that organizational standards are being followed, distinct from individual verification of specific outputs.

Data classification — Categorizing information by sensitivity level to determine which AI tools and environments are appropriate for processing that information, aligned with the confidentiality tier framework from Chapter 5.

Chapter 13: Regulation and Future-Proofing Your Practice

The ground is shifting beneath legal practice. Throughout this book, you've learned techniques for using AI effectively and responsibly within the current landscape. But that landscape is changing rapidly on multiple fronts. Regulators are developing frameworks that will impose new requirements on AI use. Courts are establishing expectations that will shape how AI-assisted work is presented in litigation. AI capabilities are advancing in ways that will create new opportunities and new risks. Practices that position themselves to navigate these changes will thrive; practices that assume today's approaches will remain adequate may find themselves scrambling to adapt. This chapter addresses the evolving external environment and how to build a practice that remains effective as that environment changes.

Regulation of AI has moved from theoretical discussion to concrete reality. The European Union's AI Act creates a comprehensive framework classifying AI systems by risk and imposing requirements accordingly. Various U.S. states have enacted legislation addressing AI in specific contexts. Federal agencies are developing guidance and enforcement approaches. Professional responsibility authorities are issuing opinions on lawyer AI use. This regulatory activity reflects growing recognition that AI creates risks warranting oversight. For legal practitioners, regulation means new compliance obligations, new client advisory opportunities, and new constraints on how tools can be used.

Technological evolution is moving from text generation to agentic behavior and multimodal capability. Current large language models generate text based on prompts. Emerging AI agents can take actions: sending emails, booking appointments, executing transactions, and navigating software interfaces. This shift from "reading and writing" to "doing" amplifies both utility and risk. Multimodal models that process text, images, audio, and video simultaneously enable new legal applications—reviewing accident photos, analyzing surveillance video, processing handwritten notes—but also introduce new challenges for authentication and evidence management.

Future-proofing your practice requires more than just adopting new tools. It requires cultivating durable skills that remain valuable regardless of technological change. Strategic judgment, client relationship management, ethical reasoning, and creative problem-solving are human capabilities that AI complements but does not replace. Focusing development on these durable skills ensures that you remain relevant even as routine tasks are increasingly automated. The lawyer of the future is not just a user of tools but a strategist who directs them.

What This Chapter Covers

- The emerging regulatory landscape for AI and what it means for law firms and clients
- Key regulatory themes including transparency, risk classification, data governance, and liability
- How courts are adapting procedural rules to address AI-assisted litigation
- Emerging capabilities: AI agents, multimodal models, and advanced reasoning
- The durable human skills that will remain central to legal practice
- A 90-day roadmap for integrating AI into your practice systematically
- Strategies for continuous learning to keep pace with rapid change

Chapter Summary

AI regulation is developing rapidly, moving from general principles to specific requirements. The EU AI Act establishes a risk-based framework that will influence regulation globally. In the U.S., a patchwork of state laws and federal agency guidance is emerging. Key regulatory themes include transparency (disclosing when AI is used), risk management (assessing and mitigating AI risks), data governance (protecting data used in AI systems), and liability (determining responsibility for AI harms). Lawyers must navigate these regulations for their own practices and advise clients on compliance. Courts are adapting rules to address AI in litigation, with disclosure and verification requirements becoming standard.

Emerging AI capabilities will transform practice further. AI agents that can execute tasks autonomously will enable new efficiencies but raise supervision challenges. Multimodal models will allow lawyers to analyze diverse evidence types directly. Advanced reasoning capabilities may enable AI to handle more complex legal analysis. These developments will require lawyers to adapt their workflows and supervision strategies. Despite these changes, core human skills remain essential: judgment, empathy, strategy, and ethics are not automated. Future-proofing involves investing in these durable skills while maintaining technological competence.

A practical roadmap helps lawyers move from understanding to action. The 90-day integration plan outlines concrete steps for adopting AI responsibly: developing policies, selecting tools, training staff, and updating workflows. Continuous learning strategies ensure that this integration remains current as technology and regulation evolve. The future of legal practice belongs to those who combine technological capability with professional excellence.

13.1 What "AI Regulation" Means for Law Offices

"AI regulation" often sounds like something that applies to tech giants developing models, not law firms using them. But regulation affects AI users as well as developers. As improved frameworks emerge, they will impose obligations on professional services firms that deploy AI tools. Understanding the scope of regulation helps you anticipate compliance requirements rather than reacting to them after they're enacted.

Regulation affects law offices in two ways: as regulated entities themselves, and as advisors to regulated clients. As regulated entities, law firms must comply with rules regarding data privacy, consumer protection, and professional responsibility. New AI regulations may add requirements for transparency, bias testing, and risk assessment when firms use AI for tasks like hiring, billing, or client service. As advisors, lawyers must understand the regulatory environment to help clients navigate it. The expertise you develop to manage your own AI use becomes a service you can offer clients facing similar challenges.

The "Brussels Effect" means the EU AI Act will likely influence global standards even for firms without European offices. The Act classifies AI systems by risk level—unacceptable, high, limited, and minimal—and imposes obligations accordingly. "High-risk" systems face stringent requirements for data governance, documentation, transparency, human oversight, and accuracy. While most legal practice tools currently fall into lower risk categories, tools used for certain sensitive tasks (like migration, employment, or administration of justice) might be classified as high-risk. U.S. firms using tools developed for the global market will encounter compliance features built to satisfy EU standards.

In the United States, regulation is emerging through a combination of state legislation, federal agency action, and executive orders. States are enacting AI-specific legislation at an accelerating pace. Colorado's AI Act (SB24-205, signed May 2024) is the first comprehensive state AI law in the United

States, imposing obligations on deployers of high-risk AI systems including impact assessments and disclosure requirements. California, New York, and other states are advancing their own legislation addressing automated decision-making, bias, and disclosure. Federal agencies like the FTC, EEOC, and DOJ are applying existing laws to AI contexts, enforcing prohibitions against discrimination and unfair practices. At the federal level, the regulatory posture has shifted. The Biden administration's Executive Order 14110 on Safe, Secure, and Trustworthy AI (October 2023) directed federal agencies to develop safety standards and reporting requirements. That order was revoked in January 2025 by Executive Order 14179, Removing Barriers to American Leadership in Artificial Intelligence, which takes a deregulatory approach emphasizing AI dominance and innovation over prescriptive safety requirements. The shift illustrates how rapidly the federal landscape can change and why lawyers must monitor executive and agency action rather than assuming any single directive is permanent. This patchwork requires lawyers to monitor developments across multiple jurisdictions.

Professional regulation creates a parallel track of obligations. Bar associations and courts regulate lawyer conduct specifically. As discussed in Chapter 4, existing duties of competence and supervision already apply to AI. Future professional regulation may become more prescriptive, potentially requiring specific disclosures, certifications, or technical standards for legal AI tools. Staying compliant requires monitoring both general AI regulation and legal-specific professional rules.

> Practice Point Treat AI regulation as a developing practice area. Even if you do not consider yourself a "regulatory lawyer," AI compliance will intersect with employment law, intellectual property, contracts, torts, and corporate governance. Basic fluency in AI regulatory concepts is becoming part of general legal literacy.

Regulatory uncertainty is a feature of the current environment. Rules are being written, interpreted, and challenged in real time. Compliance in this environment requires a principles-based approach: focusing on the underlying goals of regulation—fairness, transparency, safety, accountability—rather than just checking boxes. Practices that align with these principles are better positioned to adapt as specific rules crystallize.

13.2 Key Regulatory Themes: Transparency, Risk Tiers, Privacy, IP

While specific regulations vary, certain themes recur across jurisdictions and frameworks. Understanding these themes helps you anticipate what compliance will look like regardless of the specific rule.

Transparency is the most common regulatory requirement. Regulators want users to know when they are interacting with AI and when AI has contributed to decisions affecting them. For law firms, this reinforces the disclosure considerations discussed in Chapter 10. Clients may have a statutory right to know if AI is used in their representation, independent of professional ethics obligations. Transparency requirements may also extend to labeling AI-generated content, creating watermarking or disclosure obligations for marketing materials and work product.

Risk Tiers structure many regulatory frameworks. Not all AI is treated equally. A chatbot answering scheduling questions is low risk; an AI system screening job applicants or predicting recidivism is high risk. For law practice, this means compliance burdens will vary by tool and task. Using AI for legal research might be lower risk than using AI for jury selection or sentencing advocacy. Mapping your AI use cases to regulatory risk tiers helps you allocate compliance resources effectively.

Privacy and Data Governance remain central. AI regulation builds on existing privacy frameworks like GDPR and CCPA. The intersection of AI and privacy creates new challenges: can you "delete" data from a trained model? Does processing data to generate embeddings constitute a new use requiring consent? Law firms, as custodians of sensitive client data, must navigate these questions. Using client data to train internal models or fine-tune vendor models raises privacy issues that go beyond standard confidentiality analysis.

> **Risk Note** Data used to train models may be difficult to remove. "Machine unlearning," removing specific data points from a trained model, is technically challenging. If you use client data for training without proper rights, you may face compliance issues that are hard to remediate without retraining the model from scratch.

Intellectual Property issues are evolving rapidly. Who owns AI-generated content? Does training on copyrighted works constitute infringement? Courts and legislatures are grappling with these questions. For law firms, this creates uncertainty about ownership of AI-assisted work product. If a significant portion of a brief or contract is AI-generated, is it copyrightable? Does it matter? Clients may care about ownership of work product they pay for. Contracts with AI vendors must address IP rights clearly, allocating ownership of inputs, outputs, and any derived models.

Liability allocation is a key regulatory battleground. When AI causes harm—through defamation, discrimination, or error—who is responsible? The developer? The deployer? The user? Legal frameworks typically default to holding the user (the lawyer) responsible for professional services, but allocation of liability between firms and AI vendors will be a subject of intense negotiation. Indemnification clauses in AI vendor contracts deserve careful scrutiny as regulatory liability standards evolve.

Courts are unlikely to accept "the algorithm made me do it" as a defense against liability. Whether under professional responsibility rules or general tort principles, the human user who chooses to deploy a tool and act on its outputs retains primary responsibility. Regulatory frameworks reinforce rather than replace this human accountability.

Bias and Fairness regulations address the risk of AI replicating or amplifying historical discrimination. Laws restricting automated decision-making in employment, housing, and credit are expanding. Law firms using AI for internal operations (hiring, evaluation) or advising clients on AI implementation must test for and mitigate bias. "Algorithmic impact assessments" may become standard requirements for high-stakes AI deployments.

13.3 Courts and Procedural Rules

Courts have moved faster than legislatures in addressing AI. Where state legislation is still emerging and federal regulation continues to shift, individual judges and court systems have been writing their own rules in real time. For practitioners, this means the most immediate AI compliance obligations often come not from a state bar opinion or a statute, but from the local rules, standing orders, and case-specific directions of the court where you are filing.

The dominant pattern is the standing order. Standing orders apply to all cases before a particular judge and reflect that judge's assessment of what safeguards are needed for AI-assisted filings. Judge Brantley Starr's order in the Northern District of Texas, discussed in Chapter 4, was an early and widely cited example, requiring attorneys to certify either that no portion of a filing was drafted by generative AI or that any AI-generated language was verified for accuracy by a human being. Other judges have issued similar orders, varying in specificity and severity. Some require brief certification; others impose detailed verification procedures and explicit consequences for noncompliance. The orders share a common premise: where the rules of professional conduct require accurate filings, AI assistance changes the means but not the obligation.

Local rules represent a more durable form of court-level regulation. Where standing orders apply to one judge's cases, local rules govern an entire court. A growing number of district and appellate courts have amended local rules to address AI use, typically by adding disclosure obligations to existing certification requirements under Rule 11 or its state-law equivalents. Local rules carry more weight than standing orders because they are adopted through formal rulemaking processes and apply uniformly across the court's docket.

Case-specific directions emerge in orders judges enter in individual matters. A judge confronted with AI-generated content for the first time may issue a directive applicable only to that case, requiring counsel to disclose how AI was used and what verification was performed. These orders generally arise from one of two situations: the court has concerns about a specific filing, or the court is using the case as a vehicle to set expectations for future filings.

Disclosure requirements have become the most common court response. The threshold question is when disclosure is required: any AI use, only substantive AI use, or only AI use that affects citations or factual assertions. Practice varies. Some orders require disclosure of all AI assistance; others limit disclosure to AI use that produced specific content in the filing. The trend appears to favor broader disclosure, but the specifics remain jurisdiction-specific. The practical effect is that a single filing strategy may need to accommodate different disclosure thresholds depending on the court.

Certification requirements convert disclosure into formal commitment. Where disclosure tells the court what happened, certification confirms that the filing meets specific standards regardless of how it was produced. A certification that AI-generated content was verified for accuracy by a human being is more enforceable than a disclosure that AI was used, because the certification creates a record the court can rely on if accuracy problems later surface. Treat certifications under these orders with the same care you give Rule 11 certifications generally.

Verification standards in some courts now exceed the baseline that professional responsibility rules require. A few courts have adopted rules or standing orders requiring verification of every citation in an AI-assisted filing, regardless of whether the lawyer has independent confidence in any particular citation. This represents a meaningful departure from default practice, where lawyers typically verify citations they have specific reason to question rather than every citation in every filing. Where these heightened standards apply, the practical workflow changes: every authority cited must be checked against the underlying source before filing.

Sanctions for AI failures have set expectations even in courts without AI-specific rules. *Mata v. Avianca*, discussed in Chapter 4, is the most cited example, but similar sanctioning decisions have followed in other jurisdictions. The common pattern is consistent: a lawyer files content containing fabricated citations, the court discovers the fabrication, and the court imposes sanctions for the failure to verify. The specific sanctions vary — monetary penalties, referrals to disciplinary authorities, requirements that the lawyer notify clients — but the underlying message is the same. Where AI assistance produces errors that competent review would have caught, the court will hold the lawyer accountable regardless of whether the court has adopted any AI-specific rule.

> Practice Point Before filing in any court, check whether the court has adopted AI-related rules or standing orders. Court websites, local rules pages, and the judge's individual practices are the primary sources; some courts also publish AI policies separately. Treat AI disclosure as a procedural requirement to be verified, not as a default assumption either way.

The landscape is still developing, and the speed of development varies by jurisdiction. Some courts have moved quickly to adopt detailed rules; others are watching what happens elsewhere before committing to an approach. For lawyers practicing in multiple courts, this fragmentation requires ongoing attention. A rule that exists in one jurisdiction today may not exist in the next jurisdiction you file in tomorrow, and what exists in either may be different by the end of the year.

13.4 Emerging Capabilities: Agents, Multimodal Tools, New Risks

The AI tools discussed in this book are primarily text-based generators. The next wave of technology brings new capabilities that transform how AI can be used—and the risks it creates.

AI Agents

AI Agents move from "chat" to "action." Instead of just generating text, agents can use tools: browsing the web, sending emails, querying databases, and executing transactions. An agentic legal assistant might not just draft a contract but email it to the client, update the matter management system, and schedule a follow-up meeting.

- **Opportunity:** Massive efficiency gains in multi-step workflows.
- **Risk:** "Looping" errors where an agent takes incorrect actions repeatedly (e.g., emailing the wrong person 50 times). Supervision becomes harder when the AI is doing, not just writing. Governance must address permissions: what creates liability (sending a settlement offer) versus what is safe (drafting it).

Multimodal Models

Multimodal Models process and generate text, images, audio, and video natively. A lawyer could upload photos of an accident scene, video from a body camera, and the text of a police report, asking the AI to synthesize a narrative.

- **Opportunity:** Holistic evidence analysis. Reviewing surveillance video or audio recordings becomes as searchable as text documents.
- **Risk:** Hallucination extends to visual and auditory data. The AI might "see" something in a blurry video that isn't there. Verification requires checking the source media, not just reading text.

Advanced Reasoning and Planning

Advanced Reasoning and Planning capabilities are improving. Reasoning models have become a distinct category. They are designed to 'think' before responding by outlining steps and checking logic. This may reduce hallucinations and improve performance on complex legal analysis tasks that require multi-step reasoning.

- **Opportunity:** Handling more complex legal questions with greater reliability.
- **Risk:** Over-reliance. As models get better, the temptation to skip verification grows. The "zone of danger" shifts to more subtle, plausible-sounding errors that are harder to catch.

Personalized Models

Personalized Models fine-tuned on a firm's specific work product will become standard. Instead of a generic model, you will use a model trained on your firm's contracts, your brief bank, and your style guide.

- **Opportunity:** Higher quality drafts that match your firm's voice and standards immediately.
- **Risk:** Data privacy and security become even more critical. Protecting the firm's proprietary "knowledge base" from leaking into the public model or being accessed by competitors becomes a key IT priority.

13.5 The Durable Human Skills

In a world where AI can draft, research, and analyze, what is left for the lawyer? The answer is: the most human parts of practice. The value proposition of a lawyer shifts from "access to information" to "judgment and relationship."

Strategic Judgment: AI can list options; it cannot choose the one that aligns with a client's risk tolerance, business goals, and personal values. Deciding whether to sue is different from knowing how to sue. Strategy requires integrating legal analysis with real-world context—politics, optics, psychology, economics—that AI cannot fully model.

Client Relationships and Empathy: Clients are often stressed, angry, or frightened. They need to feel heard and understood. AI can simulate empathy, but it cannot care. The human connection—the "counselor" role—becomes the primary differentiator. A lawyer who can calm a client, manage expectations, and deliver bad news with compassion provides value no machine can replicate.

Ethical Reasoning: AI follows rules; it does not have a conscience. Legal practice involves ethical grey areas where rules conflict. Navigating these dilemmas requires moral reasoning and professional character. Deciding what is "right" rather than just what is "legal" remains a human responsibility.

Creative Problem Solving: AI generates based on patterns in past data. It struggles with truly novel situations that break patterns. Lawyers excel at creative arguments—finding a new application for an old statute, distinguishing a precedent in a novel way, or structuring a deal to bridge a seemingly unbridgeable gap. Innovation often comes from breaking the pattern, which is the opposite of what predictive models do.

Advocacy and Persuasion: While AI can draft persuasive text, the act of advocacy—standing up in court, looking a judge in the eye, negotiating face-to-face—relies on human credibility and presence. People persuade people. The trust established between advocate and decision-maker is a human dynamic.

AI commoditizes the mechanical aspects of law: finding cases, summarizing documents, drafting standard clauses. This raises the value of the non-commoditized skills: judgment, empathy, creativity, and presence. Future-proofing means doubling down on the human elements of lawyering.

13.6 Your 90-Day AI Integration Roadmap

Reading this book is the first step. Action is the next. Here is a 90-day plan to move from "interested" to "integrated." A printable version of this roadmap is available on the companion site as an implementation tracker.

Month 1: Foundation and Experimentation

- **Week 1:** Select one low-risk, general-purpose AI tool (e.g., ChatGPT Enterprise, Claude). Get an account. Read the Terms of Service to confirm data privacy.
- **Week 2:** Complete the Mini-Lab from Chapter 3. Build your first "Prompt Pack" for a recurring non-client task (e.g., summarizing legal news, drafting marketing emails).
- **Week 3:** Draft your personal "Rules of Use" (Chapter 1). Define what you will and won't use AI for.
- **Week 4:** Conduct a "Tool Audit" (Chapter 6). Inventory what you use and identify one gap where a new tool might help.

Month 2: Workflow Integration

- **Week 5:** Pick one substantive workflow (e.g., Research or Drafting). Apply the Chapter 7 or 8 framework to a closed matter (a "post-mortem" test) to see how AI would have handled it.
- **Week 6:** Use AI on a live, low-stakes task (e.g., internal memo, initial research) using the verification protocols from Chapter 7. Measure the time saved versus time spent verifying.
- **Week 7:** Develop a "Matter Playbook" for your most common case type (Chapter 11 Mini-Lab).
- **Week 8:** Review your intake and client communication templates. Use AI to rewrite one standard letter into plain English (Chapter 10).

Month 3: Governance and Scaling

- **Week 9:** Draft or update your firm's AI Acceptable Use Policy (Chapter 12). If you are a solo, write it for yourself.
- **Week 10:** Conduct a "Confidentiality Audit" (Chapter 5). Ensure no sensitive data is leaking into consumer tools.
- **Week 11:** Train one colleague or staff member on a specific AI workflow you've mastered. Teaching solidifies learning.
- **Week 12:** Assessment. Review the last 90 days. What worked? What failed? Where did you save time? Set goals for the next quarter.

Start small. Don't try to transform your entire practice in a week. Pick one task, master the AI workflow for it, and then move to the next. Consistency beats intensity.

13.7 Continuous Learning: Keeping the Book Current

A book about AI is out of date the moment it is printed. The specific tools mentioned may change, but the frameworks—verification, supervision, confidentiality, prompt engineering—are durable. To keep the specifics current, a curated list of recommended learning resources is maintained on the companion site:

Curate Your Feed: Follow 2-3 reliable sources for legal tech news (e.g., Bob Ambrogi's LawSites, Legaltech News, specific bar association tech committees). Don't try to drink from the firehose of general AI news; focus on legal applications.

Test New Models: When a major new model releases, spend an hour testing it against your standard prompt pack. Does it handle logic better? Is it more concise? Update your expectations accordingly.

Update Your Playbooks: Treat your prompt library and playbooks as living documents. Schedule a quarterly "Maintenance Day" to prune prompts that no longer work and add new ones.

Community: Join a user group or committee. Hearing how other lawyers are solving problems is the fastest way to learn.

The journey of AI integration is not a switch you flip; it is a practice you cultivate. Like the law itself, it requires study, adaptation, and a commitment to professional excellence. You are now equipped with the map. The territory is yours to explore.

> **Key Point** The future of legal practice is not "Human versus AI." It is "Human + AI" versus "Human alone." By mastering these tools, respecting their risks, and focusing on your unique human value, you secure your place in that future.

Key Takeaways

AI regulation affects law offices through multiple streams that require distinct attention. Regulation of AI systems generally imposes requirements on tools you use, with frameworks like the EU AI Act classifying legal AI as high-risk and imposing corresponding obligations on both providers and users. Regulation of lawyer conduct applies existing professional responsibility rules to AI use, with bar associations issuing guidance interpreting competence, supervision, confidentiality, and candor obligations in AI contexts. Regulation affecting client matters creates advisory opportunities as clients across industries navigate their own AI compliance obligations. The regulatory landscape is fragmented across jurisdictions, evolving rapidly, and requires ongoing monitoring rather than one-time learning. Documented good-faith compliance reasoning demonstrates appropriate attention even when regulatory requirements remain uncertain.

Key regulatory themes recur across different frameworks and jurisdictions. Transparency requirements mandate disclosure of AI use to courts, clients, and affected parties, with expectations for meaningful disclosure content rather than mere boilerplate acknowledgment. Risk-based classification assigns different regulatory treatment based on the level of risk AI systems pose, with legal AI often classified as high-risk due to its potential impact on legal rights. Privacy regulation addresses training data, inputs, and outputs, applying existing privacy frameworks to AI processing while AI-specific provisions develop. Intellectual property considerations address training data legality, output ownership, and infringement risk in AI-generated content. Liability frameworks are developing to address AI-related harms, with lawyers retaining responsibility for AI-assisted work product regardless of vendor involvement. Understanding these themes helps you navigate specific regulations by recognizing underlying concerns.

Courts are addressing AI through local rules, standing orders, case-specific directions, and sanctioning decisions. Disclosure requirements are the most common judicial response, requiring lawyers to disclose when AI assisted in preparing court filings, though specificity varies across jurisdictions. Certification requirements formalize verification obligations, creating records of lawyer commitment to accuracy that compound consequences if certifications prove false. Verification standards may exceed minimum professional responsibility requirements, with some courts requiring verification of every citation in AI-assisted filings. Sanctions for AI failures signal expectations even where AI-specific rules don't exist, putting lawyers on notice that courts will hold them accountable regardless of formal rule adoption. Requirements vary by court and continue evolving, requiring ongoing attention to developments in jurisdictions where you practice.

Emerging AI capabilities will expand what AI can do in legal practice while creating new risks requiring governance adaptation. AI agents that take autonomous actions create efficiency potential but multiply supervision challenges, as multi-step processes between human checkpoints create oversight gaps that current frameworks don't fully address. Multimodal AI that processes images, audio, and video enables analysis of content types that text-only tools cannot handle, expanding applications for document review, evidence analysis, and communication while intensifying verification challenges. Enhanced reasoning capabilities make AI useful for more sophisticated analytical tasks, improving issue spotting, contextual research, and nuanced document analysis. Real-time capabilities, integration with other systems, and personalization features are emerging to make AI assistance more contextual and seamlessly embedded in practice. Each capability expansion brings corresponding risk expansion that governance must address through adaptation rather than static approaches.

Durable human skills remain essential regardless of AI advancement and become more rather than less important as AI handles routine tasks. Legal judgment that synthesizes facts, law, context, and experience into sound conclusions requires understanding that AI cannot replicate, including understanding of clients as individuals, contexts that documents don't capture, and patterns developed through years of practice experience. Client relationships depend on human connection, empathy, and trust developed through personal interaction that AI cannot provide. Ethical reasoning navigates competing obligations and ambiguous situations through moral reflection that AI is not equipped to perform. Strategic thinking anticipates consequences and adapts to changing circumstances in ways that require forward-looking judgment beyond AI capabilities. Creativity, persuasive communication, leadership, mentorship, and wisdom all remain distinctly human contributions to excellent lawyering. Skill development and time allocation should balance AI proficiency with these durable capabilities.

The 90-day integration roadmap provides structured implementation moving from assessment through foundation to integration. Phase 1 establishes current state, maps tasks to AI opportunities, assesses risks, sets measurable goals, and engages stakeholders. Phase 2 selects and deploys tools through focused adoption, develops governance proportionate to practice size and risk, creates training materials, and begins initial deployment with lower-risk applications. Phase 3 integrates AI into regular workflows, adapts to practice-area-specific needs, monitors effectiveness against goals, refines based on findings, and plans expansion beyond initial implementation. Post-implementation continuous development maintains progress through regular reassessment, capability expansion, governance evolution, and ongoing learning. The sequence matters more than the specific timeline; deliberate, systematic implementation produces better results than ad hoc adoption that misses important steps.

Continuous learning keeps AI knowledge current as capabilities, regulations, and best practices evolve faster than any static resource can capture. Learning sources span professional associations, legal and technology publications, vendor communications, regulatory monitoring, peer communities, and direct experimentation. Learning formats include formal training and informal content consumption, with peer learning and hands-on experimentation providing practical grounding that publications alone cannot. Learning priorities should focus limited time on developments with clear relevance to your practice, with assessment questions helping evaluate which new capabilities warrant attention. Building learning into practice routines through explicit time allocation and triggers connected to existing habits makes continuous learning sustainable over a career. This book's frameworks provide foundation; your ongoing learning builds on and extends that foundation as the field evolves.

Chapter Checkpoint

Before concluding, make sure you can:

- ☐ Identify how different regulatory streams affect your practice: regulation of AI systems generally, regulation of lawyer conduct through professional responsibility, and regulation affecting client matters that creates advisory opportunities
- ☐ Apply key regulatory themes including transparency, risk-based classification, privacy, and intellectual property considerations to your AI use decisions
- ☐ Determine and comply with court-specific AI requirements in jurisdictions where you practice, including disclosure obligations, certification requirements, and verification standards
- ☐ Assess emerging AI capabilities including agents, multimodal tools, and enhanced reasoning for relevance to your practice while maintaining appropriate skepticism about claimed performance
- ☐ Identify the durable human skills that remain essential regardless of AI advancement: judgment, relationships, ethics, strategy, creativity, communication, leadership, and wisdom

- ☐ Implement the 90-day integration roadmap adapted to your circumstances, moving systematically from assessment through foundation to integration
- ☐ Establish continuous learning practices that keep your AI knowledge current through diverse sources, appropriate formats, and routines built into regular practice
- ☐ Balance AI proficiency with development of durable human capabilities, creating complete professional competence for the AI era

What's Next: The Journey Continues

This book ends, but your AI integration journey does not. The frameworks, techniques, and principles you've learned provide foundation for a practice that will continue evolving as AI capabilities advance, regulations develop, and your own experience accumulates. What remains is applying what you've learned, adapting as circumstances change, and contributing to the profession's collective understanding of how to use these powerful tools responsibly.

The transformation underway in legal practice is real but manageable. AI will not replace lawyers, but lawyers who use AI effectively will have significant advantages over those who do not. The capabilities you've developed through this book position you to capture those advantages while managing the risks that accompany them. Competence, not fear, should guide your approach. You understand how AI works, what it can and cannot do, and how to use it within professional responsibility obligations. That understanding is the foundation for confident, responsible AI use.

The profession needs lawyers who engage thoughtfully with AI. Some lawyers will resist AI entirely, missing opportunities to serve clients better. Others will adopt AI recklessly, creating problems that harm clients and the profession's reputation. The lawyers the profession needs are those who occupy the thoughtful middle: leveraging AI capabilities while maintaining the judgment, ethics, and human connection that define excellent lawyering. Your engagement with this book suggests you're committed to that thoughtful approach.

Professional responsibility remains the constant amid technological change. The specific tools will evolve, but the obligations endure: serve clients competently, maintain their confidences, deal honestly with courts and others, and conduct yourself with the integrity the profession requires. AI creates new contexts for applying these obligations but doesn't change their fundamental nature. Lawyers who ground their AI use in professional responsibility principles will navigate whatever changes come more successfully than those who treat AI as somehow separate from professional obligations.

The human dimensions of legal practice become more rather than less important. As AI handles more routine tasks, the distinctly human contributions, including judgment, relationships, ethics, and strategy, comprise a larger share of what lawyers provide. Investing in these capabilities alongside AI proficiency creates professionals who offer more value than either capability alone could provide. The lawyers who thrive will be those who combine technological facility with deeply human skills.

Your clients will benefit from what you've learned. Faster, more thorough research. Higher-quality documents produced more efficiently. Better-prepared litigation strategy. Clearer communication that helps clients understand their situations. These benefits translate into better outcomes and better service. The effort you've invested in developing AI competence ultimately serves the clients who depend on you.

The regulatory and judicial landscape will continue developing. Requirements that seem novel today will become routine. New requirements will emerge that current guidance doesn't anticipate. The monitoring and adaptation skills you've developed will serve you as the landscape evolves. Stay engaged with developments, participate in professional discussions about appropriate standards, and contribute to the profession's collective wisdom about AI use.

Emerging capabilities will create opportunities you cannot currently imagine. The AI tools available in five years will be substantially more capable than today's tools. Applications that seem futuristic now may become routine. The foundation you've built, including understanding of AI principles, governance frameworks, and professional responsibility integration, positions you to adopt new capabilities thoughtfully rather than scrambling to catch up.

Share what you learn with colleagues and the profession. The lawyers who engage seriously with AI have insights worth sharing. Your experiences, both successes and failures, contribute to collective understanding. Participate in professional communities, mentor colleagues who are earlier in their AI journeys, and contribute to the guidance that will shape professional standards. The profession's successful integration of AI depends on practitioners sharing what they learn.

Return to this book as a reference, not just a text to read once. The frameworks, checklists, and prompt packs are designed for ongoing use. When you face a new AI challenge, the relevant chapter may provide guidance. When you need to develop governance for your organization, the templates and examples offer starting points. When you train others, the concepts and structures provide curriculum foundation. A resource used repeatedly provides more value than one read and shelved.

The practice of law has always required adaptation to changing circumstances. New statutes, new case law, new client needs, and new technologies have always demanded that lawyers continue learning throughout their careers. AI is the current frontier of that adaptation, but it is not the last frontier. The learning skills and adaptive mindset you've developed will serve you for whatever comes next, not just for AI.

Begin where you are with what you have. You don't need to implement everything in this book immediately. Start with the applications most relevant to your practice. Build from there as confidence and capability develop. Perfect is the enemy of good; competent AI use that starts now provides more value than perfect AI use that never begins. Take the first step, then the next, then the next.

The future of legal practice belongs to lawyers who embrace both technological capability and enduring professional values. You've invested in developing both. Now the work is applying what you've learned, continuing to learn as the field evolves, and contributing to a profession that serves clients and society with the integrity that has always defined excellent lawyering. The tools are new, but the mission endures.

Key Terms

Risk-based regulation — A regulatory approach classifying AI systems by potential harm, imposing stricter requirements on higher-risk systems and creating a tiered compliance landscape.

Algorithmic discovery — Litigation over the design, function, and output of AI systems used by parties, analogous to e-discovery but focused on algorithms rather than documents.

AI Agents — AI systems capable of autonomous action, using software tools to complete multi-step tasks without human intervention at each step. Unlike chatbots, agents can change the state of the world.

Multimodal AI — AI systems that process and generate multiple content types — text, images, audio, and video — natively within a single model, enabling analysis of diverse evidence types that text-only tools cannot handle.

Brussels Effect — The tendency of EU regulations, including the AI Act, to set global standards because companies building products for the European market incorporate EU compliance requirements into their default offerings worldwide.

Machine unlearning — The technically challenging process of removing specific data points from a trained AI model, relevant when data was used without proper rights or when deletion is required under privacy regulations.

Durable human skills — Professional capabilities that remain valuable regardless of AI advancement, including strategic judgment, client relationships, ethical reasoning, creative problem-solving, and persuasive advocacy.

Algorithmic impact assessment — A structured evaluation of the potential effects of an AI system on individuals and groups, increasingly required by regulation for high-stakes AI deployments including those affecting legal rights. # Appendices

Appendix A: The Prompt Blueprint + Debugging Cheatsheet

This appendix condenses the core prompting framework from Chapter 3 into a practical reference you can use while working. Its purpose is speed. When a prompt fails, underperforms, or produces something that feels vague, generic, or unreliable, the fastest way to improve it is usually not to start over from scratch. It is to diagnose which element is missing, weak, or misaligned.

The Prompt Blueprint

Every strong legal prompt contains five elements. When one is absent or underdeveloped, output quality declines in predictable ways.

1 Role

Role tells the model who it should be for purposes of the task. This does not mean pretending the model is licensed or authoritative in a way it is not. It means assigning a functional posture that shapes tone, depth, and analytical orientation.

WEAK

> "Summarize this."

BETTER

> "Act as a careful junior associate preparing an internal summary for a supervising attorney."

Role is especially useful when you want the model to adopt a specific perspective: cautious reviewer, opposing counsel, client educator, or contract redliner.

2 Context

Context gives the model the background it needs to produce relevant output. Context can include the type of matter, jurisdiction, audience, procedural posture, factual background, and any source material the task depends on.

WEAK

> "Draft a response."

BETTER

> "This is a response to a counterparty's markup of an indemnification clause in a software services agreement. Our client is the vendor and wants to narrow downstream liability while preserving the commercial relationship."

Context reduces generic output. It also helps the model understand what matters and what does not.

3 Task

Task is the specific thing you want the model to do. The clearer the task, the better the result. A vague task invites filler. A precise task invites focused work.

WEAK

> "Help me with this contract."

BETTER

> "Compare the original clause and the counterparty's proposed revision, identify every substantive change, explain the practical effect of each change, and suggest a response position from our client's perspective."

Task instructions should be action-oriented and concrete. If the work has multiple parts, list them explicitly.

4 Format

Format tells the model how to organize the output. This is one of the most underused but highest-value prompt elements. Many disappointing outputs are not wrong in substance — they are simply delivered in a form that is difficult to use.

WEAK

> No format specified.

BETTER

> "Return the answer in four sections: (1) key issue, (2) governing rule, (3) application to facts, (4) practical recommendation. Use short paragraphs and bullet points only where necessary."

Format can specify headings, table structure, paragraph length, order of analysis, and level of detail.

5 Guardrails

Guardrails tell the model what not to do. They prevent the most common failure modes: overstatement, fabrication, false confidence, and scope creep.

WEAK

> No guardrails specified.

BETTER

> "Do not cite any authority you have not been given. Do not state legal conclusions with more confidence than the provided materials support. If something is unclear or ambiguous, say so rather than resolving it yourself."

Guardrails are especially important in legal work because the default failure modes — invented citations, overstated holdings, fabricated facts — carry professional responsibility consequences.

> **The Blueprint at a Glance** Role: Who is the model for this task? — Context: What does it need to know? — Task: What exactly should it do? — Format: How should the output be structured? — Guardrails: What must it avoid or limit?

Debugging Cheatsheet

When a prompt fails, the failure usually falls into one of a few predictable categories. Use this section to diagnose the problem and identify the fix.

Problem: The output is generic

CAUSE The prompt lacks enough context or the task is too broad.

FIX Add matter-specific background, identify the audience, and narrow the task.

> "Draft a client letter explaining a denied motion to dismiss to a first-time litigant who is anxious and prefers plain-English explanations."

Problem: The output is disorganized

CAUSE No format was specified.

FIX Add explicit format instructions: sections, headings, ordering, length.

> "Organize the answer under clear headings. Use numbered sections. Put the recommendation at the end."

Problem: The output fabricates authorities

CAUSE The model was asked to produce legal citations without source material.

FIX Use source-bounded prompts. Provide the authorities you want analyzed.

> "Use only the cases I have provided. Do not cite any authority not in the attached materials."

Problem: The output overstates confidence

CAUSE No guardrails were specified.

FIX Add explicit uncertainty instructions.

> "Do not provide a definitive conclusion where the authorities are mixed. Flag where additional facts would change the answer."

Problem: The output misses the point

CAUSE The task instruction was too vague or the role was absent.

FIX Clarify the task, add a role, and specify what a successful output looks like.

> "Act as a careful associate preparing a memo for a partner. The partner needs to know: can we win this motion? State your conclusion first."

Problem: The output is too long

CAUSE No length constraint was specified.

FIX Specify the desired length or format explicitly.

> "Limit your answer to three paragraphs. Do not repeat the facts back to me."

Problem: The output is too short

CAUSE The task was underspecified or the model inferred a simpler request than intended.

FIX Be explicit about depth: number of issues, length, level of analysis.

> "Address each of the five factors in the test separately. For each factor, explain the rule, apply it to the facts, and state your conclusion."

Problem: The output ignores key constraints

CAUSE Constraints were mentioned in passing rather than stated as guardrails.

FIX Move constraints to a dedicated guardrails section at the end of the prompt.

```
"Do not discuss attorneys' fees. Do not address claims we have
already settled. Limit your analysis to the contract claims only."
```

Remember Do not treat a weak prompt as evidence that AI is useless. Do not treat a strong output as evidence that verification is unnecessary. If a prompt fails, rebuild it. If an output succeeds, review it anyway. That discipline, more than any single tool or technique, is what makes AI usable in legal practice.

Appendix B: Prompt Bank: Ready-to-Use Prompts by Task

This appendix is for working lawyers, legal professionals, and law students who want prompts they can actually use, not just admire. The prompts below are meant to be copied, adapted, and improved in light of your own practice. They are not magic formulas. They work best when you customize them for your matter, your audience, your jurisdiction, and your risk tolerance.

These prompts also assume the discipline developed throughout this book. If a prompt asks for legal analysis, that does not eliminate the need for verification. If a prompt helps draft a client update, that does not eliminate your responsibility to ensure the communication is accurate, appropriately toned, and suited to the client's needs. If a prompt helps organize facts or authorities, that does not mean the organization is correct just because it is clean.

Use these prompts as starting points. Tighten them. Add your own examples. Save the ones that work. Revise the ones that do not. Over time, the strongest prompt bank is not the one printed in a book. It is the one you build from actual use.

How to Use This Appendix

These prompts are deliberately written in a practitioner voice. They are meant to be workable on a busy day, when you need a reliable starting point faster than you can build one from scratch. That means they favor clear instructions, source control, and practical outputs over clever phrasing. If a prompt produces something that feels polished but wrong, trust your verification process, not the polish. A prompt is a workflow tool, not a substitute for legal judgment.

You should also feel free to turn strong prompts into repeatable templates. If a prompt works well for a certain kind of motion, agreement, client update, or internal memo, save it with light placeholders and reuse it. Over time, your best prompt bank will become matter-tested, style-consistent, and aligned with how you actually practice rather than how someone else says you should. That is the standard this appendix is trying to support.

Each prompt is written to be practical rather than ornamental. Most are designed to do one specific job well. Some are source-bounded and assume you will paste in facts, authorities, contract language, or excerpts from the record. Others are designed for lower-risk tasks such as brainstorming, organization, or translation into plain English. Adapt them as needed by changing:

- jurisdiction
- audience
- document type
- factual constraints
- source restrictions
- formatting instructions
- caution instructions

1. General Prompting and Setup

Use these to establish baseline AI behavior before any substantive task.

1 **General role setup for legal work**

```
Act as a careful legal assistant helping me think through a task for
attorney review. Be accurate, practical, and explicit about
uncertainty. Do not invent facts, authorities, or procedural
requirements. If information is missing, identify the gap instead of
filling it in.
```

2 **General source-bounded setup**

```
Use only the material I provide below. Do not add authorities,
facts, or assumptions that are not in the supplied material. If the
supplied material is incomplete, say so directly.
```

3 **General output-structure setup**

```
Return the answer in this structure:
1. Core issue
2. Governing rule or principle
3. Application / analysis
4. Risks or uncertainties
5. Recommended next step
Keep each section concise and practical.
```

4 **Tight uncertainty-control prompt**

```
Distinguish clearly between:
- what is established by the materials provided
- what is likely but not certain
- what remains unknown or requires verification
Do not blur these categories.
```

5 **Opposing-counsel mode**

```
Act as capable opposing counsel trying to identify the weaknesses,
ambiguities, and vulnerabilities in the position set out below.
Focus on the strongest attacks, not weak or contrived ones.
```

6 **Client-facing translation mode**

```
Act as a lawyer preparing a clear explanation for a client with no
legal training. Use plain English, short paragraphs, and practical
examples where useful. Avoid legal jargon unless you define it.
```

7 **Executive-summary mode**

```
Assume the reader is a busy decision-maker. Put the most important
conclusion first, then the key reasons, then the recommended next
step. Do not bury the answer.
```

8 **Checklist mode**

```
Turn the material below into a practical checklist. Use clear,
action-oriented steps. Sequence the steps in the order a lawyer
would actually perform them.
```

2. Research Question Framing and Search Planning

Use these to break down a research problem before diving into authorities.

9 **Break a broad research question into sub-issues**

```
I need to research the following question:
[insert question]

Break this into specific legal issues and sub-issues I should
research. For each one, tell me what kind of authority is most
likely to matter, such as statute, case law, regulation, contract
language, local rule, or agency guidance.
```

10 **Generate search terms**

```
I am researching this issue:
[insert issue]

Generate search terms, synonyms, related concepts, and alternative
phrasing that courts or lawyers might use. Include variations that
could matter across jurisdictions or practice areas.
```

11 **Frame the issue for a memo**

```
Help me turn this broad problem into a properly framed internal
research question for a legal memo:
[insert problem]

Give me:
- a narrow issue statement
- a broader backup issue statement
- key factual questions that affect the analysis
```

12 Identify threshold questions first

```
Before I dive into the merits of this issue, identify the threshold
questions I should resolve first, such as jurisdiction, timing,
standing, exhaustion, enforceability, waiver, or applicable law.
Issue:
[insert issue]
```

13 Spot hidden research angles

```
Based on the facts below, what legal angles or issue clusters might
I miss if I focus only on the obvious claim?
[insert fact summary]
```

14 Generate research plan

```
Create a step-by-step research plan for this issue:
[insert issue]

Sequence the plan in the order a lawyer should actually research it,
from threshold matters through the core merits questions to possible
defenses and remedies.
```

3. Research with Provided Authorities

Use these when you have specific authorities and need AI to analyze, compare, or synthesize them.

15 Summarize provided authorities

```
Using only the authorities I provide below, summarize what they say
about the following issue:
[insert issue]

For each authority, give me:
- the key holding or rule
```

```
- why it matters
- any important limitation or qualification
```

16 **Compare authorities**

```
Compare the authorities below on this issue:
[insert issue]

Identify:
- where they agree
- where they differ
- whether the difference is a true conflict or just different fact
patterns
```

17 **Extract holdings carefully**

```
For each authority below, extract the actual holding relevant to my
issue. Stay close to the source language. Do not paraphrase broadly.
Preserve any qualifications or limiting language.
```

18 **Build a rule statement from authorities**

```
Using only the authorities below, draft a rule statement suitable
for an internal memo on:
[insert issue]

Then identify where the rule is uncertain, split, or fact-dependent.
```

19 **Identify weak spots in my rule statement**

```
Here is my current rule statement:
[insert rule statement]

Using only the authorities below, tell me where my statement
overstates, understates, or mischaracterizes the law.
```

20 **Build a synthesis paragraph**

```
Using only the authorities below, draft one synthesis paragraph
explaining how the cases fit together and what practical rule
emerges.
```

21 **Flag unsupported propositions**

```
Review the analysis below and identify any proposition that is not
adequately supported by the authorities provided.
[insert analysis]
[insert authorities]
```

22 Build a research memo outline

```
Using only the authorities and facts below, build an outline for a
research memo. Organize it as:
- Issue
- Brief answer
- Facts
- Analysis
- Conclusion
```

4. Argument Mapping and Counterarguments

Use these to build, stress-test, and defend your legal arguments.

23 Build my argument map

```
Based on the authorities and facts below, build an argument map for
my position. Identify:
- the core conclusion
- the supporting premises
- the authorities for each premise
- the factual assumptions doing the work
```

24 Build the opposing argument map

```
Based on the same authorities and facts, build the strongest
opposing argument map. Show how the other side would structure the
issue and what authorities they would emphasize.
```

25 Steelman the other side

```
Present the strongest possible version of the opposing side's
argument. Do not give me weak points or strawman arguments. I want
the best version I need to beat.
```

26 Identify my weak links

```
Here is my current argument:
```

```
[insert argument]

Identify the weakest links in the reasoning. Tell me where a judge
or opposing counsel is most likely to push back.
```

27 Pre-mortem my motion

```
Assume the judge denies this motion. What is the most likely reason
for the denial based on the arguments outlined below?
[insert argument outline]
```

28 Generate responses to counterarguments

```
For each of the following likely counterarguments, draft the
strongest response I can make using the authorities and facts below:
[insert counterarguments]
[insert authorities/facts]
```

5. Drafting and First-Draft Generation

Use these to generate initial drafts from instructions or source materials.

29 Draft from instructions

```
Draft [document type] using the following parameters:

Purpose:
[insert purpose]

Context:
[insert matter context]

Constraints:
[insert constraints]

Structure:
[insert structure]

Output requirements:
[insert style, length, format, audience]

Do not invent facts or authorities. If something is missing,
identify the gap.
```

30 Draft from source materials

```
Using only the materials below, draft a first version of [document
type]. Build from the supplied content rather than generating from
general patterns.
[insert source materials]
```

31 Draft a litigation section from verified inputs

```
Draft the [statement of facts / argument section / conclusion] using
only the record citations and authorities below. Do not add any
facts or authorities I did not provide.
```

32 Draft a client letter from an internal memo

```
Based on the internal analysis below, draft a client-facing letter
that explains the issue in plain English, gives the bottom line, and
clearly identifies the next steps.
```

33 Draft multiple alternatives

```
Generate three versions of this clause, paragraph, or section:
- one conservative
- one balanced
- one aggressive
After each version, briefly explain what makes it conservative,
balanced, or aggressive.
```

34 Tighten a draft

```
Revise the text below to make it clearer, tighter, more direct, and
more professional. Do not change the substantive meaning unless you
flag the proposed change separately.
```

6. Contract Clause Work

Use these to generate, compare, improve, and risk-spot contract provisions.

35 Generate clause options

```
Draft three versions of a [clause type] for the following deal
context:
[insert context]
```

```
For each version, explain what risk it addresses, what it leaves
exposed, and which side it favors.
```

36 Improve a one-sided clause

```
Analyze the clause below from my client's perspective. Identify
every provision that is unfavorable to my client, explain the risk
it creates, and then draft a revised version that better balances
the parties' interests.
[insert clause]
```

37 Compare two clauses

```
Compare these two clause versions and identify every substantive
difference. For each difference, explain:
- what changed
- why it matters
- which party benefits
```

38 Spot hidden risks in a clause

```
Review the clause below and identify hidden risks, ambiguities,
loopholes, or practical enforcement problems. Prioritize the risks
from highest to lowest.
```

39 Generate fallback positions

```
My preferred language is below. If the counterparty rejects it,
generate three fallback positions that still protect my client's
core interests. Explain what each fallback gives up.
```

40 Analyze interaction effects

```
Read these two provisions together and tell me whether they create
any unexpected interaction, inconsistency, or gap in protection.
[insert provision 1]
[insert provision 2]
```

41 Explain a clause to a client

```
Explain the following clause to a client in plain English. Focus on
what it means in practice, what risks it creates, and what the
client is giving up or gaining.
```

42 Draft negotiation comments

```
Draft short, professional markup comments explaining why we are
proposing the revisions below and why the counterparty's language is
not acceptable.
```

7. Redlining and Negotiation Support

Use these to analyze markups, plan responses, and prepare for negotiation.

43 Analyze the markup

```
Compare our original draft and the counterparty's redline. Identify
every substantive change. Categorize each as:
- economic
- liability / risk allocation
- procedural
- cosmetic or stylistic

Then rank the substantive changes by importance to my client.
```

44 Read the negotiation pattern

```
Looking across this markup, what does the counterparty appear to
care most about? What negotiating pattern do you see?
```

45 Draft a response position memo

```
Based on the redline analysis below, draft a short internal memo
identifying:
- must-hold positions
- flexible points
- possible trades
- fallback options
```

46 Identify issue trades

```
Looking at both parties' apparent priorities, suggest possible issue
trades that could help move the negotiation without sacrificing my
client's core objectives.
```

47 Prepare for negotiation call

```
Prepare a one-page negotiation sheet from the following issues.
Organize it under:
- non-negotiables
- preferred outcomes
- fallback positions
- trade opportunities
- talking points
```

48 Test whether a concession is acceptable

```
If we accept the following concession, what risks remain? What
scenarios could still leave my client underprotected?
[insert concession]
```

8. Litigation Drafting

Use these for demand letters, pleadings, motions, and briefs built from verified inputs.

49 Draft a demand letter

```
Draft a demand letter using only the facts and legal theories below.
Keep the tone firm but professional. Do not add facts, dates, or
legal claims I did not provide.
```

50 Draft a complaint from a fact sheet

```
Using only the facts below, draft a complaint asserting the
following claims:
[insert claims]

Use only the facts I provide. If any required factual element is
missing, identify the gap rather than inventing it.
```

51 Draft an answer from a complaint and response chart

```
Using the complaint and my paragraph-by-paragraph response notes
below, draft an answer that admits, denies, or states insufficient
knowledge consistently with my notes.
```

52 Draft a statement of facts with record cites

```
Draft a statement of facts using only the facts and record citations
below. Every factual sentence must remain tied to a provided
citation.
```

53 Draft an argument section from verified authorities

```
Draft the argument section using only the legal standards,
authorities, and facts below. Do not add cases or propositions I did
not provide.
```

54 Draft a reply brief response

```
The opposition brief raises the following arguments:
[insert summary]

Using only my authorities and facts below, draft a reply section
responding directly to those arguments.
```

55 Tighten a brief introduction

```
Revise this brief introduction to make it clearer, more forceful,
and more direct without overstating the law or the facts.
```

56 Check a brief for overclaiming

```
Review the draft below and flag any sentence that sounds more
certain, broader, or more aggressive than the authorities and facts
actually support.
```

9. Discovery and Review

Use these to draft requests, evaluate objections, and build review protocols.

57 Draft document requests

```
Based on the claims, defenses, and factual issues below, draft
targeted requests for production designed to obtain the most
important documents first.
```

58 Draft interrogatories

```
Draft interrogatories designed to force the opposing party to
identify the people, events, documents, and explanations central to
the dispute described below.
```

59 Evaluate incoming requests for objections

```
Review these discovery requests and identify which are potentially
objectionable as overbroad, vague, unduly burdensome, privileged, or
irrelevant. For each, suggest a specific objection and any narrower
response position.
```

60 Draft a proportionality argument

```
Draft a proportionality argument opposing the request below. Use the
burden facts I provide and focus on why the likely burden outweighs
the likely benefit.
```

61 Build a review protocol

```
Create a document review protocol for this matter. Include relevance
criteria, privilege criteria, hot-document criteria, and escalation
rules for uncertain documents.
```

62 Spot privilege candidates

```
Review the document descriptions below and flag which are likely
privilege candidates, explaining why they should be escalated for
lawyer review rather than auto-produced.
```

63 Draft privilege log descriptions

```
Using the document metadata below, draft concise privilege log
descriptions that identify the sender, recipients, date, and general
subject matter without revealing the substance of the legal advice.
```

64 Identify hot documents

```
Based on the summaries below, identify which documents are likely to
be the most strategically important for depositions, motion
practice, or settlement leverage. Explain why.
```

10. Depositions, Hearings, and Witness Prep

Use these to build outlines, plan impeachment, and prepare witnesses.

65 Build a deposition outline

```
Create a deposition outline for the following witness. Organize by
topic, objective, key documents, and likely follow-up questions.
[insert witness role and goals]
```

66 Draft a question module

```
Draft a focused line of questioning designed to establish:
[insert target fact or admission]

Include questions that lock the witness into a position before
showing the key exhibit.
```

67 Identify impeachment points

```
Based on the documents and prior statements below, identify the best
impeachment points if the witness denies knowledge of the issue.
```

68 Organize exhibits strategically

```
Review the potential exhibits below and suggest the most effective
order for using them in the deposition or hearing. Explain what
narrative or strategic effect that order produces.
```

69 Prepare a witness for deposition

```
Act as opposing counsel and ask hard but realistic deposition
questions on the topics below. After each question, explain what the
witness needs to understand about the risk in answering poorly.
```

70 Summarize a deposition transcript

```
Summarize the transcript below by identifying:
- key admissions
- harmful testimony
- inconsistencies
- follow-up investigation points
- testimony useful for motions or impeachment
```

11. Investigations and Fact Development

Use these to build timelines, map relationships, and identify gaps in the record.

71 Build a master timeline

```
Using the materials below, create a chronological timeline of
events. For each entry, include the date, actors involved, event
description, and source.
```

72 Build an entity / relationship map

```
Identify all people and entities mentioned in the materials below
and map their relationships, reporting lines, and roles in the
events.
```

73 Identify factual gaps

```
Review the timeline and factual summary below. Identify the most
important gaps in the evidence, the periods of silence, and the
questions that need investigation next.
```

74 Identify patterns or anomalies

```
Analyze the records below and identify unusual patterns,
inconsistencies, or anomalies that may warrant deeper investigation.
```

75 Draft interview questions

```
Based on the timeline and documents below, draft interview questions
for [witness name / role]. Prioritize questions that clarify gaps,
test explanations, and challenge inconsistencies.
```

12. Client Communication and Intake

Use these to process intake, explain legal concepts, and draft client updates.

76 Summarize an intake conversation

```
Based on the intake notes or transcript below, extract and organize:
- key facts
- parties
- dates
- documents mentioned
- possible legal issues
- follow-up questions
```

77 Spot legal issues from a client narrative

```
Based on the client narrative below, identify possible legal claims,
defenses, procedural issues, and timing concerns that warrant closer
review.
```

78 Explain a legal concept in plain English

```
Explain the following legal concept to a client with no legal
training. Use plain English, practical examples where helpful, and
avoid jargon unless you define it.
[insert concept]
```

79 Draft a client update

```
Draft a client update using the following structure:
- current status
- what happens next
- timing
- what I need from you

The client prefers [detailed / concise] communication and has no
legal training.
```

80 Draft a difficult-news communication

```
Draft a client communication explaining the following adverse
development. Acknowledge the setback, explain what it means, and
explain the path forward without overpromising.
```

81 Explain options without steering

```
Draft a plain-English explanation of the following options for the
client. Present them fairly, with pros, cons, risks, and practical
implications, without pushing the client toward one choice.
```

13. Marketing, Thought Leadership, and Public-Facing Writing

Use these to draft content that demonstrates expertise while complying with advertising rules.

82 Draft a blog post

```
Draft a blog post for [audience] on [topic]. Keep it practical,
accurate, and readable. Do not sound promotional or generic. End
with a simple note that the content is general information, not
legal advice.
```

83 Turn a legal development into client-facing content

```
Explain the practical significance of the following legal
development for [target audience]. Focus on what changed, why it
matters, and what actions readers should consider.
```

84 Draft a LinkedIn post from a longer article

```
Turn the article below into a concise LinkedIn post that highlights
the key takeaway, sounds professional, and invites further reading
without sounding salesy.
```

85 Generate thought-leadership topics

```
Suggest ten article or presentation topics for a lawyer in [practice
area]. Focus on topics potential clients actually care about and
questions they are likely to search for.
```

86 Check marketing content for rule-risk

```
Review the content below for potentially problematic advertising
language, including superlatives, implied guarantees, unsupported
expertise claims, or wording that sounds misleading.
```

14. Practice Management, Governance, and Internal Systems

Use these to build playbooks, improve billing, draft policies, and design training.

87 Draft a matter playbook outline

```
Create a matter playbook outline for [matter type]. Include phases,
key documents, common issues, deadlines, trigger events, lessons
learned, and prompts worth preserving.
```

88 Create a reusable prompt template

```
Build a reusable prompt template for the following recurring task:
[insert task]

The template should follow Role, Context, Task, Format, and
Guardrails, and should require only minimal matter-specific
customization.
```

89 Improve a billing narrative

```
Improve the time-entry narrative below so it clearly and
professionally describes the work performed without exaggerating or
changing the substance of what was actually done.
```

90 Draft an internal AI policy section

```
Draft a short policy section for a law firm addressing [acceptable
use / confidentiality / verification / disclosure / incident
response]. Keep it specific enough to guide behavior without
becoming unnecessarily bureaucratic.
```

91 Draft training objectives

```
Create training objectives for a one-hour internal session on [AI
topic]. The objectives should be practical, specific, and tied to
actual lawyer workflows.
```

92 Build an audit checklist

```
Create a short audit checklist for reviewing whether AI use on a
matter complied with internal policy, confidentiality requirements,
and verification expectations.
```

15. Prompt Repair and Quality Control

Use these when a prompt underperforms and you need to diagnose and fix it.

93 Diagnose why a prompt failed

```
Here is the prompt I used:
[insert prompt]

Here is the output I received:
[insert output]

Diagnose why the prompt underperformed. Identify what is missing or
weak in the Role, Context, Task, Format, or Guardrails. Then rewrite
the prompt.
```

94 Rewrite a weak prompt

```
Take the weak prompt below and rewrite it into a professional legal
prompt using the Prompt Blueprint. Then explain what changed and
why.
[insert prompt]
```

95 Reduce hallucination risk

```
Rewrite the following prompt to reduce hallucination risk and
increase controllability. Add source restrictions, uncertainty
instructions, and any necessary output limits. [insert prompt]
```

96 Improve a prompt for a specific audience

```
Rewrite this prompt so the output is better suited for [audience:
client / supervising partner / judge / in-house counsel /
colleague].
[insert prompt]
```

97 Convert a one-shot prompt into a chained workflow

```
Break the following big prompt into a better sequence of smaller
prompts, with each stage feeding into the next and allowing review
before moving forward.
[insert prompt]
```

98 Stress-test a draft before use

```
Review the draft below and identify:
- unsupported assertions
- overbroad conclusions
- places where the tone is wrong
- places where a reasonable reader would want more support
Then suggest specific revisions.
```

16. High-Value Closing Prompts

Use these as final checks before relying on or sending any AI-assisted work.

99 "What am I missing?" prompt

```
Based on the materials below, what important issue, risk, factual
gap, or practical concern have I not asked about but should be
thinking about before I finalize this work?
```

100 "What would a skeptical reviewer attack?" prompt

```
If a skeptical judge, client, partner, or opposing counsel reviewed
this work, what are the three strongest criticisms they would likely
make?
```

101 "What needs verification?" prompt

```
Review the text below and identify every statement that requires
verification before I can rely on or send this work.
```

102 "Condense without oversimplifying" prompt

```
Condense the material below into a shorter version for a busy
reader, but preserve any qualifications, uncertainty, or limiting
language necessary for accuracy.
```

103 "Turn analysis into action" prompt

```
Based on the analysis below, turn it into an action-oriented next-
steps list for the lawyer handling the matter.
```

104 "Convert into checklist" prompt

```
Take the workflow or analysis below and convert it into a practical
checklist that a lawyer could actually use in sequence.
```

105 "Teach it back" prompt

```
Explain the following issue as if teaching it to a junior lawyer who
is intelligent but unfamiliar with the subject. Focus on what they
need to understand to avoid mistakes.
```

Final Note

The value of a prompt bank is not that every prompt will be perfect for every matter. The value is that it shortens the distance between "I know what I need" and "I know how to ask for it well."

The best prompts in your practice will be the ones you actually use, revise, and keep because they helped you do better work. Let this appendix be the beginning of that process, not the end of it.

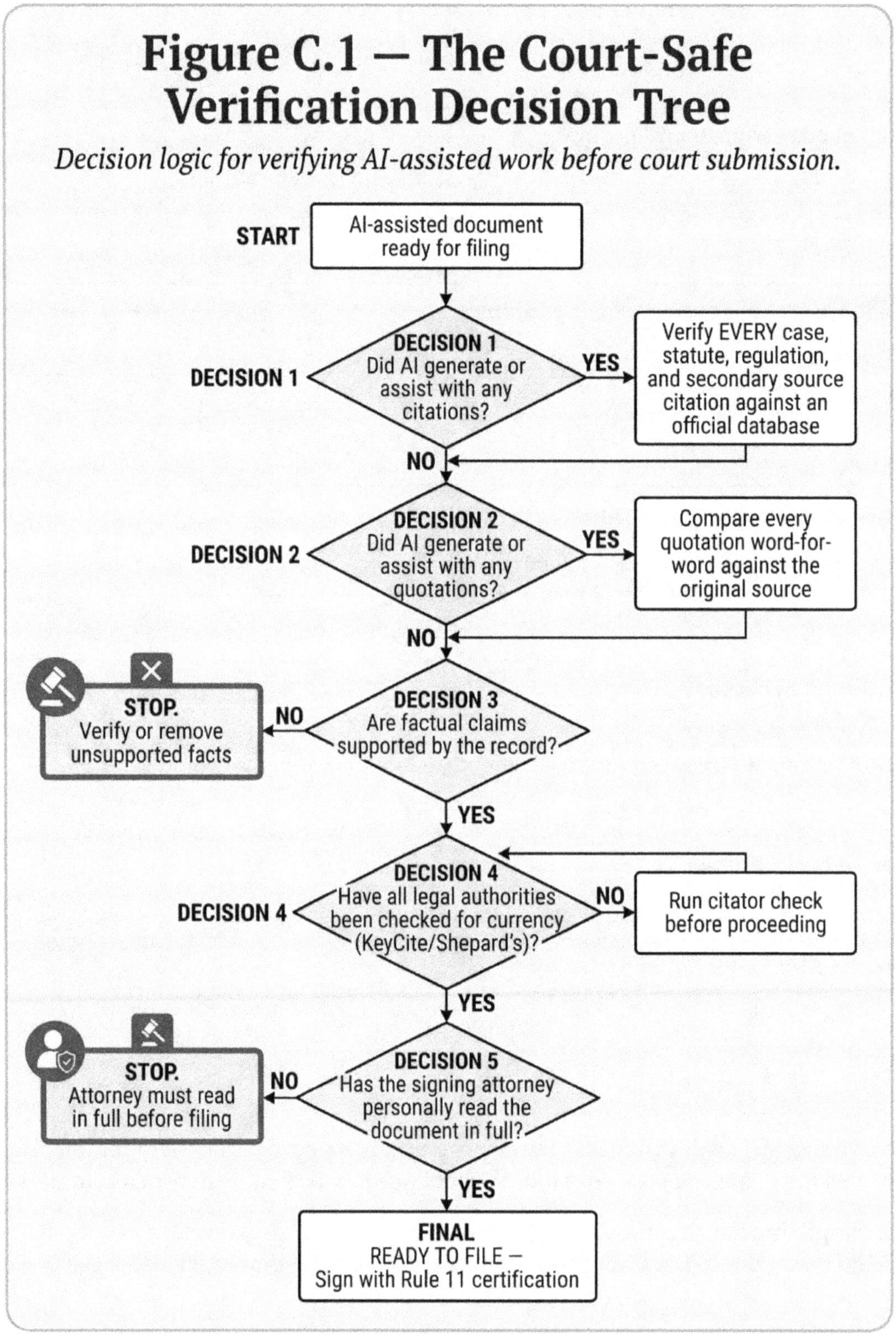

Figure C.1 — The Court-Safe Verification Decision Tree

This appendix provides verification tools specifically designed for court-facing work. The materials address the heightened accuracy obligations that apply when AI-assisted work product will be submitted to courts, responding to the professional responsibility requirements discussed in Chapter 4 and the verification techniques from Chapter 7. These tools are designed for practical use: print the one-page checklist and keep it at your desk, use the citation procedure for every court filing, and maintain the workflow log as documentation of your verification process.

1. Verification Checklist (One-Page)

Print this checklist and complete it for every AI-assisted court filing.

AI-Assisted Court Filing Verification Checklist

Matter: ______________________

Court: ______________________

Document: ______________________

Filing Date: ______________

Attorney: ______________________

Date Completed: ______________

Section 1: AI Use Documentation

- ☐ AI tool(s) used: ______________

AI was used for (check all that apply):

- ☐ Legal research
- ☐ Drafting text
- ☐ Summarizing authorities
- ☐ Generating arguments
- ☐ Citation formatting
- ☐ Editing/proofreading
- ☐ Other: ______________
- ☐ Court disclosure requirements reviewed for this jurisdiction
- ☐ Required AI disclosure included in filing (if applicable)

Section 2: Citation Verification

☐ Every case citation verified against official reporter or reliable database

Number of case citations: ____ Number verified: ____

☐ Every statutory citation verified against current code

Number of statutory citations: ____ Number verified: ____

☐ Every regulatory citation verified against current CFR or state equivalent

Number of regulatory citations: ____ Number verified: ____

☐ Every secondary source citation verified

Number of secondary citations: ____ Number verified: ____

☐ Citation format checked against Bluebook or local rules

☐ No citations were generated by AI without independent verification

Section 3: Quotation Verification

☐ Every quotation checked against original source

Number of quotations: ____ Number verified: ____

☐ Quotations are word-for-word accurate

☐ Ellipses and alterations properly indicated

☐ Quotations are not taken out of context

☐ Pinpoint citations provided for all quotations

Section 4: Factual Accuracy

☐ Every factual statement traced to record or verified source

☐ Record citations provided for all factual claims

☐ No facts were invented or assumed by AI

☐ Facts are characterized accurately and fairly

☐ Opponent's positions stated accurately

Section 5: Legal Accuracy

☐ Holdings distinguished from dicta

☐ Legal standards accurately stated

☐ Authorities support propositions for which they are cited

- ☐ Adverse authority identified and addressed
- ☐ Jurisdiction confirmed for all cited authorities
- ☐ Currency checked: no overruled, superseded, or withdrawn authorities

Section 6: Compliance

- ☐ Page/word limits satisfied
- ☐ Formatting requirements met
- ☐ Filing deadlines confirmed
- ☐ Certificate of service accurate
- ☐ Local rules reviewed and followed
- ☐ Signature block complete and accurate

Section 7: Final Review

- ☐ Document read in full by signing attorney
- ☐ Arguments are sound and supportable
- ☐ No overstatements or mischaracterizations
- ☐ Document represents attorney's own work product (AI-assisted, not AI-authored)
- ☐ Signing attorney takes full responsibility for all content

Certification

I have completed the verification steps above. I have personally reviewed this document and verified its accuracy. I am prepared to certify this filing under Rule 11 or equivalent.

Signature: ______________________________

Date: _______________

2. Citation and Quote-Check Procedure

This procedure provides step-by-step guidance for verifying citations and quotations in AI-assisted documents. Follow this procedure for every court filing where AI assisted with research, drafting, or citation work.

Citation Verification Procedure

Purpose

AI tools may generate citations that appear legitimate but do not exist, cite authorities that exist but don't support the stated proposition, or contain errors in citation format. This procedure ensures every citation in a court filing has been independently verified before submission.

Scope

This procedure applies to:

- All case citations
- All statutory citations
- All regulatory citations
- All citations to secondary sources
- All citations to record materials

Procedure

Step 1: Extract All Citations

Before beginning verification, create a complete list of every citation in the document.

- Method A (Manual): Read through the document and list each citation.
- Method B (AI-Assisted): Use this prompt to extract citations:

"Review the following document and list every citation to legal authority, including cases, statutes, regulations, and secondary sources. For each citation, provide: 1. The citation as it appears; 2. The page number where it appears; 3. The proposition it supports. Do not verify the citations; only extract them. [Paste document]"

- Review the AI-generated list against the document to ensure completeness.

Step 2: Verify Case Citations

For each case citation:

- ☐ Confirm existence: Look up the case in Westlaw, Lexis, or official reporter. Confirm the case exists with the cited name, citation, court, and year.
- ☐ Confirm holding: Read the cited portion of the case. Confirm it supports the proposition for which you cited it.
- ☐ Check currency: Run a citator check (KeyCite, Shepard's, or equivalent). Confirm the case has not been overruled, reversed, or superseded on the relevant point.
- ☐ Verify pinpoint: If a pinpoint citation is provided, confirm the specific page contains the relevant material.
- ☐ Check format: Confirm citation format complies with Bluebook and any local rules.

> **Red flags requiring additional attention:** Case you've never heard of that perfectly supports your position. Unpublished or hard-to-find decisions. Very old cases cited for current propositions. Cases from unexpected jurisdictions.

Step 3: Verify Statutory Citations

For each statutory citation:

- ☐ Confirm existence: Look up the statute in the current official code or reliable database.
- ☐ Confirm current text: Verify the statutory language matches what you've represented. Check for recent amendments.
- ☐ Confirm applicability: Verify the statute applies to the relevant time period and circumstances.
- ☐ Check related provisions: Review surrounding sections for context that might affect interpretation.
- ☐ Verify format: Confirm citation format complies with Bluebook and local rules.

Step 4: Verify Regulatory Citations

For each regulatory citation:

- ☐ Confirm existence: Look up the regulation in the current CFR or state equivalent.
- ☐ Confirm current text: Verify the regulatory language is current. Check Federal Register or equivalent for recent amendments.
- ☐ Confirm effective date: Verify the regulation was in effect during the relevant period.

☐ Check enabling statute: Confirm the regulation is within the agency's statutory authority if relevant.

☐ Verify format: Confirm citation format complies with requirements.

Step 5: Verify Secondary Source Citations

For each secondary source citation:

☐ Confirm existence: Verify the treatise, law review article, or other source exists.

☐ Confirm content: Verify the source says what you've represented.

☐ Check edition/date: Confirm you're citing a current edition if currency matters.

☐ Assess authority: Consider whether the source carries appropriate weight for your purpose.

Step 6: Document Verification

For each citation verified, document:

- Citation verified
- Source used for verification
- Date verified
- Any issues identified and how resolved
- Verifier initials

Maintain this documentation in the matter file.

Quotation Verification Procedure

Purpose

Quotations must be word-for-word accurate. AI tools may paraphrase, truncate, alter, or fabricate quotations. This procedure ensures every quotation has been verified against the original source.

Scope

This procedure applies to:

- All direct quotations from cases
- All direct quotations from statutes or regulations
- All direct quotations from secondary sources
- All direct quotations from the record
- All direct quotations from any other source

Procedure

Step 1: Identify All Quotations

Review the document and identify every passage in quotation marks, every block quotation, and every phrase attributed to a source with language suggesting direct quotation (e.g., "the court stated," "the statute provides").

Step 2: Locate Original Source

For each quotation, obtain the original source document. Do not rely on AI's representation of what the source says. Do not rely on quotations of quotations. Go to the original.

Step 3: Word-for-Word Comparison

Compare the quotation in your document to the original source word by word. Check for:

- ☐ Accuracy: Every word matches exactly
- ☐ Completeness: No words omitted without ellipsis indication
- ☐ Proper ellipses: Omissions properly indicated with ellipses
- ☐ Proper alterations: Any changes in capitalization, verb tense, or other alterations indicated with brackets
- ☐ Proper emphasis: Added emphasis indicated as such; original emphasis preserved or noted

Step 4: Context Verification

Verify that the quotation, as used, does not misrepresent the source's meaning. Check for:

- ☐ Fair representation: The quotation fairly represents the source's point
- ☐ No distortion: Surrounding context doesn't change the meaning of the quoted language
- ☐ Complete thought: The quotation doesn't truncate in a way that changes meaning
- ☐ Accurate attribution: The quotation is attributed to the correct source (majority opinion vs. dissent, holding vs. dicta)

Step 5: Pinpoint Verification

Verify that pinpoint citations are accurate.

- ☐ Page number correct: The quoted language appears on the cited page
- ☐ Paragraph number correct: If citing by paragraph, the quotation appears in the cited paragraph
- ☐ Section correct: If citing a statute or regulation by section, the quotation appears in the cited section

Step 6: Document Verification

For each quotation verified, document:

- Quotation location in your document (page, line)
- Original source consulted
- Page/location in original source
- Any discrepancies identified and how resolved
- Verifier initials

Common Citation and Quotation Errors

Citation Errors to Watch For

Error Type	Example	How to Catch
Fabricated case	"Smith v. Jones, 523 F.3d 456 (9th Cir. 2018)" that doesn't exist	Database lookup confirms no such case
Wrong citation	Correct case name but wrong reporter, volume, or page	Database lookup shows different citation
Wrong court	Case exists but from different court than cited	Database lookup shows actual court
Wrong year	Case exists but from different year	Database lookup shows actual year
Overruled	Case exists but has been overruled on cited point	Citator check shows negative treatment
Wrong proposition	Case exists but doesn't say what's claimed	Reading the case shows different holding

Quotation Errors to Watch For

Error Type	Example	How to Catch
Fabricated quotation	Quotation doesn't appear in cited source	Word search in original source finds nothing
Paraphrase as quotation	Similar language but not exact	Word-by-word comparison shows differences
Truncated quotation	Missing words change meaning	Reading full passage shows different meaning
Combined quotations	Parts of different sentences combined	Checking original shows discontinuous text

Error Type	Example	How to Catch
Wrong attribution	Quotation from dissent attributed to majority	Checking original shows different source
Missing alteration marks	Changed words without brackets	Comparison shows undisclosed alterations

Verification Standards by Document Type

Document Type	Citation Standard	Quotation Standard	Documentation
Supreme Court brief	100% verification	100% verification	Full documentation
Appellate brief	100% verification	100% verification	Full documentation
Dispositive motion	100% verification	100% verification	Full documentation
Other motion	100% verification	100% verification	Standard documentation
Opposition/Reply	100% verification	100% verification	Standard documentation
Trial brief	100% verification	100% verification	Standard documentation
Letter to court	100% verification	100% verification	Standard documentation

There is no document type submitted to a court where less than 100% verification is acceptable.

3. AI Workflow Log Template

This log documents AI use in legal work, creating an audit trail that demonstrates appropriate process and enables reconstruction of how work was produced if questions arise.

AI Workflow Log — Matter Information

Field	Entry
Matter Name/Number	
Client	
Matter Type	
Responsible Attorney	
Log Maintained By	
Log Period	

Entry Format

For each significant AI use, create a log entry with the following information:

Log Entry #: ________

Date/Time: __________________________

Attorney: __________________________

Task Description: ____

AI Tool Used and Configuration

AI Tool Used: ____

Tool Configuration:

- ☐ Consumer version
- ☐ Enterprise version (organization account)
- ☐ Private/self-hosted deployment
- ☐ Other: ___________

Data Input Summary

- ☐ Public information only
- ☐ Confidential client information (describe category): ___________

☐ Privileged information
☐ Sanitized/anonymized information
☐ Information from provided documents only (source-bounded)

Prompt Summary: ____

(Brief description of what was requested; full prompts may be attached or maintained separately)

Output Summary: ____

(Brief description of what AI produced)

Review and Verification

Verification Step	Completed	By	Date
Initial review for responsiveness	☐		
Accuracy review	☐		
Citation verification (if applicable)	☐		
Quote verification (if applicable)	☐		
Factual verification (if applicable)	☐		
Supervisor review (if required)	☐		

Modifications Made: ____

(Describe significant changes made to AI output before use)

Final Use

☐ Used substantially as generated
☐ Used with significant modification
☐ Used as starting point only; substantially rewritten
☐ Not used; generated alternative approach
☐ Other: __________

Work Product Destination

☐ Internal working document
☐ Client communication
☐ Court filing
☐ Opposing party communication
☐ Other: ___________

Issues or Concerns: ____

(Note any problems identified, verification failures, or other issues)

Attachments

☐ Full prompt text
☐ Full AI output
☐ Verification documentation
☐ Final work product
☐ Other: ___________

Simplified Log for Routine Tasks

For lower-risk, routine AI uses, a simplified log entry may be sufficient:

Date	Attorney	Task	Tool	Data Type	Review	Use

Column Guide

- Date: Date of AI use
- Attorney: Attorney using AI
- Task: Brief description (e.g., "draft email," "research summary")
- Tool: AI tool used

- Data Type: P (public), C (confidential), S (sanitized), B (source-bounded)
- Review: Self (S), Supervisor (Sup), Citation check (CC)
- Use: As-is (A), Modified (M), Starting point (SP), Not used (N)

Log Retention and Access

Retention Period

AI workflow logs should be retained for the same period as matter files, or longer if required by:

- Firm policy
- Client requirements
- Regulatory requirements
- Anticipated litigation hold

Access

Logs should be accessible to:

- Responsible attorney
- Supervising attorneys
- Firm management (for governance purposes)
- Client (upon request, subject to work product considerations)

Logs containing detailed prompts and outputs may contain work product and should be protected accordingly.

Storage

Store logs:

- With matter file for matter-specific logs
- In governance records for aggregate reporting
- Securely, with access limited to authorized personnel

Aggregate Reporting

For governance purposes, logs can be aggregated to report:

Metric	Purpose
AI uses by tool	Track tool adoption and usage patterns

Metric	Purpose
AI uses by task type	Identify common applications
AI uses by data sensitivity	Monitor confidential data handling
Verification completion rates	Assess policy compliance
Issues identified	Track AI-related problems
Use rates (as-is vs. modified)	Assess AI output quality

Sample quarterly summary format: AI Use Summary [Quarter/Year]

Category	Count	Notes
Total logged AI uses		
By tool: [Tool A]		
By tool: [Tool B]		
Research tasks		
Drafting tasks		
Analysis tasks		
Court filings with AI assistance		
Verification issues identified		
Policy exceptions requested		

Sample Completed Log Entries

Example 1: Research Task

Log Entry #127 — Research Task

Date/Time: March 15, 2026, 2:30 PM

Attorney: J. Smith

Task Description: Research on enforceability of non-compete agreements for low-wage employees under recent state law developments.

AI Tool Used: [Legal Research AI Tool]

Tool Configuration: Enterprise version

Data Input Summary: Public information only

Prompt Summary: Requested summary of state law developments affecting non-compete enforceability for employees earning below specified wage thresholds. Specified jurisdictions of interest.

Output Summary: AI provided summary of developments in 8 states with citations to statutes and recent cases. Identified trend toward restricting enforcement for lower-wage workers.

Review and Verification:

- Initial review: Completed by JS, 3/15
- Accuracy review: Completed by JS, 3/15
- Citation verification: Completed by JS, 3/16
- Quote verification: Completed by JS, 3/16
- Supervisor review: Completed by MP, 3/17

Modifications Made: Corrected one statutory citation that had wrong section number. Added two additional recent cases identified through independent research. Removed one case AI cited that didn't actually address wage thresholds.

Final Use: Used with significant modification

Work Product Destination: Internal working document (research memo)

Issues or Concerns: One citation was incorrect (wrong section number). One case didn't support the proposition. Reminder to verify all AI-generated citations.

Example 2: Drafting Task for Court Filing

Log Entry #128 — Drafting for Court Filing

Date/Time: March 18, 2026, 10:00 AM

Attorney: J. Smith

Task Description: Draft argument section for motion to compel discovery responses, specifically section on proportionality analysis.

AI Tool Used: [General AI Assistant - Enterprise]

Tool Configuration: Enterprise version

Data Input Summary: Sanitized/anonymized information; information from provided documents only (source-bounded)

Prompt Summary: Provided sanitized summary of discovery dispute (party names removed, specific facts generalized). Provided text of three cases on proportionality to use as authority. Requested draft argument section applying proportionality factors.

Output Summary: AI produced approximately 1,200-word draft argument section applying proportionality factors from provided cases to generalized fact pattern.

Review and Verification:

- Initial review: Completed by JS, 3/18
- Accuracy review: Completed by JS, 3/18
- Citation verification: Completed by JS, 3/18
- Quote verification: Completed by JS, 3/18
- Factual verification: Completed by JS, 3/18
- Supervisor review: Completed by MP, 3/19

Modifications Made: Replaced generalized facts with actual case facts. Revised characterization of one holding that AI overstated. Added additional argument point not in AI draft. Adjusted tone to be more assertive. Verified all three case citations and two quotations against original sources.

Final Use: Used with significant modification

Work Product Destination: Court filing

Issues or Concerns: AI characterized one holding too broadly; original case was more limited. Reminder that even source-bounded prompting requires verification of how AI characterizes provided authorities.

Attachments: Full prompt text, full AI output, verification documentation, final work product

Example 3: Client Communication

Log Entry #129 — Client Communication

Date/Time: March 20, 2026, 3:00 PM

Attorney: J. Smith

Task Description: Draft client update letter explaining recent motion ruling and next steps.

AI Tool Used: [General AI Assistant - Enterprise]

Tool Configuration: Enterprise version

Data Input Summary: Confidential client information (case status, strategy)

Prompt Summary: Provided summary of motion ruling, current case status, and next steps. Requested draft client letter explaining ruling in accessible terms with recommended actions.

Output Summary: AI produced draft client letter approximately 500 words explaining ruling, implications, and recommended next steps.

Review and Verification:

- Initial review: Completed by JS, 3/20
- Accuracy review: Completed by JS, 3/20
- Tone/clarity review: Completed by JS, 3/20

Modifications Made: Adjusted explanation of ruling to be clearer on specific implications. Strengthened recommendation section. Added specific deadline for client decision. Personalized opening and closing.

Final Use: Used with significant modification

Work Product Destination: Client communication

Issues or Concerns: None. AI draft was helpful starting point but required personalization and clarity improvements typical for client communications.

Appendix D: AI Governance Issues to Resolve with Your Own Counsel

This appendix is intentionally different from the other appendices in this book.

It does not provide a model firm policy for adoption. It does not provide client disclosure language for direct use in engagement letters or communications. And it is not intended to function as legal advice for any specific lawyer, firm, legal department, jurisdiction, or client relationship.

Instead, this appendix is a structured framework for identifying the questions that a responsible practice should resolve before adopting, scaling, or formalizing AI use. Those questions should then be answered in light of the rules, ethics opinions, client requirements, court orders, and practical realities that apply to the reader's own practice. In some organizations, that means consultation with in-house ethics counsel, general counsel, outside counsel, risk management personnel, or practice leadership. In smaller practices, it may mean a deliberate self-audit followed by consultation with appropriate outside advisors where needed.

The goal is not to avoid making decisions. The goal is to avoid making them casually.

A firm that copies generic policy language without understanding what it means may create a false sense of safety. A lawyer who pastes disclosure language into an engagement letter without thinking through what the language commits them to may create avoidable risk. Good governance requires more than forms. It requires judgment, internal alignment, and clear understanding of what the practice is actually prepared to do and supervise.

Use this appendix to identify what your practice still needs to decide.

1. Internal AI Policy Design Questions

Any internal AI policy should answer, clearly and specifically, the questions below.

A. Scope and Coverage

- ☐ Who does the policy apply to?
- ☐ Does it apply only to lawyers, or also to staff, contract professionals, interns, vendors, and consultants?
- ☐ Does it apply only to client work, or also to internal operations, marketing, billing, hiring, and training?
- ☐ Does it apply to all AI tools, or only generative AI systems?
- ☐ What counts as "AI use" for purposes of the policy?

A policy that does not clearly define its scope will create confusion immediately. People cannot follow rules that do not clearly state who is covered and what conduct is covered.

B. Approved Tools and Access Control

- ☐ Who decides which tools are approved?
- ☐ Who may approve a new tool?
- ☐ Is approval centralized, distributed, or hybrid?
- ☐ Are different tools approved for different practice groups or workflows?
- ☐ Are personal accounts prohibited for work purposes?
- ☐ Are there controls preventing access to unapproved tools from firm systems?

The key question is not simply "What tools do we like?" It is "What tools are we willing to stand behind, supervise, and defend as appropriate for our work?"

C. Prohibited Uses

- ☐ What tasks should be prohibited outright?
- ☐ Are there categories of work where AI may assist but may not draft the final output?
- ☐ Are final privilege determinations prohibited from AI completion without human lawyer review?
- ☐ Are lawyers prohibited from using AI to generate final legal conclusions in certain categories of matters?
- ☐ Are there client, court, or practice contexts where AI use should be categorically disallowed?

The point of a prohibited-use section is not to sound strict. It is to identify the work that the practice is not prepared to entrust to AI because the risks are too high or the required supervision cannot realistically be maintained.

D. Data Classification and Input Rules

- ☐ What types of information may be entered into which types of tools?
- ☐ Is there a data-tier framework the policy should follow?
- ☐ What must be anonymized, abstracted, or redacted before use?
- ☐ Are there categories of information that may never be entered into any AI system?
- ☐ What client-specific restrictions must be observed?

☐ How should lawyers handle outside counsel guidelines that prohibit or limit AI use?

This is one of the most important policy areas because data handling choices often create the most serious downstream risk.

E. Verification and Review Gates

☐ What level of verification is required for each type of AI-assisted work?

☐ What review is required before AI-assisted work can go to a client, court, opposing counsel, regulator, or third party?

☐ Who is responsible for that review?

☐ Are there mandatory review checklists for specific work types?

☐ What is the review expectation for court filings versus internal work product?

If verification standards are left vague, the policy will fail where it matters most.

F. Documentation and Logging

☐ What AI use should be documented?

☐ When is a workflow log required?

☐ What should be logged: tool used, prompt summary, output summary, verification steps, final reviewer?

☐ Who can access those logs?

☐ How long are logs retained?

☐ Can the practice protect everything it chooses to log?

Auditability is valuable, but over-logging can create its own confidentiality and security exposure.

G. Disclosure and Transparency

☐ When, if ever, should AI use be disclosed to clients?

☐ When, if ever, is disclosure required by courts, standing orders, or local rules?

☐ How should disclosures be calibrated for different audiences?

☐ What processes ensure those disclosures are accurate?

☐ Who approves disclosure practices?

This is not just a drafting question. It is a policy question about transparency, consistency, and defensibility.

H. Training and Competency

☐ What baseline training is required before someone may use approved AI tools?

☐ Is access to certain tools conditioned on training completion?

☐ Who trains new hires?

☐ How often should refresher training occur?

☐ How will the practice keep people current as tools and risks change?

A policy that assumes competence without investing in it is not a governance solution.

I. Incident Response and Escalation

☐ What counts as an AI incident?

☐ Who must be notified if one occurs?

☐ What are the first containment steps?

☐ When should the client be informed?

☐ When does the issue need to be escalated to ethics counsel, general counsel, or leadership?

☐ What post-incident review is required?

Incident response should not be improvised at the moment of failure.

2. Questions to Resolve Before Discussing AI Use with Clients

Not every practice will approach client disclosure the same way. But before you say anything to a client about AI use, your practice should answer several threshold questions.

A. What Are You Actually Telling the Client?

- ☐ Are you disclosing that AI may be used in a general way, or are you describing specific categories of use?
- ☐ Are you making commitments about review, confidentiality, or tool restrictions that your practice can actually honor?
- ☐ Are you describing present practice, or are you describing a future aspiration?

The language used with clients must match actual practice. Anything else creates avoidable risk.

B. What Level of Detail Is Appropriate?

- ☐ Does the client need only a general notice?
- ☐ Does the matter call for more specific explanation because of data sensitivity, client sophistication, or client policy?
- ☐ Would too much detail confuse rather than inform?
- ☐ Would too little detail fail to meet the client's expectations or rights?

This is not a one-size-fits-all decision.

C. Is the Client Entitled to Restrict AI Use?

- ☐ Does the client's outside counsel guidance address AI?
- ☐ Does the client have internal policy requirements that affect use?
- ☐ Has the client requested restrictions?
- ☐ Can the practice operationally comply with those restrictions?
- ☐ If the client prohibits AI use, how will that be documented and implemented?

A promise to avoid AI is easy to make and sometimes harder to operationalize than lawyers assume.

D. What Will Happen If the Client Asks Follow-Up Questions?

- ☐ Who answers questions about specific tools?
- ☐ Who answers questions about data handling?
- ☐ Who answers questions about internal supervision or verification practices?
- ☐ Is there a consistent internal position on what the practice is and is not willing to say?

A vague initial statement often leads to more specific follow-up questions. The practice should be prepared for that.

3. Escalation Triggers: When to Stop and Check with Your Own Counsel

The following situations should generally trigger a pause and further review rather than casual judgment in the moment:

- A client requests or proposes specific contractual language about AI use
- A client prohibits AI use and the matter is already using AI-assisted workflows
- A court rule, standing order, or judge-specific requirement appears to require AI disclosure or certification
- The practice is considering training or fine-tuning a system on internal or client-derived materials
- The proposed AI use touches sensitive health, employment, financial, biometric, or other regulated information
- The matter involves cross-border data issues
- An AI output appears to contain fabricated authority, invented facts, or compromised confidential information
- A lawyer or staff member used an unapproved tool with client-sensitive material
- The practice is considering allowing an AI system to take actions rather than merely generate text
- A vendor changes terms, data handling practices, or model behavior in ways that affect prior approval assumptions

The more novel, sensitive, or externally consequential the issue is, the less appropriate it is for casual internal improvisation.

4. A Practical Policy-Design Workflow

A safe and usable policy usually emerges from a sequence like this:

- Identify actual workflows where the practice is already using or wants to use AI.
- Map the risks in those workflows: accuracy, confidentiality, supervision, disclosure, billing, and client expectations.
- Decide which uses are approved, conditionally approved, or prohibited.
- Translate those decisions into plain-language rules.
- Build the corresponding training, verification, and documentation habits.

- Test whether the policy can actually be followed in live matters.
- Revise the policy based on real use, incidents, feedback, and changes in the tools or the regulatory environment.

A policy written in abstraction and never tested against actual workflows will usually fail in practice.

5. Questions for Smaller Practices

Small firms and solo practices often assume governance is something only large organizations need. That is a mistake.

Even if your governance structure is simple, you still need to answer the same basic questions:

- ☐ What tools are you using?
- ☐ What data may go into them?
- ☐ What review is mandatory?
- ☐ What is prohibited?
- ☐ What do you do if something goes wrong?

A short, clear policy that you actually follow is better than a sophisticated policy packet that never shapes behavior.

6. Questions for Larger Organizations

Larger firms and legal departments have an additional governance challenge: consistency across many people, practice groups, and workflows.

Questions to resolve include:

- ☐ Who owns AI governance?
- ☐ Which decisions are central and which are local?
- ☐ How are exceptions approved?
- ☐ How are clients with specific AI restrictions handled across teams?
- ☐ How are audit trails and review expectations implemented consistently?
- ☐ How are incidents escalated across legal, IT, security, and leadership functions?

At scale, ambiguity becomes expensive very quickly.

7. Final Rule

Do not adopt policy language because it sounds modern, careful, or comprehensive.

Adopt policy positions only when your practice understands what they mean, can actually follow them, and is prepared to supervise the work that follows from them.

That is the difference between governance language and governance.

Appendix E: Vendor/Tool Intake Checklist + Risk Scorecard

This appendix is designed to help you evaluate AI tools before they become embedded in your practice. The goal is not to produce a false sense of certainty. The goal is to force the right questions early, document the answers, and make tool adoption a reasoned decision rather than a reaction to a polished demo.

Throughout this book, tool selection has been framed as a professional judgment problem rather than a shopping problem. A tool may look impressive and still be a poor fit for your practice. A vendor may have strong marketing language and weak contractual commitments. A platform may perform well in one narrow use case and badly in another. This appendix translates the tool-selection framework from Chapter 6 into something operational: a repeatable intake process, a scoring model, and sample evaluations you can adapt to your own practice.

The checklist is organized around the four criteria that matter most in legal AI adoption: accuracy, security, cost, and governance fit.

Those criteria should be assessed together. A highly accurate tool with weak security may still be unacceptable. A secure tool with weak governance fit may never be used well enough to justify its cost. A low-cost tool with unclear data handling may create downstream risk that outweighs its pricing advantage. The point of the scorecard is not to pretend these tradeoffs can be solved by math alone. The point is to make the tradeoffs visible.

How to Use This Appendix

This appendix has three parts:

- Vendor / Tool Intake Checklist — use this when evaluating a specific AI tool or vendor before approval.
- Risk Scorecard — use this to turn your checklist findings into a documented recommendation.
- Sample Completed Scorecards — use these examples as models for how to document approval, rejection, and conditional approval.

This appendix works best when paired with Chapter 6. The chapter explains why these questions matter. This appendix gives you a format for asking them consistently.

1. Vendor / Tool Intake Checklist

Use this checklist before approving any AI tool for use in a law practice, legal department, clinic, or practice group.

A. Basic Identification
☐ Tool name recorded

- ☐ Vendor name recorded
- ☐ Product version / tier identified
- ☐ Intended user group identified
- ☐ Intended use cases identified
- ☐ Decision owner identified
- ☐ Date of evaluation recorded
- ☐ Renewal / review cycle identified

B. Accuracy Review

Use this section to determine whether the tool is actually good at the tasks you want it to perform.

- ☐ The proposed use case is clearly defined
- ☐ The task category is identified (research, drafting, review, summarization, intake, contract analysis, etc.)
- ☐ The vendor has explained how accuracy was measured
- ☐ Accuracy claims are tied to a methodology rather than marketing language alone
- ☐ The tool has been tested on at least five real or closely simulated tasks from your practice
- ☐ The testing included at least one harder or edge-case scenario
- ☐ Known limitations have been identified and documented
- ☐ The tool's likely failure modes are understood
- ☐ The tool's performance is acceptable for the intended use case
- ☐ The tool is not being approved for uses outside the tested scope

Notes

Tested use cases: ______________________________

What the tool did well: ______________________________

What the tool did poorly: ______________________________

Failure modes observed: ______________________________

Tasks the tool should not be used for: ______________________________

C. Security and Data Handling Review

Use this section to determine whether the tool is appropriate for the kinds of information your practice would place into it.

- ☐ Data handling terms were reviewed in writing
- ☐ The vendor clearly states whether inputs are used for model training
- ☐ The vendor clearly states retention periods for inputs and outputs
- ☐ The vendor clearly states who can access customer data
- ☐ Encryption at rest is documented
- ☐ Encryption in transit is documented
- ☐ Security certification or independent audit information is available (for example SOC 2, ISO 27001, or equivalent)
- ☐ Subprocessors are disclosed
- ☐ Incident response / breach notification terms are available
- ☐ Contract terms match the vendor's marketing claims
- ☐ The tool is appropriate for the intended data tier
- ☐ The tool is not being approved for data types it cannot safely handle

Notes

What data types may be entered into the tool:

____

What data types must stay out of the tool:

____

Training / retention concerns: ______________________________

Security certifications / evidence reviewed:

____

Open security questions: ______________________________

D. Cost Review

Use this section to assess total cost of ownership rather than subscription price alone.

- ☐ Subscription or license cost documented
- ☐ Per-user or usage-based pricing documented
- ☐ Implementation cost estimated

- ☐ Training cost estimated
- ☐ Ongoing administration cost estimated
- ☐ Integration cost estimated if applicable
- ☐ Exit or switching cost considered
- ☐ Expected efficiency gains documented realistically rather than optimistically
- ☐ The likely value to the practice justifies the total cost

Notes

Direct cost: ______________________________

Implementation / training cost: ______________________________

Expected use level: ______________________________

Expected return or practical value: ______________________________

Cost concerns: ______________________________

E. Governance Fit Review

Use this section to determine whether your organization can actually deploy and supervise the tool responsibly.

- ☐ There is a clear internal owner for the tool
- ☐ Policies can support the intended use case
- ☐ Training can be delivered to the users who need it
- ☐ Review expectations for the tool's outputs are clear
- ☐ Supervisors understand how to oversee work produced with this tool
- ☐ Audit / logging expectations are clear enough for the intended use
- ☐ The tool fits with existing workflow rather than forcing risky workarounds
- ☐ The organization has the capacity to maintain the tool over time
- ☐ A review date has been scheduled to reassess fit after deployment

Notes

Who owns this tool internally: ______________________________

Who approves use cases: ______________________________

Training needed before launch: ______________________________

Review / supervision requirements: ______________________________

Governance concerns: ______________________________

F. Final Intake Questions

Before approval, answer these directly.

- ☐ What exactly are we approving this tool to do?
- ☐ What exactly are we not approving it to do?
- ☐ What data may go into it?
- ☐ What data must stay out?
- ☐ What human review remains mandatory?
- ☐ What event would trigger re-evaluation or suspension of use?

2. Risk Scorecard

The scorecard below is not a substitute for judgment. It is a way of documenting that judgment.

Score each category on a 1–5 scale.

- 5 = strong / low concern / clearly acceptable
- 4 = acceptable with minor concerns
- 3 = mixed / conditional / requires mitigation
- 2 = serious concern
- 1 = unacceptable without major change

Scoring Guide — Accuracy

Score	Meaning
5	Strong performance on tested tasks, clear limitations, acceptable failure profile for intended use
4	Generally reliable for intended use, with modest limitations that are manageable through review

Score	Meaning
3	Mixed performance, acceptable only with strong guardrails and narrow use case restrictions
2	Weak or inconsistent performance on realistic tasks
1	Not reliable enough for the intended use

Scoring Guide — Security

Score	Meaning
5	Strong contractual and technical protections, clear data handling, no training on customer data, good evidence of security controls
4	Generally sound protections with limited open questions
3	Usable only with significant restrictions or mitigation steps
2	Major uncertainty or weakness in data handling or security posture
1	Inappropriate for any meaningful legal use involving sensitive information

Scoring Guide — Cost

Score	Meaning
5	Strong value relative to total cost of ownership
4	Reasonable value with manageable implementation cost
3	Unclear or moderate value relative to cost
2	Weak value proposition or difficult cost justification
1	Cost clearly outweighs likely value

Scoring Guide — Governance Fit

Score	Meaning
5	Clear internal ownership, easy policy integration, realistic training and supervision burden
4	Good fit with modest governance demands
3	Fit is possible but only with additional process, training, or oversight
2	Difficult to govern responsibly in the current environment
1	Governance burden exceeds organizational capacity

Risk Scorecard Form

Tool: ______________________________________

Use Case: ______________________________________

Evaluator: ______________________________________

Date: ______________________

Scores (1–5):

Accuracy Score: ____

Security Score: ____

Cost Score: ____

Governance Fit Score: ____

Total Score: ________

Recommended Outcome:

- ☐ Approve
- ☐ Approve with conditions
- ☐ Pilot only
- ☐ Do not approve

Required Conditions / Mitigations: ______________________________________

Review Date: ______________________

Rationale: ____

Suggested Interpretation

Total Score	Suggested Outcome
17–20	Approve, assuming no hidden red flags outside the scorecard
13–16	Approve with conditions or limited pilot
9–12	Pilot only or defer pending mitigation
8 or below	Do not approve

Use judgment even when the total score looks favorable. A tool with a high total score but a weak security score may still be unacceptable for client-confidential use.

3. Sample Completed Scorecards

Example 1: Clear Approval

Field	Entry
Tool	Enterprise Contract Review Platform
Use Case	Contract summary and clause comparison for commercial agreements
Evaluator	J. Smith
Date	April 13, 2026
Accuracy Score	5
Security Score	5
Cost Score	4
Governance Fit Score	4

Field	Entry
Total Score	18 / 20
Recommended Outcome	Approve
Required Conditions / Mitigations	Use only for contract review and comparison tasks. Do not use for final legal conclusions without lawyer review.
Review Date	October 13, 2026
Rationale	Tested well on real contract-review tasks, uses enterprise protections, aligns with current workflows, and has manageable training demands.

Example 2: Conditional Approval

Field	Entry
Tool	General AI Assistant
Use Case	Internal brainstorming, outline generation, and first-pass reorganization of non-sensitive text
Evaluator	J. Smith
Date	April 13, 2026
Accuracy Score	3
Security Score	4
Cost Score	5
Governance Fit Score	3
Total Score	15 / 20

Field	Entry
Recommended Outcome	Approve with conditions
Required Conditions / Mitigations	No client-confidential facts. No citation generation. No final legal analysis. Use only with source-bounded prompts for internal support tasks.
Review Date	July 13, 2026
Rationale	Useful and inexpensive for low-risk drafting support, but requires strict guardrails and clear review expectations.

Example 3: Rejection

Field	Entry
Tool	Consumer AI Chatbot (Free Tier)
Use Case	General legal drafting and client communication support
Evaluator	J. Smith
Date	April 13, 2026
Accuracy Score	2
Security Score	1
Cost Score	5
Governance Fit Score	1
Total Score	9 / 20
Recommended Outcome	Do not approve

Field	Entry
Required Conditions / Mitigations	None. The proposed use case is not acceptable for this tool.
Review Date	N/A
Rationale	Weak security posture for legal work, unclear data handling, no enterprise controls, and an unacceptable governance burden if used for client matters.

4. Final Rule

Do not approve a tool because it is impressive. Approve it because you can explain, defend, and supervise how it will be used.

The point of intake is not to create a paper trail for its own sake. The point is to slow the decision down enough that the organization asks the right questions before adoption creates momentum that is hard to reverse.

That is what a good vendor and tool intake process should do.

Appendix F: Tool Cards: Evaluating Categories of AI Tools

This appendix is the ongoing-reference companion to Appendix E.

Appendix E helps you decide whether a tool should be approved in the first place. Appendix F helps you document, monitor, and manage that tool after approval. The distinction matters. An intake checklist is a decision document. A tool card is a living reference document. One helps you decide whether to bring a tool into the practice. The other helps you remember what the tool is, what it is allowed to do, what it is not allowed to do, who owns it internally, and when it should be reviewed again.

A good tool card should let a lawyer, supervisor, knowledge professional, or risk manager answer a few practical questions quickly.

The point is not bureaucratic documentation for its own sake. The point is operational clarity. Once a practice starts using multiple AI tools across multiple workflows, memory becomes unreliable. People forget what was approved, under what conditions, and for what uses. Tool cards solve that problem by creating a stable internal reference for each approved tool category or tool instance.

1. What a Tool Card Is

A tool card is a short, structured internal reference document for an AI tool or tool category. It captures the most important practical and governance information in one place so that lawyers and staff do not have to reconstruct it from contract files, vendor emails, committee notes, or institutional memory.

Tool cards are most useful after a tool has already passed initial review. They are not procurement scorecards. They are operational reference records. In other words:

- Appendix E asks: Should we approve this tool?
- Appendix F asks: Now that it is approved, how do we describe it, govern it, and keep track of it?

This distinction prevents duplication and helps the practice treat approval as the beginning of governance rather than the end of it.

2. What a Tool Card Should Include

At minimum, a tool card should include:

- Tool name
- Vendor
- Tool category
- Deployment type
- Date approved
- Internal owner

- Approved uses
- Prohibited uses
- Allowed data tier
- Known limitations
- Mandatory supervision or verification requirements
- Review cycle
- Next review date
- Notes on incidents, updates, or special conditions

Some practices will want more detail. Others will want less. But if the card does not answer the practical questions that working lawyers actually have, it will not get used.

3. Blank Tool Card Template

Use the following template as a starting point.

Field	Entry
Tool Name	
Vendor	
Tool Category	
Deployment Type	
Version / Tier	
Date Approved	
Internal Owner	
Practice Group(s) Using It	
Approved Uses	
Prohibited Uses	

Field	Entry
Allowed Data Tier	
Required Guardrails	
Required Human Review	
Known Limitations	
Known Failure Modes	
Related Policy / Procedure	
Training Required Before Use	
Review Cycle	
Next Review Date	
Notes / Incidents / Updates	

4. Category-Level Tool Cards

The examples below are not endorsements of any specific vendor. They are category-level illustrations showing what a completed tool card can look like.

A. General AI Assistant

Field	Entry
Tool Name	General AI Assistant
Vendor	Varies

Field	Entry
Tool Category	General-purpose generative AI
Deployment Type	Enterprise or secure managed deployment preferred
Version / Tier	Approved enterprise tier only
Date Approved	[Insert date]
Internal Owner	Innovation lead / practice management lead
Practice Group(s) Using It	Cross-practice, subject to policy
Approved Uses	Brainstorming, plain-English translation, internal summarization, drafting from source-bounded inputs, internal organization tasks
Prohibited Uses	Final legal advice without lawyer review, unsupervised citation generation, final privilege calls, entering client-confidential data into unapproved environment
Allowed Data Tier	Tier 1 or Tier 2 by default; higher only if enterprise controls specifically allow it
Required Guardrails	Source-bounded prompting when accuracy matters; explicit instruction not to invent facts or authorities
Required Human Review	Mandatory review before any external use; heightened review for client-facing or court-facing work
Known Limitations	Hallucination, overconfidence, generic drafting, subtle factual drift
Known Failure Modes	Fabricated authorities, invented facts, overclaiming, tone mismatch
Related Policy / Procedure	AI Acceptable Use Policy; Client Data Handling Policy

Field	Entry
Training Required Before Use	Foundational AI training
Review Cycle	Quarterly
Next Review Date	[Insert date]
Notes / Incidents / Updates	Use only through approved environment; no personal accounts for work purposes

B. Legal Research AI Platform

Field	Entry
Tool Name	Legal Research AI Platform
Vendor	Varies
Tool Category	Research / authority retrieval
Deployment Type	Vendor-hosted legal research platform
Version / Tier	Approved research tier
Date Approved	[Insert date]
Internal Owner	Research lead / library / litigation support
Practice Group(s) Using It	Litigation, appellate, advisory practices
Approved Uses	Initial issue framing, authority identification, authority summarization, synthesis support, memo outlining
Prohibited Uses	Blind reliance on summaries, filing without independent authority review

Field	Entry
Allowed Data Tier	Tier 1 and Tier 2; no unnecessary client-identifying detail
Required Guardrails	Verify every authority independently before relying on it
Required Human Review	Mandatory authority verification and quote-checking for any cited proposition
Known Limitations	Can mischaracterize holdings, miss better authorities, flatten distinctions across cases
Known Failure Modes	Overbroad synthesis, incomplete treatment of split authority, weak jurisdiction sensitivity
Related Policy / Procedure	Research Verification Procedure
Training Required Before Use	Foundational AI training plus research-platform training
Review Cycle	Semiannual
Next Review Date	[Insert date]
Notes / Incidents / Updates	Useful for acceleration, not a substitute for reading the cases

C. Drafting Assistance Tool

Field	Entry
Tool Name	Drafting Assistance Tool
Vendor	Varies
Tool Category	Drafting / document generation
Deployment Type	Enterprise deployment or secure drafting environment

Field	Entry
Version / Tier	Approved drafting tier
Date Approved	[Insert date]
Internal Owner	Document automation / KM / practice group lead
Practice Group(s) Using It	Transactional, advisory, selected litigation uses
Approved Uses	First drafts, source-bounded drafting, clause variation generation, internal revision support
Prohibited Uses	Final filing-ready or client-ready document generation without human revision and verification
Allowed Data Tier	Depends on deployment; default to lower-risk data unless expressly approved
Required Guardrails	Use defined instructions, source materials, and output limits; prohibit invented facts and authorities
Required Human Review	Mandatory substantive review, completeness review, and verification before release
Known Limitations	Fluent but generic language, hidden omissions, overconfident phrasing
Known Failure Modes	Missing required provisions, weak tailoring, unsupported assertions
Related Policy / Procedure	Drafting Workflow Policy; Style Guide
Training Required Before Use	Foundational training plus drafting workflow training
Review Cycle	Quarterly

Field	Entry
Next Review Date	[Insert date]
Notes / Incidents / Updates	Best used as a starting point, not an endpoint

D. Contract Analysis Tool

Field	Entry
Tool Name	Contract Analysis Tool
Vendor	Varies
Tool Category	Extraction / comparison / contract review
Deployment Type	Enterprise contract-review platform
Version / Tier	Approved review tier
Date Approved	[Insert date]
Internal Owner	Transactional lead / KM / legal ops
Practice Group(s) Using It	Corporate, commercial, procurement, internal legal ops
Approved Uses	Key term extraction, deviation spotting, document comparison, obligation summaries
Prohibited Uses	Final risk acceptance decisions without lawyer review
Allowed Data Tier	As approved by security review and client obligations
Required Guardrails	Verify flagged changes and summaries against the actual contract text

Field	Entry
Required Human Review	Required for all material risk judgments and negotiation decisions
Known Limitations	Can identify text differences without understanding commercial significance
Known Failure Modes	False positives, false negatives, weak prioritization of actual deal risk
Related Policy / Procedure	Contract Review Protocol
Training Required Before Use	Tool-specific training plus contract workflow training
Review Cycle	Semiannual
Next Review Date	[Insert date]
Notes / Incidents / Updates	Most valuable for extraction and comparison, not for final business judgment

E. Discovery / Review Tool

Field	Entry
Tool Name	Discovery / Review Tool
Vendor	Varies
Tool Category	Document review, classification, triage
Deployment Type	Review platform / hosted e-discovery environment
Version / Tier	Approved litigation support tier
Date Approved	[Insert date]

Field	Entry
Internal Owner	Litigation support / e-discovery lead
Practice Group(s) Using It	Litigation, investigations, regulatory response
Approved Uses	Relevance review, prioritization, hot document identification, privilege candidate flagging, clustering
Prohibited Uses	Final privilege determination without human review; final production decisions without lawyer oversight
Allowed Data Tier	As approved under litigation support security controls
Required Guardrails	Review protocol must define responsiveness, privilege, escalation, and QC rules
Required Human Review	Mandatory privilege review and QC sampling
Known Limitations	Can misclassify edge-case documents and over-read ambiguous content
Known Failure Modes	Privilege false negatives, overinclusive responsiveness, clustering that obscures nuance
Related Policy / Procedure	Review Protocol; Privilege Review Procedure
Training Required Before Use	Litigation support training; tool-specific review training
Review Cycle	Quarterly
Next Review Date	[Insert date]
Notes / Incidents / Updates	Works best when combined with documented review criteria and ongoing QC

5. How to Maintain Tool Cards

Tool cards only work if they are kept current.

A stale tool card can be as misleading as no tool card at all. If the vendor changes terms, a new feature introduces a new risk, an incident occurs, or internal policy changes, the card should be updated. At minimum, every card should have:

- An internal owner
- A review cycle
- A next review date
- A place for updates or incidents

Typical triggers for updating a tool card include:

- Vendor terms or privacy changes
- New feature release
- Incident or near miss
- Expansion of approved use
- Policy revision
- Annual or quarterly governance review

6. Final Rule

A tool card should answer the question, "What do we need everyone in this practice to know about this tool right now?"

If it cannot do that quickly and clearly, it is not functioning as a tool card.

The point is not to create paperwork. The point is to reduce ambiguity, preserve institutional memory, and make responsible use easier than improvisation.

Appendix G: Glossary of AI Terms for Legal Professionals

This appendix provides a condensed, book-wide reference glossary. The chapter-end Key Terms sections remain the primary place for fuller definitions in context; this appendix is designed for quick lookup.

Acceptable Use Policy (AUP) for AI — A document outlining the permissible ways employees can use artificial intelligence tools within an organization, specifically addressing approved tools, prohibited activities, and data security requirements.

ABA Formal Opinion 512 — The ABA Standing Committee on Ethics and Professional Responsibility's July 2024 opinion addressing generative AI in legal practice. Covers competence, confidentiality, communication, candor, supervision, and fees under the Model Rules as applied to generative AI tools.

AI Agents — AI systems capable of autonomous action, using software tools to complete multi-step tasks without human intervention at each step.

AI Governance — The framework of decision rights, policies, and processes that an organization uses to manage the risks and opportunities of artificial intelligence.

Algorithmic discovery — Litigation over the design, function, and output of AI systems used by parties, analogous to e-discovery but focused on algorithms rather than documents.

Algorithmic impact assessment — A structured evaluation of the potential effects of an AI system on individuals and groups, increasingly required by regulation for high-stakes AI deployments including those affecting legal rights.

Anchoring — The model's tendency to commit to an early interpretation and develop it at the expense of alternatives.

Argument mapping — Creating a structured representation of how legal arguments connect premises, authorities, facts, and conclusions.

Assertion-source matrix — A verification tool mapping each factual assertion in a document to its supporting source in the record or investigation materials.

Audience instructions — Specifications about who will receive the AI output, allowing the model to calibrate vocabulary, formality, assumed knowledge, and emphasis for the intended reader.

Audit Trail (in AI governance) — The chronological record of system activities, including user logins, prompts entered, and outputs generated.

Audit-ready documentation — Records sufficient to demonstrate, if later questioned, that AI-assisted work was performed in compliance with professional obligations.

Automation implementation checklist — A checklist ensuring chatbots and intake automation are accurate, compliant, accessible, and appropriately limited to avoid unauthorized practice issues.

Brussels Effect — The tendency of EU regulations, including the AI Act, to set global standards because companies building products for the European market incorporate EU compliance requirements into their default offerings worldwide.

Candor to the tribunal (Model Rule 3.3) — The duty prohibiting lawyers from making false statements of fact or law to a court.

Case theory — The coherent narrative explaining why your client should prevail, integrating facts and law into a persuasive story.

Classify — Sorting items into defined categories according to a classification scheme. Reliability depends on category clarity: well-defined categories with objective criteria produce consistent results, while fuzzy categories that require judgment produce inconsistent ones.

Clause generation — Using AI to produce draft contract language for a specific provision.

Clause library — An organized collection of contract provisions preserved for reuse, including the clause language, annotations explaining its use and history, and metadata for retrieval.

Client communication checklist — A structured review ensuring communications address client needs, maintain appropriate tone, and include necessary information.

Client comprehension verification — Checking that a client actually understood an explanation, using techniques beyond "does that make sense?"

Conditional logic (in intake forms) — Showing or hiding questions based on previous answers, creating customized form experiences that ask only relevant questions.

Confidentiality (Model Rule 1.6) — The duty not to reveal information relating to the representation of a client.

Contemporaneous time recording — Capturing time entries as work is performed or immediately after, rather than reconstructing time at the end of a day or week.

Content repurposing — Adapting a single piece of content for use in multiple formats and channels, extending the value of content creation efforts.

Context packaging — The process of selecting, structuring, and presenting information in a prompt so the model has what it needs to produce useful output.

Context window — The maximum amount of text that an AI model can process in a single interaction.

Contract analysis tools — AI tools that extract, compare, and summarize contract provisions.

Controlled reuse — Verifying that retrieved work product fits the current context before incorporating it, including checking for legal changes, jurisdictional differences, and factual distinctions that might make past work inapplicable or misleading.

Critique — Evaluating existing work to identify weaknesses, gaps, errors, or areas for improvement. Critique is useful because AI approaches the work without the cognitive biases its author developed while creating it, though it will also raise false alarms and miss issues requiring deeper contextual understanding.

Data classification — Categorizing information by sensitivity level to determine which AI tools and environments are appropriate for processing that information, aligned with the confidentiality tier framework from Chapter 5.

Data handling practices — How an AI system processes, stores, retains, and uses information entered into it.

De-identification and abstraction — De-identification removes or obscures identifying information, while abstraction goes further by transforming specific facts into general analytical questions.

Deliverable-first prompting — An approach that begins with a detailed specification of the desired output, then works backward to provide the role, context, and task instructions needed to produce it.

Discovery planning — Determining what evidence you need, what evidence the opponent needs, and how to obtain or protect that evidence efficiently.

Document template — A standardized starting point for a recurring document type, capturing preferred structure, standard language, formatting conventions, and quality standards.

Drafting checklist — A standardized list of steps for AI-assisted document creation across the workflow stages of instructions, first draft, revision, and verification.

Drafting instructions — Guidance to AI defining what a document must accomplish, what context applies, what constraints limit the draft, how it should be structured, and what output format is expected.

Drafting verification — The final review process confirming that a document accomplishes its intended purpose and that all factual and legal assertions are accurate.

Durable human skills — Professional capabilities that remain valuable regardless of AI advancement, including strategic judgment, client relationships, ethical reasoning, creative problem-solving, and persuasive advocacy.

Enterprise AI deployment — AI services provided under business contracts with negotiated terms, typically including confidentiality provisions, data handling commitments, and restrictions on using inputs for model training.

Extract — Pulling specific information out of a larger body of source material. Extraction is concrete and verifiable because the information either appears in the source or it does not.

Fact constraint — The requirement that every factual assertion in litigation documents must trace to a verifiable source: the evidentiary record, good-faith allegations with expected support, or matters subject to judicial notice.

Fact invention — AI generating factual assertions that do not come from your inputs, based on patterns in training data about what facts typically appear in similar documents.

Foundational AI training — Baseline education that all lawyers using AI should receive, covering how AI works, professional responsibility obligations, and fundamental prompt engineering skills.

General AI assistants — Large language models trained on broad datasets and accessible through chatbot interfaces or embedded in productivity software.

Generate — Creating new content from instructions rather than reshaping existing material. Generation draws on patterns from training data rather than verified sources, making it the most powerful and the least reliable of the task types.

Governance model — The organizational structure determining who has authority over AI-related decisions, ranging from centralized committees to distributed practice group leadership to hybrid approaches combining central standards with local implementation.

Grounded AI — AI systems connected to authoritative data sources that constrain outputs to information actually present in those sources.

Hallucination — Output that is fabricated rather than grounded in fact.

Holding extraction — Capturing what a court actually decided by using language close to the opinion's own words and preserving the scope and limitations of the ruling.

Human fallback — The availability of human assistance when automated systems cannot meet user needs, ensuring automation supplements rather than replaces human interaction.

Human-in-the-loop (HITL) — The requirement that a human professional review and approve AI outputs before they are finalized, filed, or sent to clients. The fundamental safety mechanism for professional AI use.

Impeachment preparation — Identifying specific prior statements or documents that contradict the witness's likely testimony.

Incident response protocol — A predefined plan for responding to AI-related failures, including steps for containment, investigation, notification of affected parties, and remediation to prevent recurrence.

Inconsistent reasoning — A failure mode in which the model's conclusions do not follow from its analysis, or in which different parts of the output contradict each other.

Information relating to the representation — The scope of Rule 1.6 protection.

Intake — Gathering information from new clients or about new matters: client narrative, relevant documents, parties, and potential issues.

Intake automation — Using technology, including chatbots, online forms, and automated workflows, to gather preliminary information from potential clients and route inquiries before human involvement.

Integrated prompt pack — A collection of prompts designed for a specific matter type, organized to align with the matter's phase structure and key tasks, allowing lawyers to access relevant prompts at each stage of the matter.

Interaction effects — How multiple contract provisions work together to create outcomes that neither provision creates alone.

Issue spotting — Identifying potential legal claims, defenses, or complications a client's factual situation raises, including issues the client may not have recognized.

Issue trading — Exchanging concessions across different contract provisions, where each party gives ground on issues less important to them in exchange for gains on issues more important.

Jurisdiction blending — A failure mode in which the model's output mixes legal principles from different jurisdictions without flagging the substitution.

Knowledge management — The systematic capture, organization, and reuse of valuable work product and expertise across a practice, preventing the waste of recreating work that has already been done.

Large language model (LLM) — A type of AI trained on massive text datasets to predict what words should follow a given input.

Machine unlearning — The technically challenging process of removing specific data points from a trained AI model, relevant when data was used without proper rights or when deletion is required under privacy regulations.

Matter playbook — A structured guide documenting the practical workflow for handling a specific matter type, capturing what experienced practitioners know about effective approaches, common issues, and lessons learned.

Motion strategy — Deciding which motions to file, when to file them, and how to frame the arguments to advance the overall case theory.

Multimodal AI — AI systems that process and generate multiple content types — text, images, audio, and video — natively within a single model, enabling analysis of diverse evidence types that text-only tools cannot handle.

Omission — A failure mode in which important information that should appear in AI output is missing, even though the output otherwise looks complete. Omissions are especially dangerous because the reader cannot see what was left out.

Overconfidence — A failure mode in which the model presents uncertain or contested propositions in the same confident register as well-established rules. The tone of the output does not track the underlying certainty of the content.

Phase structure — Documentation of how matters of a specific type typically progress, identifying stages, milestones, and transitions that help lawyers understand where they are in a matter and what comes next.

Pin cite — A citation identifying the specific page or pages where supporting material appears within a longer source.

Plain-English communication — Explaining legal concepts in everyday language clients can understand without legal training, replacing jargon with accessible terms while maintaining accuracy.

Practice checklist — A standardized list of steps to complete for a recurring task type, ensuring consistent quality and preventing omission of important elements.

Prebill review — Reviewing invoices before sending them to clients, verifying accuracy, appropriateness, and professionalism of all entries.

Procedural explanation — Describing to clients what will happen in their legal matter: steps, timeline, and what will be expected of them.

Prompt Blueprint — A five-element structure for constructing effective AI prompts: Role (who the AI should be), Context (information it needs), Task (what to produce), Format (how to structure it), and Guardrails (what to avoid).

Prompt chaining — Using the output of one prompt as input for a subsequent prompt, building complex analysis through a sequence of simpler steps.

Prompt library — A personal or organizational collection of effective prompts saved for reuse, customized for specific practice needs and refined based on experience.

Prompt template — A standardized starting point for recurring AI-assisted tasks, capturing effective instructions, output specifications, and quality requirements.

Prompting standardization — Organizational practices that promote consistent, effective prompting across users: shared templates, prompt libraries, training programs, quality benchmarks, and supervision that includes prompting feedback.

Quality assurance sampling — Periodic review of a random selection of AI-assisted work product to verify that organizational standards are being followed, distinct from individual verification of specific outputs.

Question decomposition — Breaking a broad research question into specific, discrete legal issues that can be researched independently and then synthesized into a complete answer.

Reasonable fees (Model Rule 1.5) — Fees determined by multiple factors including time, difficulty, skill required, results obtained, and customary charges.

Record citation — A specific reference to evidentiary materials in the case file that supports a factual assertion.

Redline analysis — Systematic review of counterparty markup to identify, categorize, and assess proposed changes.

Relevance review — Determining which documents in a collection are responsive to discovery requests and relevant to the claims or defenses.

Reliability ladder — A framework for matching AI tasks to appropriate supervision levels.

Research memo — A written document presenting legal analysis of a specific question, typically including issue, brief answer, facts, analysis, and conclusion.

Research question refinement — Transforming vague or broad research questions into precise legal issues suitable for systematic investigation.

Risk framing — How potential negative developments are presented to clients, balancing the need to inform against the risks of creating unnecessary anxiety.

Risk framing in client communications — Presenting uncertainties in ways that inform without unnecessarily alarming, acknowledging genuine risks while maintaining appropriate confidence.

Risk memo — A document advising a client about legal risks in a transaction, contract language, or course of action.

Risk-based regulation — A regulatory approach classifying AI systems by potential harm, imposing stricter requirements on higher-risk systems and creating a tiered compliance landscape.

Rule 11 certification — The implicit certification made when a lawyer signs a court filing, representing that the filing has legal and factual support and is not filed for improper purpose.

Shadow AI — Unauthorized use of AI tools that have not been approved by the organization's governance process, often driven by restrictive policies that push users toward unsanctioned alternatives.

Signal (in legal citation) — Indicates the relationship between a cited authority and the proposition it supports.

Source-bounded drafting — Providing AI with specific source materials (precedents, facts, legal standards) that constrain the draft to verified content rather than free generation.

Source-bounded prompting — Constraining the model to work only with authorities you provide, prohibiting it from generating additional citations.

Standing order — A directive issued by an individual judge that applies to all cases on that judge's docket.

Status-next-timeline-action framework — A structure for client updates organized around the questions clients actually have: Where do things stand? What happens next? When? What do I need to do?

Staying current — The ongoing effort to maintain accurate, up-to-date knowledge about AI tools, practices, and regulatory requirements as the landscape evolves.

Steelmanning — Presenting an opposing position in its strongest possible form rather than attacking a weakened version.

Style system — A coordinated approach to maintaining consistent voice, tone, and formatting conventions when AI assists drafting across multiple drafters, documents, or matters.

Style verification — Reviewing documents specifically to confirm they meet organizational style standards.

Supervision (under the Model Rules) — The obligation of lawyers with managerial or supervisory authority to ensure that conduct by those under their supervision conforms to professional rules.

Synthesis (in legal research) — Analyzing authorities collectively to identify principles, patterns, and frameworks that address the research question.

Technological competence — The obligation under Comment 8 to Model Rule 1.1 to understand the benefits and risks of technology used in legal practice.

Temporal confusion — A failure mode in which the model conflates current law with superseded law, or fails to recognize that a statute was amended or a case was overruled.

Terminology consistency — Using the same words and phrases to describe the same concepts throughout a document, avoiding variations that suggest unintended distinctions or ambiguity.

Thought leadership — Content positioning you as a knowledgeable authority in your field through articles, guides, presentations, and other materials.

Tool card — A structured reference document capturing key information about a specific AI tool: capabilities, data handling practices, appropriate uses, limitations, and current status.

Tool-specific training — Education focused on the particular AI tools a practice uses, covering features, capabilities, limitations, and best practices specific to those tools as distinct from general AI skills that transfer across platforms.

Tool-task matching — Deploying specific AI tools for specific task types based on alignment between task requirements and tool capabilities.

Total cost of ownership — All costs associated with adopting and operating a tool: licensing/subscription, implementation, training, ongoing administration, and productivity impact during transition.

Training bias — Systematic patterns in AI output reflecting imbalances in training data.

Transformation — Converting content from one form to another while preserving its substance.

Triage (in verification) — Allocating review effort based on consequence.

Trigger events — Occurrences during a matter that require specific responses or activate particular obligations, as distinguished from calendar deadlines that arrive on fixed dates.

Two-pass verification — Separating content generation from content critique into distinct prompts.

Update prompts — Prompts that specify client characteristics so AI can calibrate tone, detail level, and approach.

Update schedule — A predetermined rhythm for communicating with clients about matter status, ensuring regular communication regardless of whether significant developments have occurred.

Voice preservation — Ensuring AI-assisted content reflects your authentic voice, perspective, and expertise rather than generic AI style.

About the Author

Carlos A. Baradat is an attorney, legal technology advisor, and educator focused on the intersection of law, artificial intelligence, and digital risk. He is the founder of Legal Tech Dynamics, a consulting firm that advises law firms, businesses, and professionals on eDiscovery, digital forensics, and the responsible use of emerging technologies in legal practice.

Carlos has more than a decade of experience working at the intersection of law and technology. His work includes advising on data preservation, electronic discovery strategy, and technology-driven workflows, as well as helping legal professionals understand and implement artificial intelligence tools in a way that aligns with professional responsibility obligations.

In addition to his practice, Carlos serves as an adjunct professor at multiple universities, where he teaches courses in legal technology, eDiscovery, and artificial intelligence in the legal profession. His teaching focuses on translating complex technical concepts into practical, real-world applications that students and practitioners can use immediately.

Carlos is also the creator of The Virtual Jurist, a platform dedicated to educating lawyers and legal professionals on the evolving role of technology in law, including artificial intelligence, automation, and digital evidence. Through his writing, teaching, and consulting, he emphasizes a disciplined, practical approach to innovation—one that prioritizes accuracy, confidentiality, and professional judgment over speed or novelty.

AI for Legal Practice reflects his core philosophy: that artificial intelligence can significantly improve legal work when used thoughtfully, but that it must be approached with structure, skepticism, and a strong understanding of the risks involved.

www.ingramcontent.com/pod-product-compliance
Lightning Source LLC
LaVergne TN
LVHW060021210926
848481LV00013B/3757